KT-545-819

Alan Grainger is Lecturer in Physical Geography at the University of Salford. He has been researching tropical land use and forestry issues since 1977, and his first book on this subject, *Desertification*, was published by Earthscan in 1983. He is currently writing a book on the human impact on the tropical rain forests.

WITHDRAWN FROM
THE LIBRARY

KA 0129921 2

UNIVERSITY OF
STER

THE THREATENING DESERT

Controlling Desertification

by Alan Grainger

Earthscan Publications Ltd, London

in association with

United Nations Environment Programme, Nairobi

First published 1990 by
Earthscan Publications Ltd
3 Endsleigh Street, London WC1H 0DD

Copyright © 1990 International Institute
for Environment and Development (IIED).

All rights reserved

British Library Cataloguing in Publication Data
Grainger, Alan
  The threatening desert: controlling desertification
  1. Desertification. Control
  I. Title
  333.73'13

  ISBN 1-85383-041-0

Production by David Williams Associates 01-521 4130
Typeset by DP Photosetting, Aylesbury, Bucks.
Printed and bound in Great Britain by
Guernsey Press Ltd, Guernsey, C.I.

Earthscan Publications Ltd is a wholly owned and editorially
independent subsidiary of the International Institute for
Environment and Development.

Cover photo of Kalsaka Village, Yatenga Province, in Burkina Faso by
Mark Edwards/Still Pictures

KING ALFRED'S COLLEGE
WINCHESTER

574.
52652   01299212
GRA

Contents

Figures and Tables

FIGURES

TABLES

Acronyms

AVHRR	Advanced very high resolution radiometer (instrument used in satellite monitoring)
CARE	A voluntary agency based in the USA
CIDA	Canadian International Development Agency
CILSS	Comité Permanent Inter-états de lutte contre la Sécheresse dans le Sahel (Permanent Interstate Committee for Drought Control in the Sahel)
DESCON	Consultative Group for Desertification Control
EEC	European Economic Community
FAO	UN Food and Agriculture Organization
GCM	Global circulation model
GEMS	Global Environment Monitoring System (a programme of UNEP)
IAWGD	Interagency Working Group on Desertification
IBPGR	International Board for Plant Genetic Resources
ICRAF	International Council for Research in Agroforestry
ICRISAT	International Crop Research Institute for the Semi-Arid Tropics
IIASA	International Institute for Applied Systems Analysis
IITA	International Institute for Tropical Agriculture
ILCA	International Livestock Centre for Africa
MSS	Multi-spectral scanner (instrument used in satellite monitoring)
NAS	US National Academy of Sciences
NASA	US National Aeronautic and Space Administration
NCWK	National Council of Women of Kenya
NGLS	UN Liaison Service for Non-Governmental Organizations

NGO	Non-governmental organization
NOAA	US National Oceanographic and Atmospheric Administration
NRC	US National Research Council
OECD	Organization for Economic Co-operation and Development
OPEC	Organization of Petroleum Exporting Countries
ORSTOM	Office de la Recherche Scientifique et Technique Outre-Mer
OXFAM	A voluntary agency based in the UK, originally the Oxford Committee for Famine Relief
SIDA	Swedish International Development Agency
UN	United Nations Organization
UNCOD	UN Conference on Desertification
UNDP	UN Development Programme
UNEP	UN Environment Programme
UNESCO	UN Educational, Scientific and Cultural Organization
UNICEF	UN Children's Fund
UNIDO	UN Industrial Development
UNSO	UN Sudano-Sahelian Office
USAID	US Agency for International Development
WFP	World Food Programme
WHO	World Health Organization
WMO	World Meteorological Organization

Preface

I had two main aims in writing this book: first, to provide a more comprehensive treatment of the nature, causes and scale of desertification than was possible in my previous book, *Desertification*, published in 1983; second, to identify the key strategies which may be used to control desertification, and illustrate these with examples of projects, both successful and otherwise.

Satisfying the first aim was made difficult by: (a) a continuing lack of reliable data on the scale of desertification; (b) our still woefully inadequate understanding of desertification processes, owing to the small number of scientific studies which have been carried out in this field. There is still a vigorous debate about the respective roles of human impact and climate in causing desertification, and Chapter 1 should therefore be seen as a contribution to this debate. To avoid confusion with my previous book, my publishers advised me to use *The Threatening Desert* as the main title for this one. I hope that it is apparent from the text that I do not subscribe to view that desertification mainly refers to the expansion of natural deserts. The real "threat" lies in our inability to manage drylands in a sustainable way, and our unwillingness to acknowledge just how serious a global problem desertification has become.

That I was not able to find many examples of successful projects in support of my second aim is indicative of the lack of action taken to control desertification since the United Nations Conference on Desertification in 1977. Many of the projects which are implemented fail to achieve their objectives because they focus primarily on land-management techniques and neglect the social and economic dimensions of the problem.

I have tried to satisfy the needs of two main types of reader: the general reader or student who is interested in the whole subject of desertification and wants a concise overview; and the member of staff of a government agricultural department, international aid agency or non-governmental organization who is more concerned with the design and implementation of projects to bring desertification under control. Government policy-makers should find something of interest here too, since the role of policy is referred to throughout the text.

This book would not have been possible without the help of many people. I am grateful to the UN Environment Programme (UNEP) for funding the book and to my commissioning editor Lloyd Timberlake, formerly with Earthscan, who encouraged me to take a wholly fresh approach to the subject. Richard Sandbrook, of the International Institute for Environment and Development (IIED), and Neil Middleton, of Earthscan Publications, were patient and understanding during the process. UNEP's Desertification Control Branch provided information from its database of projects intended to control desertification, and Alex Forbes of IIED's Washington DC office helped to track down suitable projects for possible inclusion in the book. I would also like to thank the staff members of the US Agency for International Development, the World Bank, the Canadian International Development Agency, CARE, Lutheran World Relief and all the other agencies who assisted in my search for projects.

The book was mainly written while I was the holder of a Gilbert F. White Fellowship at Resources for the Future in Washington DC. I am deeply indebted to Resources for the Future for the fellowship and for the excellent facilities that were invaluable in allowing me to complete the manuscript. Thanks are also due to the staff of the libraries of the Oxford Forestry Institute and the former Institute of Agricultural Economics, both at the University of Oxford, where much of the initial literature research was carried out.

Many colleagues, by kindly agreeing to comment on all or part of various drafts of the book, have made it much better than it would otherwise have been. It is impossible to thank them all here, but the contributions of Harold Dregne, Michael Glantz, Andrew Goudie, Jean Gorse, Dick Grove, Mike Hulme, Peter Lamb, Jack Mabbutt, Tim Resch, Poul Sihm and Jeremy Swift cannot go unacknowledged. Needless to say, I am responsible for any inaccuracies which remain.

I wish to thank the following people for allowing me to reproduce illustrations from their publications: Harold Dregne (Figs 1.1, 1.6 and 4.2); Dick Grove (Fig. 1.2); Peter Jones (Fig. 1.11); Peter Lamb (Fig. 1.7); Mike Hulme (Fig. 1.8); and Nick Middleton (Fig. 1.9). I am also very grateful to Gustav Dobrzynski of the Department of Geography, University of Salford, for drawing Figs. 1.4, 1.5 and 4.1. I am responsible for Figs. 1.4 and 1.5, but am indebted to the authors of previous maps (acknowledged in footnotes) upon which these were based.

Finally, although UNEP funded the preparation of this book, the views which it expresses are entirely my own. I hope that the book properly represents the dedicated efforts of the many people who are engaged in trying to control desertification, and will inspire others to follow in their footsteps.

Alan Grainger
Salford
January 1989

The International Institute for Environment and Development (IIED), the author and the publishers would like to thank the United Nations Environment Programme (UNEP) whose generous funding made the research for and the writing of this book possible.

For more information on the work of UNEP, write to: UNEP, PO Box 30552, Nairobi, Kenya

Introduction

Desertification is one of the most serious problems facing the world today. Large parts of the dry areas that cover more than one-third of the earth's land surface are being degraded, with serious effects on the environment, food production, and the lives of millions of people. Desertification, characterized by the degradation of soil and vegetative cover, can occur in any dry area, not just on the fringes of natural deserts. It is a global phenomenon, affecting both developed and developing nations, and is a particular problem in Africa, the Middle East, India and Pakistan, China, Australia, the USSR, the USA, Latin American countries such as Brazil and Chile, and European countries such as Greece, Spain and Portugal.

This book has three main parts: the first (Chapters 1–4) looks at the nature, causes and extent of desertification; the second (Chapters 5–9) describes ways by which it can be brought under control and gives examples of projects which have aimed to do this; and the third (Chapter 10) evaluates progress made so far in controlling desertification.

Desertification has been occurring for millennia, but became a matter of worldwide concern in the early 1970s when a major drought in the Sahel region of West Africa, which in the opinion of most experts continues to this day, killed between 50,000 and 250,000 people, about 3.5 million cattle, and countless sheep, goats and camels (Caldwell 1984; Kates *et al.* 1977). This prompted the United Nations to convene a Conference on Desertification (UNCOD) in Nairobi in 1977 to agree on a Plan of Action to combat desertification and bring it under control by the year 2000.

The extensive scientific investigations which preceded UNCOD showed that the Sahel tragedy was not just a natural disaster caused by lack of rainfall, but the result of a chronic process of land degradation in which people had a key role. The four main direct causes of desertification, described in detail in Chapter 2, were identified as overcultivation, overgrazing, deforestation and the mismanagement of irrigated cropland. However, while poor land use can simply be the result of bad management, it is greatly influenced by periods of drought, during which cropping and grazing become more intensive in order to maintain overall food production; by poverty and other aspects of economic underdevelopment, which make it difficult for farmers to manage their lands in a sustainable way; and by misguided government policies, which are often biased against the maintenance or improvement of traditional farming systems. The underlying social, economic and political causes of desertification are discussed in Chapter 3.

There is much debate about the relative contributions of human impact and drought to desertification, a subject discussed in Chapter 1. Annual rainfall totals are highly variable in dry areas and so droughts (periods of below-average rainfall) occur quite frequently. UNCOD considered that human impact was the main cause of desertification, and that the role of drought was rather like a catalyst, merely speeding up the long-term process of degradation that had been occurring before the drought began. Normally, droughts are relatively short-term phenomena, lasting for only a few years at most. Whereas droughts in other dry regions have come and gone since UNCOD, the drought in the Sahel has continued despite several years of relatively good rainfall (such as 1988). It is therefore possible that the region is actually experiencing a long-lasting decline in rainfall. The exact cause of this is not known, although a number of explanations have been proposed. Some experts claim that the drought has been prolonged by a change in the reflective properties of the ground surface caused by desertification in the region, while others see it as a consequence of a much wider change in global climate caused by such mechanisms as the "greenhouse effect".

Soil erosion and the removal of vegetative cover, the two main physical characteristics of desertification, are also described in Chapter 1. They are actually found throughout the tropics, not just in dry areas, and present a major challenge to governments concerned with finding a balance between economic development and the conservation of natural resources. Thus, although there are similarities between land degradation in the Sahel and that occurring in mountain areas like the Himalayas and tropical rain forest areas like the Amazon Basin, desertification in the drylands is distinguished by the involvement of drought, and by the very severe effects which it has on the environment and on human and animal populations.

One way in which desertification affects human beings is by undermining food production and contributing to malnutrition and famine. However, as will be discussed in Chapter 3, famine need not inevitably follow drought or desertification. We have seen in Ethiopia, Sudan and elsewhere that it can occur even in their absence when poverty, war, misguided government food policies and other factors prevent food shortages in one area from being compensated by supplies from another. UNCOD was certainly prompted by the famine that occurred in the Sahel region in the early 1970s, but it focused on the way in which famine and other forms of human misery were the result of a long-term process of degradation that was exacerbated by drought. UNCOD argued that short-term relief measures during and immediately after the drought were not enough to prevent such tragedies from recurring in the future. Also needed was a long-term commitment to sustainable agricultural development and environmental rehabilitation. That message is as true today as it was in 1977. Controlling desertification will not guarantee an end to famine, but it will make an important contribution towards that end.

The differing views on the role of climate in desertification referred to above are a predictable consequence of our quite limited knowledge of the subject, which allows a high degree of subjectivity to enter into the opinions of experts on dryland issues. Further scientific studies of the causes, effects and scale of desertification are therefore required so that judgements in the future can be based on fact rather than intuition. Current estimates of the overall extent and rate of increase of desertification are presented in Chapter 4. According to one estimate, the area suffering from at least moderate desertification is at the most 20 million square kilometres (sq km); but another gives an area of 32 million sq km, almost a quarter of the earth's land surface. The only estimate we have of the rate of desertification is 202,460 sq km per annum. All of these estimates are known to be very inaccurate, and lack of good data is a major constraint on the willingness of governments and international agencies to allocate funds for the control of desertification. It is vital to improve the monitoring of desertification by using the many sophisticated remote-sensing techniques, such as satellite imaging, which are at our disposal. Before this can be done, however, desertification needs to be defined much more rigorously in terms of measurable ground characteristics (called desertification indicators). This would have the additional benefit of helping to secure greater agreement on what desertification is and how it is caused.

UNCOD was much more than a forum at which representatives from ninety-five countries, fifty UN agencies and offices, eight intergovernmental organizations and sixty-five non-governmental organizations vowed to bring desertification under control. It agreed on a Plan of Action giving detailed guidelines to countries prone to drought and desertification to help them ensure that their agricultural production would be sustainable in the face of further climatic variations and would not lead to environmental degradation. The Plan also contained a list of priority recommendations for immediate action by governments and international agencies, and a set of ambitious transnational programmes in which governments of countries in a number of regions would work together to halt desertification.

The major contribution of the UNCOD Plan of Action undoubtedly lay in emphasizing that the key to controlling desertification was not to erect physical barriers against desert encroachment but to make land use more sustainable. It suggested various techniques and approaches by which rainfed cropping, irrigated cropping, livestock management and forest resource management could each be improved in order to prevent continued overcultivation, salinization and waterlogging of irrigated lands, overgrazing and deforestation respectively. The general aim was to improve both the productivity and the sustainability of each land use, while at the same time ensuring that it was only practised on the types of land appropriate to it. Thus, it was hoped that making rainfed cropping more productive on the better lands would prevent its expansion on to less fertile lands that were highly prone to soil erosion.

Chapters 5–8 outline the wide range of techniques that have been advocated (many of them at UNCOD) for improving the four major types of land use, and these are illustrated by examples from actual projects. Chapter 9 looks at techniques for soil conservation and sand dune stabilization that can be used to reclaim degraded land or to prevent degradation from ever becoming serious. The projects included in these chapters were identified in the course of a survey made by the author with the assistance of the UN Environment Programme (UNEP). The dominance of forestry and soil conservation projects does not reflect a bias on the part of the author, but merely the fact that a much larger number of projects appear to have been undertaken in these fields than in cropping and livestock raising. Most of the projects mentioned here were completed before 1987, when this survey took place. Ongoing projects have generally not been included because their eventual success or failure was difficult to predict.

The number of projects which have been successful in controlling desertification is remarkably small, and Chapter 10 shows how little has been done to implement the priority recommendations of the UNCOD Plan of Action or the co-operative, transnational programmes which it proposed. Why so little progress? The easiest answer, that the governments of both developed and developing nations have lacked the interest and the will to take action, is nevertheless quite a valid one. UNCOD came about because of a relatively short-term problem – the Sahel drought and its aftermath – but it proposed a long-term solution, attacking causes rather than symptoms. As often happens, when the symptoms appeared to go away, interest in a long-term cure waned. Another reason was that the Plan was very radical, challenging many traditional notions of agricultural development held by officials in governments and international agencies. Implementing certain of its recommendations would therefore require major shifts in policy. Furthermore, parts of the Plan were politically naïve, for while the large-scale international collaborative programmes which it proposed were highly principled and fully appropriate to the magnitude and seriousness of the problem, some of them were impractical because of the poor relations, or even enmity, which existed between countries that were required to work together.

Yet another reason for the lack of progress is that if techniques for improving land use are to be widely adopted they must be economically attractive to people living in the affected areas and compatible with their cultures. The Plan did include recommendations dealing with social needs, but these were not integrated with the techniques recommended for improving land use. As numerous examples in Chapters 5–9 demonstrate, when projects do not take into account the needs and wishes of local people they run into tremendous obstacles and are likely to fail. Some experts may claim that we have at our disposal all the techniques necessary to control desertification, but experience has proved otherwise. The social and policy components of projects are not optional extras or even supplementary to the basic techniques for resource management. They are absolutely essential, and need to be fully integrated with resource management if projects are to succeed. An element of "social engineering" is therefore required, as well as the introduction of new agricultural technologies. Arousing people's awareness and enthusiasm is a lot more difficult than performing fairly mechanical processes like sowing seeds or planting trees. Therein lies the challenge for controlling desertification and the reason why so many projects have failed. Despite the importance of social aspects, it was felt to be more convenient in this book to structure desertification control techniques and projects in the conventional way, by type of land use. The social components and policy priorities desirable for successful projects are therefore discussed in the context of each land use, although they are also brought together in summary form in Chapter 10.

Desertification is a problem that will not go away. It transcends the boundaries of nations, scientific disciplines and land uses. It challenges us to look afresh at the relationship between environment and development. It exposes the limitations of our ability to manage natural resources for human sustenance in some of the world's harshest climes. It reveals our woeful ignorance of a problem of global proportions. This book aims to describe a phenomenon about which we know surprisingly little, although we recognize it to be serious, widespread and complex, and to document the earnest efforts of those who are trying to bring it under control. In many instances questions are asked which cannot yet be answered, and obstacles are identified which we do not know how to overcome. If desertification is to be controlled, there is clearly much to be done.

1. What Is Desertification?

Desertification is the degradation of lands in dry areas. This chapter opens with a review of some key definitions of desertification, briefly describes the general distribution of drylands and the reasons for their occurrence, distinguishes between desertification and natural desert, and then introduces the two main physical characteristics of desertification – the degradation of soil and vegetation. Desertification is not the desert expansion of popular imagination. Instead it is essentially a subtle, dispersed and continuous process which mainly occurs far away from desert fringes, with the outright conversion of fertile land into desert only taking place in extreme cases.

The direct cause of desertification is poor land use. A long-term change in climate could make an area more desert-like without human intervention, but so far there is no firm evidence that this is taking place. Poor land use is not unique to dry areas, and consequently soil erosion and the removal of vegetative cover are major problems throughout the tropics. However, it is the severe environmental and human effects and the influence of low and erratic rainfall which make desertification such a distinctive and worrying problem. There are four main types of poor land use in the drylands: overcultivation, overgrazing, mismanagement of irrigated cropland and deforestation. These are briefly reviewed here as a prelude to a more detailed discussion in Chapter 2, and some examples are also given of desertification in the civilizations of the ancient world. The role of climate is then examined in detail. At UNCOD, with no evidence of long-lasting climatic change in the Sahel or other seriously affected areas, desertification was viewed as a long-term process of degradation which merely accelerated during occasional periods of intense drought. Since then, however, the drought in the Sahel has continued, and there is a growing recognition that global climate could be changing or about to change. So it is time for a reappraisal of the role of climate in desertification.

Although Chapter 1 focuses on the physical aspects of desertification, this does not detract from the importance of the human repercussions (such as famine) or the underlying social and economic causes (such as poverty), both of which are discussed in Chapter 3. Indeed, identifying the indirect causes of desertification (whether human or climatic) is extremely important if effective strategies are to be devised to prevent continued famine. Regrettably, we still have only partial understanding of many aspects of desertification. The major areas of uncertainty are referred to in the course of this chapter and are summarized at the end.

DESERTIFICATION DEFINED

UNCOD defined desertification as "the diminution or destruction of the biological potential of the land, and can lead ultimately to desert-like conditions", and called it "an aspect of the widespread deterioration of ecosystems under the combined pressure of adverse and fluctuating climate and excessive exploitation" (UN 1977). Desertification was also implicitly understood at UNCOD to be restricted to dry areas. Harold Dregne of Texas Tech University later expanded this definition by referring to desertification as "the impoverishment of terrestrial ecosystems under the impact of man ... the process of deterioration in these ecosystems that can be measured by reduced productivity of desirable plants, undesirable alterations in the biomass and the diversity of the micro and macro fauna and flora, accelerated soil deterioration, and increased hazards for human occupancy" (Dregne 1985).

Both these definitions stress that desertification consists of a decline in biological productivity or production potential, characteristic of a long-term process of degradation (or change in climate), rather than simply a short-term decline in the level of production that commonly occurs during a drought. Although crop yields will fall sharply in the course of a drought, the decline should only be temporary and is normally reversed when the drought ends. Both UNCOD and Dregne also placed desertification in the broader context of the "widespread deterioration of ecosystems", thereby preserving the intentions of the French forester Aubréville, who first used the term desertification in 1949 to refer to a general process of ecological degradation that started with deforestation, not necessarily in dry areas, and ended with land being turned into desert (Aubréville 1949). In emphasizing the role of human impact, rather than climate, Dregne's definition was more exclusive than UNCOD's, but simply stated more explicitly the understanding reached at UNCOD of the relative strengths of the roles of human impact and climate in causing desertification.

These definitions are too general to be used in estimating the actual extent and degree of desertification in a particular area. Until now, assessments of the scale of desertification have had to rely upon fairly simple criteria to distinguish between land which is "slightly", "moderately" or "severely" affected. We badly need a more rigorous definition of desertification in terms of measurable characteristics called "desertification indicators". Various sets of indicators, including such parameters as the annual rate of soil erosion, have been proposed and they will be reviewed in Chapter 4.

THE DISTRIBUTION OF DRYLANDS

The world's drylands are mainly found in two belts approximately centred on the Tropics of Cancer and Capricorn (23.5° north and south of the Equator respectively), although the width of each belt in degrees of latitude can be quite large (Fig. 1.1). Drylands constitute almost all of the northern half of Africa, southwest Africa, the Middle East, parts of India and Pakistan, Mexico, North America, the western coast and southern tip of South America, and a large part of Australia.

Many of these regions are dry as a result of global patterns of atmospheric circulation. Simply put, warm air rises at the Equator and then moves towards the cooler Poles to redistribute the surplus of solar energy received at the Equator. As part of this process the two sets of air currents subside slightly to the poleward side of the two Tropics at about latitude 30°. For rain to form, warm moist air has to rise from the ground and be condensed to water in the cool upper atmosphere. Since air in the vicinity of the subtropics is

Figure 1.1: Distribution of Arid Region Climates

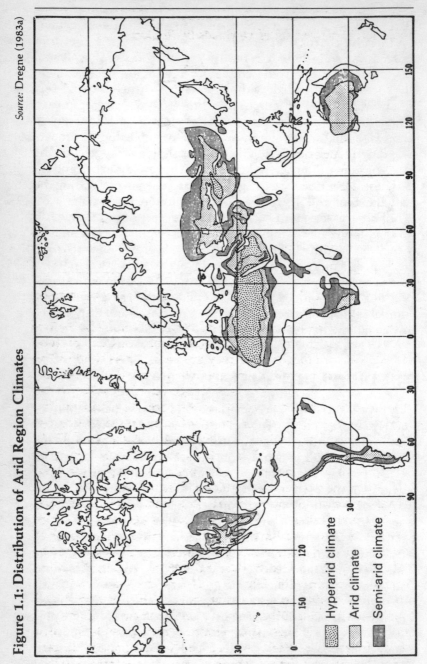

Source: Dregne (1983a)

Hyperarid climate

Arid climate

Semi-arid climate

Table 1.1: Distribution of Drylands by Region

	Area (million sq km)	% total
Africa	17.3	37
Asia	15.7	33
Australia	6.4	14
North America and Mexico	4.4	9
South America	3.1	7
Europe	0.2	0
Total	47.1	100

Note: Includes hyper-arid, arid and semi-arid zones.
Source: Dregne (1983a)

subsiding rather than rising, however, these regions receive relatively little rain.

Dry areas also occur because of other factors. The North American Great Plains and Prairies, and parts of Central Asia, for example, are dry because they are in the "rain shadow" of nearby mountains: rain-bearing winds deposit their moisture before crossing the mountains to the plains. Huge areas in the interior of the continents of Africa, Asia and Australia are dry because the moisture content of rain-bearing winds decreases the further they travel from the oceans. Coastal areas in Chile and Peru are dry because a cold northward ocean current nearby disturbs the movement of air and prevents precipitation.

The drylands cover more than a third of the earth's land surface, but are not evenly distributed (Table 1.1). More than 80% of their total area is found in just three continents: Africa (37%), Asia (33%) and Australia (14%). In contrast, the shares of North America and Mexico, South America and Europe are 9%, 7% and 0.4% respectively (Dregne 1983a). Africa, Asia and Australia account for fifty-five of the sixty-six countries affected by aridity in some way and for all the thirty-four countries with 75–100% of their national land area arid or semi-arid (Table 1.2). Of these thirty-four countries, eighteen are in Africa, fourteen in West Asia (the Middle East), one (Pakistan) in South Asia, and the remaining country is Australia (Paylore

Table 1.2: Distribution of Drylands by Country

Group	Description	Number	Percent of nation arid/semi-arid	Countries
1.	Arid	11	100	Bahrain, Djibouti, Egypt, Kuwait, Mauritania, Oman, Qatar, United Arab Emirates, Saudi Arabia, Somalia, South Yemen
2.	Predominantly arid	23	75-99	Afghanistan, Algeria, Australia, Botswana, Burkina Faso, Cape Verde, Chad, Iran, Iraq, Israel, Jordan, Kenya, Libya, Mali, Morocco, Namibia, Niger, North Yemen, Pakistan, Senegal, Sudan, Syria, Tunisia
3.	Substantially arid	5	50-74	Argentina, Ethiopia, Mongolia, South Africa, Turkey
4.	Semi-arid	9	25-49	Angola, Bolivia, Chile, China, India, Mexico, Tanzania, Togo, USA
5.	Peripherally arid	18	<25	Benin, Brazil, Canada, Central African Republic, Ecuador, Ghana, Lebanon, Lesotho, Madagascar, Mozambique, Nigeria, Paraguay, Peru, Sri Lanka, USSR, Venezuela, Zambia, Zimbabwe

Source: Paylore and Greenwell (1979)

and Greenwell 1979). Africa, Asia and Australia also account for about four-fifths of the total area that is at least moderately desertified (Mabbutt 1984; Dregne 1983b). Since Australia's drylands are so lightly populated the human impact of desertification is most severe in Africa and Asia, which contain more than four-fifths of all people affected by at least moderate desertification (Table 1.3), (Mabbutt 1984). This book therefore focuses mainly on Africa and Asia, where drylands and desertification are concentrated and the effects of desertification are most strongly felt by human beings.

Table 1.3: Areas and Numbers of People Affected by At Least Moderate Desertification by Region

	Affected area (1,000 sq km)	% Total	Affected population (millions)	% Total
Africa	7,409	37	108.00	38
Asia	7,480	37	123.00	44
Australia	1,123	6	0.23	0
Med. Europe	296	1	16.50	6
N. America	2,080	10	4.50	2
S. America & Mexico	1,620	8	29.00	10
Total	20,008	100*	281.23	100

Source: Mabbutt (1984)
*Due to rounding, the sum of these percentages is not 100.

THE ARID ZONES

The drylands of the world are divided into three climatic zones, termed the hyper-arid, arid and semi-arid zones in decreasing order of aridity. The simplest way to define the limits of each zone, and therefore to categorize aridity, is by the average amount of rain received each year. According to one scheme, hyper-arid areas receive less than 25 mm per annum; arid areas receive between 25 and 200 mm; and semi-arid areas receive between 200 and 500 mm (Heathcote 1983). For the purpose of rough comparison, most areas in western Europe receive on average between 500 mm and 1,000 mm per annum of total precipitation (as rain, snow, etc.), while areas near the Equator where the typical vegetation is tropical rain forest receive an annual rainfall of 1,800–4,000 mm and above. (To avoid confusion with the specifically defined arid zone, this book uses the term "drylands", not "arid lands", to refer collectively to all dry areas.)

Other schemes for classifying climatic zones use different rainfall limits. The UN Food and Agriculture Organization (FAO), for example, has used the following scheme: arid from 80–150 mm to 200–350 mm; semi-arid from 200–250 mm to 450–500 mm with winter rains but from 300–400 mm to 700–800 mm with summer rains (FAO 1985b). This takes account of the fact that rainfall in the drylands is generally not distributed uniformly throughout the year. In some drylands, such as those to the north of the Sahara Desert and in parts of West Asia, rainfall is concentrated in a winter rainy season; other drylands, such as those in India (the Thar Desert) and to the south of the Sahara, have a summer rainy season. FAO also includes within its definition of drylands part of another climatic zone, the sub-humid lands, which receive up to 800 mm of rainfall, although this zone cannot be easily defined in terms of rainfall alone (FAO 1985b). Before UNCOD, desertification was assumed to be confined to arid and semi-arid areas, but it was agreed in the course of the conference to include sub-humid areas, covering another 13 million sq km (Table 4.4). Desertification estimates prepared for UNCOD excluded sub-humid areas, and are not therefore directly comparable with later estimates.

The Sahel is the African region most closely associated with desertification. The name Sahel derives from a local word meaning "edge of the desert" (Grove 1978) and, although there is a diversity of usage, strictly speaking it refers to the semi-arid zone in West Africa that receives 200–400 mm of rain per annum, including parts of Senegal, Mauritania, Mali, Burkina Faso, Niger and Chad. The zone immediately to the south of the Sahel, called the Sudan Savanna, receives 400–1,000 mm of rain per annum and besides the six countries mentioned above it also includes parts of Gambia, Benin, Nigeria and Cameroon (Fig. 1.2). These two climatic zones extend eastward from Chad into Sudan and Ethiopia, two countries which, like the Sahel, have experienced many problems due to drought and desertification since the end of the 1960s. In this book the term Sahel refers only to the six West African countries in the 200–400 mm zone, and when a more general statement is required covering the whole semi-arid belt traversing Africa to the south of the Sahara, the phrase "and its extension into northeast Africa" is used to include Sudan and Ethiopia. (Somalia's climate, in this context, appears to be more closely associated with Kenya than with Sudan and Ethiopia.) The United Nations uses the term "Sudano-Sahelian region" to refer to the nineteen countries associated with the UN Sudano-Sahelian Office (UNSO). Besides those countries already mentioned it includes the Cape Verde Islands, Djibouti, Guinea, Guinea-Bissau, Uganda,

Figure 1.2: Arid Zones of West Africa

Source: Grove (1978)

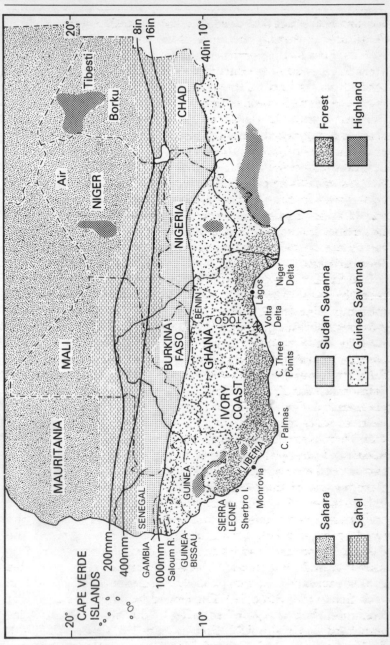

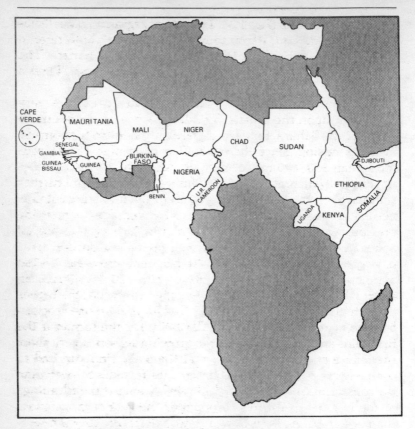

Figure 1.3: The Sudano-Sahelian Region

Source: Berry (1984a)

Kenya and Somalia (Fig. 1.3). Two other regional terms, "sub-Saharan Africa" and "Africa south of the Sahara", both refer to all African countries that lie to the south of the Sahara. The North African region includes Morocco, Algeria, Libya, Tunisia and Egypt.

The types of agriculture practised in dry areas are very dependent upon the climate. The driest lands on the southern fringe of the Sahara, for example, are only suitable for nomadic livestock raising, since the cultivation of rainfed crops needs more than 350–400 mm of rainfall. Intensive agriculture does not displace pastoralism as the dominant land use until further south where rainfall exceeds 600 mm. Even more rainfall is needed to ensure reliable crop production. Freeman, using previous work by FAO, classified Africa into agro-ecological zones according to the length of the growing season: arid areas have growing seasons of up to 75 days; semi-arid areas 75–180 days; and sub-humid tropical uplands 180–270 days (Freeman 1986; FAO 1978a). The longer the growing season, the higher the potential yield, the likelier it is that more than one crop can be grown, and the greater the flexibility for the farmer if the first rains are delayed. Where the growing season is very short the failure of the rains to arrive at the usual time can lead to complete crop failure. Irrigation enables farmers to overcome the constraints imposed by lack of rainfall and to take advantage of the high temperatures throughout the year to grow more than one crop. Egypt, although almost all its territory is hyper-arid, was able to build a great civilization based on irrigated farming that used water from the Nile.

Rainfall alone is a very inexact way of classifying climatic zones because it ignores the other factor that plays a big part in determining aridity: temperature. The higher the temperature, the more rain evaporates back into the atmosphere and the lower effective moisture is received by the land. A better definition of the boundaries of different climatic zones is therefore given by indices which combine both precipitation and temperature (as determined by the intensity of solar radiation received by a region). A number of different indices have been devised, such as those of Thornthwaite (1948), Meigs (1953) and Budyko (1958, 1974). Budyko's dryness ratio was defined as the number of times the mean net radiation at the earth's surface (R) in a year can evaporate the mean precipitation (P); the higher the ratio, the more arid the area. The dryness ratio D is easily calculated as $D = R/LP$, where L is the latent heat of vaporization of water. (Measuring L, the amount of heat required to evaporate a unit mass/volume of water, is a popular experiment in school science classes.) Putting the dryness ratio to practical use requires scientists to make arbitrary decisions about how it relates to ground conditions. In a map of the world's arid areas prepared for UNCOD, upon which Figure 1.1 is based, an area was designated desert if D was more than 10; arid, or desert margin, if D was between 7 and 10; semi-arid if D was between 2 and 7; and sub-humid if D was less than 2 (Henning and Flohn 1977).

NATURAL DESERTS AND DESERTIFICATION

At the heart of the drylands are five major zones of natural desert:

1. the Afro-Asian Desert, a great belt of desert from the Atlantic Ocean to China, including the Sahara Desert, the Arabian Desert, the Iranian Desert and the Touranian Desert in the southwest USSR (which includes the Kara Kum and Kyzyl Kum Deserts), the Thar Desert in Pakistan and India, and the Takla Makan and Gobi Deserts in China and Mongolia;
2. the North American Desert of the southwest United States and northwest Mexico, comprising the Great Basin, Mojave, Sonoran and Chihuahuan Deserts;
3. the Atacama Desert, a thin coastal strip between the Andes and the Pacific Ocean, running from southern Ecuador to central Chile; and the Patagonian Desert in Argentina to the east of the Andes;
4. the Namib and Kalahari Deserts in southwestern Africa;
5. the Australian Desert.

There is actually no strict definition of the term "desert", in terms either of average rainfall or any other characteristics. According to Jack Mabbutt the term carries at least three related connations: "empty of life", "waterless" and "unproductive" (Mabbutt 1985). John Cloudsley Thompson defined desert as a hyper-arid area receiving 25 mm of annual rainfall or less

Figure 1.4: Major Deserts of the World

Source: Based on Goudie (1984) and Dregne (1983)

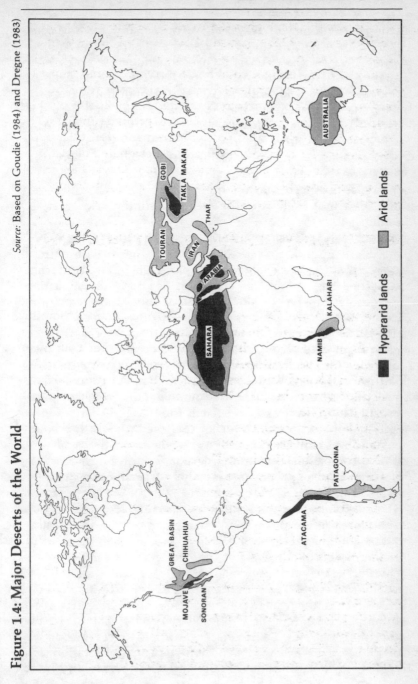

Arid lands

Hyperarid lands

(Cloudsley Thompson 1977). Only the Sahara, Takla Makan, Namib, Atacama, and Mojave Deserts, and part of the Arabian Desert, satisfy this criterion. The remainder are arid zones. Henri le Houérou, on the other hand, considered typical desert boundaries to coincide with the 100 mm rainfall limit, except for misty coastal deserts like the Atacama in Chile and Peru. He described the Sahara, Sinai, Arabian, Central Asian and Atacama Deserts as "true deserts" because of their characteristic rocky pavements and sandy plains with very limited perennial vegetation, but claimed that many so-called deserts in Central Australia, northern Mexico, the Kalahari and the southwest USA were not really true deserts but simply arid areas (le Houérou 1977).

One problem in using the term "desertification" to described land degradation is that it can lead to confusion with natural desert, creating misleading impressions such as: "land once desertified has to look like a natural desert"; "desertification irrevocably transfers productive land at some definite point in time into desert"; "desertification can only happen on the fringes of deserts"; "climatic change is necessary for desertification to occur"; and "deserts can expand of their own volition". Although reduced biological productivity is a key feature of desertified land the result may not necessarily look like the desert of popular imagination. Furthermore, desertification was clearly viewed in the UNCOD definition as a gradual process, rather than a single event in which productive land was converted into desert; thus desertified lands will show a complete spectrum of degradation, of which desert is but one extreme. This implies that desertification can be reversed as long as the process is still continuing and the extreme point (desert) has not been reached. Harold Dregne's experience with arid lands worldwide has led him to conclude that despite the presence of severe erosion in some areas "very little land has been irreversibly desertified as a result of man's activities" (Dregne 1985). Salinization of irrigated cropland is usually reversible (see p. 32), and increasing attention is now being given to the need to rehabilitate degraded irrigated areas. Human intervention may not be needed in all cases, for desertification may be reversed naturally once the factors contributing to it have been removed and vegetation can regenerate (Glantz and Orlovsky 1986). But the most important factor determining whether land is brought back into some form of productive use is economic feasibility. The reversibility of desertification is therefore probably best expressed as a function of technology, and of the prospective economic returns from the use of the reclaimed land relative to the cost of reclaiming it (Adams 1975).

Land on the desert fringe could in theory be desertified in various ways. First, desert sands could be carried on to adjacent land by winds, without any human involvement. Second, short-term adverse climatic conditions, such as drought, could lead to overcultivation, overgrazing and the consequent degradation of drylands on the fringes of deserts. These degraded lands could later be overrun by sand blown from the adjacent desert. Human impact is dominant in this mechanism, but is influenced by climate. Third, the short-term adverse climatic conditions could become so prolonged that a long-term reduction in rainfall occurs and, even without any human involvement, areas which were previously arid would become hyper-arid and in time take on the ecological characteristics of natural deserts. In practice the change in climate would also probably lead to intensified land use and consequent degradation, and so the roles of human impact and climate would again overlap.

The notion of deserts expanding of their own accord, by a combination of high winds and some kind of innate force within the deserts' rolling sand dunes (where these occur), was once widely held (Stebbing 1935), but is no longer regarded as important, although significant areas of what appear to be natural desert are thought to have been caused by human misuse. Some evidence for expansion due to overuse of land on the desert fringe came from measurements made in Sudan's Northern Kordofan province by Hugh Lamprey during a series of reconnaissance flights and field studies. These appeared to show that the southern border of the Sahara in 1975 was about 90–100 km south of its position in 1958, as indicated by a map made in that year, and therefore that the Sahara Desert was moving south at a rate of more than 5 km per annum (Lamprey 1975). Lamprey's study received much publicity at the time of UNCOD but proved controversial, receiving considerable criticism from other scientists who studied the area later and found no lasting changes (Helldén 1984, 1988; Olsson 1984).

The third mechanism, a long-term reduction in rainfall, was

not favoured at UNCOD, which took the view that because, on the evidence then available, drylands like the Sahel were not experiencing any long-term shifts in climate, there was no natural reason for desert boundaries to change; consequently, the paramount cause of desertification was human impact promoted in some way by adverse climatic conditions. The second and third mechanisms appear similar, but UNCOD stated quite categorically that desertification could happen anywhere in the drylands where human pressure led to degradation, not just on the fringes of deserts. A widely used metaphor at the time was that the deserts were not so much expanding as being "pulled out" by human actions far away from the boundaries of natural deserts. In recent years the continuing drought in the Sahel and growing concern about possible future trends in global climate (see p. 47) have given grounds for reconsidering the UNCOD position on the relevance of the third mechanism, for if the decline in rainfall does prove to be long-lasting in some areas of the drylands, then it could well cause the boundaries of natural deserts to shift.

THE TWO MAIN CHARACTERISTICS OF DESERTIFICATION

The two main characteristics of desertification are the degradation of soil and the degradation of vegetation. Dryland soils, because of their inherently low fertility, are particularly susceptible to erosion, especially when their vegetative cover has been removed or degraded. This section introduces the major types of vegetation and soils in the drylands and some of the mechanisms by which they are degraded. More detailed examples of degradation are given in Chapter 2.

Main Vegetation Types in the Drylands

Lack of water is a major constraint on the growth of plants. This is evident in the drylands, where low rainfall, often exacerbated by high temperatures, imposes limits on available soil moisture. Another important factor is seasonality, with rainfall being concentrated in one or two periods of the year separated by long dry seasons. The vegetation as a whole adapts to the general aridity by adjusting its density to the amount of water available.

Figure 1.5: Distribution of World Vegetation

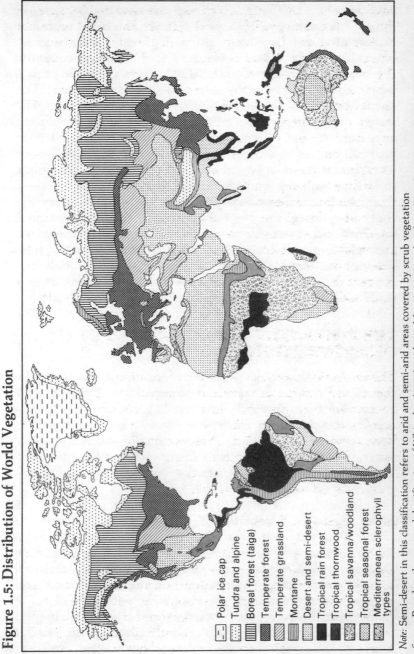

Polar ice cap
Tundra and alpine
Boreal forest (taiga)
Temperate forest
Temperate grassland
Montane
Desert and semi-desert
Tropical rain forest
Tropical thornwood
Tropical savanna/woodland
Tropical seasonal forest
Mediterranean sclerophyll types

Note: Semi-desert in this classification refers to arid and semi-arid areas covered by scrub vegetation

Sources: Based on the map and classification of Whittaker (1975), with modifications using data from Eyre (1968), Goudie (1984), Grainger (1986a) and Walter (1973).

The drier the area, the further apart plants grow. The types of plants which comprise dryland vegetation also show a number of adaptations which enable them to cope with the general lack of water and long periods without rainfall. One type, called the ephemeral plants, restrict their life cycle to the rainy period. These annual plants spring up when the rains come, then quickly produce large numbers of flowers, fruit and seeds. The cycle is completed within a few weeks. The seeds then lie dormant in the soil until the rains return the following year, when the process is repeated. Other types are perennial plants that survive by a variety of physiological adaptations. These include: growing long tap roots to reach supplies of water deep underground; having small leaves with hairy and waxy surfaces that reduce water loss by transpiration; adopting a deciduous habit – shedding leaves in the dry season to reduce water loss; and storing water in roots, stems and leaves (an adaptation shown by the cacti and other succulent plants).

Since the growth of vegetation is so strongly influenced by the climate, and in particular by average temperature and rainfall and the degree of seasonality, on a broad continental scale the distribution of the world's major vegetation types (or biomes) follows that of the major climatic zones (Fig. 1.5) (Eyre 1968; Walter 1973). Such a distribution assumes that there is complete freedom everywhere for vegetation to develop to the "climatic climax", the type most appropriate to a particular climatic zone. The actual distribution of vegetation at local level is more complex, being modified by site factors – such as soil, drainage, relief and altitude – and by human needs for settlements, agriculture, etc. With these reservations in mind, there are six main vegetation types in the drylands:

Deserts. Vegetation is sparse in true (hyper-arid) tropical deserts such as the Sahara, Arabian, Namib and Atacama Deserts, parts of Australia and southwest USA, and (in temperate latitudes) the Takla Makan Desert. Landscapes are dominated instead by sand, gravel, stony desert pavements, salt crust or naked rock. Nevertheless, some vegetation may be found in valleys, isolated depressions and gullies, where water collects preferentially. Away from the desert core, in arid zones (and in arid deserts such as the Iranian, Thar and Kalahari), the vegetative cover increases, typically comprising various kinds of grasses, low scrub (e.g., sagebrush – *Artemisia* spp.), succulent plants, and shrubs (e.g., *Acacia* spp. and *Tamarix* spp.).

Thorn woodlands. Semi-arid zones in the topics are covered by a wide range of vegetation types, characterized by thorny trees with small deciduous leaflets (commonly species of *Acacia* and *Prosopis*). Thorn woodlands vary in density and height from a scrubby mixture of stunted thorn trees, perennial shrubs and grasses to woodlands with an almost closed canopy. Succulent plants are common in the drier areas. Thorn woodlands are found extensively in Africa (e.g., in the Sahel), South America (e.g., the Brazilian *caatinga*), and India.

Savannas. The wetter semi-arid and sub-humid areas in the tropics are covered by various combinations of grasslands and trees or shrubs. The trees are 6–12 m high and have flattened crowns. Common genera include *Acacia* and *Eucalyptus*. In the driest areas the trees are widely scattered, but tree density increases (and grasses become less dominant) as annual rainfall increases and the length of the dry season decreases. Savannas are most extensive in Africa, but also occur in Australia, South America and Southern Asia. Large areas of savannas in Africa, however, are thought to have resulted from the clearance and burning of closed forest and open woodlands for agricultural purposes (see p. 28). Savannas with a high tree density, such as the *cerrado* of Brazil and the *miombo* woodlands of southern Africa, are also called "savanna woodlands" or "open woodlands". "Open" woodlands are distinguished here from "closed" forests (such as the tropical rain forests) where the canopy is relatively closed and covers a high proportion of the ground surface. Two-thirds of all open woodlands in tropical regions are found in Africa where they account for 69% of the total forest area, compared with 24% in Latin America and only 9% in Asia (Table 1.4) (Lanly 1982).

Temperate grasslands. Grasslands, generally without trees, cover large areas in the semi-arid zones in the interiors of the continents of North America (the Great Plains and Prairies) and Eurasia (where the grasslands are called "steppes"). They also occur in the southwest of South America (the "pampas" of Argentina and Uruguay).

Tropical seasonal forest. Closed forests of deciduous trees are found in the most humid tropical drylands. In Asia and Latin America these tropical dry deciduous forests then give way, with increasing rainfall, to tropical moist deciduous forests. Tropical seasonal forests are rare in Africa, presumably because of large-scale clearing and burning. African savannas therefore are found in close proximity to tropical rain forests, often with a distinct boundary between the two vegetation types.

Mediterranean sclerophyll vegetation. Areas in North Africa and Southern Europe that are close to the Mediterranean are covered by open woodlands of evergreen oak (*Quercus ilex*). The trees have small leathery leaves to reduce moisture loss in the hot dry summer. Beeches, pines, cedars and firs grow on the sides of mountains. As a result of clearance, grazing and burning over a long period, most of the oak woodlands have now been degraded to a low shrubby vegetation called *maquis* or *garrigue*. Other areas with a Mediterranean climate (winter rains and summer drought) are found in California, southwest Australia and South Africa. The last two areas have evolved distinct vegetation types. Evergreen oak forests are found in California but *chapparal*, a shrubby type of vegetation similar to *maquis*, is found as the natural vegetation in drier areas.

The Degradation of Vegetation

Degradation of vegetation occurs in the early stages of the

Table 1.4: Areas of Tropical Open Woodlands, 1980 (million hectares)

	Area (million ha)
Africa	486
Asia	31
Latin America	217
Total	734

Source: Lanly (1982)

desertification process, e.g. when deforestation makes soil more susceptible to wind and water erosion, but it also continues later in response to the decline in soil fertility and structure that follows overcultivation, overgrazing and poor irrigation management. The vegetative cover of an area may be said to be degraded when it becomes inferior to: (a) what the land could be expected to support, taking into account the climate, site conditions and historical experience; (b) what the area needs for the purposes of environmental protection.

Degradation of vegetation takes two main forms. The first involves a reduction in the overall density of vegetation cover, as represented by the biomass (the amount of vegetative material per unit area) and the proportion of land covered by vegetation. This reduction takes place when trees are cleared for cropping and grazing, cut down for fuelwood or fodder, or overbrowsed by livestock; and also when rangelands are overgrazed. The second form of degradation involves a change to a less productive type of vegetative cover, involving a modification in species composition, and possibly also in the general types of plants growing in an area. On overgrazed rangelands, for example, perennial grasses may be replaced by less palatable annual grasses and thorny, stunted shrubs, both of which are characteristic of the less productive ecosystems of drier climates. Both forms of vegetation degradation can also occur on overcultivated croplands, e.g. when the average density of vegetative cover falls owing to declining crop yields and shorter fallow periods, and (on irrigated croplands) when more saline-tolerant crop species have to be grown because of waterlogging and salinity problems.

The above definition of degradation is a pragmatic, but still inadequate, one. It recognizes that in the drylands, as in most other parts of the world, vegetative cover has been considerably changed by human activities and is found in a mixture of "natural" and agricultural environments. Actually quantifying the degree of degradation in the drylands, however, will not be an easy matter, and presents two major challenges. First, when formulating a system of desertification indicators (see Chapter 4), ways will have to be found to assess the relative merits of trees, shrubs, grasses and cultivated crops as vegetative covers, and develop an indicator (or group of indicators) which encompasses these main types of vegetation and accommodates the differences between them, for example the fact that although grasslands have a lower biomass per unit area than woodlands they also have a higher biological productivity (Whittaker 1975). Second, in order to estimate the degree of degradation it will be necessary to decide upon a "baseline" vegetative cover with which to compare the present vegetation. In natural environments the proper baseline would logically be the climatic climax vegetation. This is what would occur under conditions in which the vegetation in each area was free to develop to its ideal state, subject only to the constraints of climate and other natural influences. It corresponds to the general pattern of vegetation types described in the previous section. The present vegetative cover in the drylands, however, is quite different from the climatic climax, for large areas have been converted to crop cultivation, and the remainder, though consisting nominally of wild vegetation, has been greatly changed in the course of thousands of years of human intervention. In sub-Saharan Africa large areas of open woodlands were cleared and burned long ago for hunting, grazing and cropping, and since then the annual burning of grasslands to improve pasture growth has effectively prevented tree regeneration, and resulted in an artificially adapted vegetative cover of savanna grasslands with only a sparse covering of trees. The full development of vegetation to the climatic climax of open woodland is therefore prevented. Those trees which remain are predominantly fire-resistant and thorny to repel grazing animals, and their drought resistance is actually a secondary character (Monnier 1981; Eyre 1968). The barren hills around the Mediterranean are covered with a similarly degraded type of scrubby drought-resistant vegetation called *maquis*, the result of a long exploitation of the formerly rich evergreen forests which once covered the hills from sea to mountain top (Eyre 1968).

Main Soil Types in the Drylands

More than three-quarters of the world's arid and semi-arid lands consist of just two main types of soils: sandy soils (or entisols); and aridisols with a high mineral content, a low organic matter content, and a hard pan of mineral accumulation close to the surface (Fig. 1.6, Table 1.5). Aridisols developed

Figure 1.6: Distribution of Main Soil Types in the Drylands

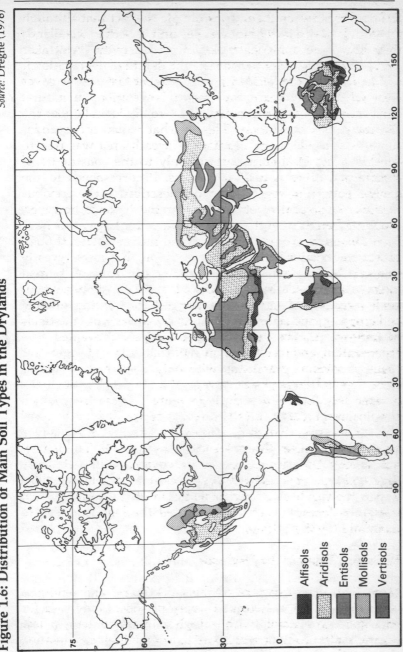

Source: Dregne (1976)

Table 1.5: Distribution of Main Soil Types in the Drylands by Region

	Africa %	Asia %	Australia %	N. America %	S. America* %
Alfisols	12	–	7	4	13
Aridisols	28	41	44	45	28
Entisols	58	34	37	8	41
Mollisols	1	20	–	41	17
Vertisols	1	5	12	2	–
Total	100	100	100	100	100

Source: Dregne (1976)
* Due to rounding up, the sum of these percentages is not 100.

where there was insufficient rainfall to leach soil constituents and deposit them in different bands at lower levels, as happens with most kinds of soils (Dregne 1976). Instead, the nutrients in aridisols remain fairly close to the surface. Because aridisols are rich in unleached nutrients and entisols are well drained, they have potential for cropping but their poor structure and low organic matter content make them susceptible to erosion (Heathcote 1983). Aridisols and entisols account for 86% of all drylands in Africa, the highest proportion of any region (Dregne 1976). Soils in sub-Saharan Africa are generally deficient in nitrogen and phosphorus, and have a low capacity to hold nutrients (Freeman 1986; Dregne 1982).

The three other types of soil found in the drylands, and particularly in semi-arid areas, are the alfisols, vertisols and mollisols. Alfisols are reddish-brown soils which are easily compacted and eroded, and are susceptible to surface crust formation. Vertisols, which are especially common in the drylands of India and Australia, have a high clay content and, though agriculturally valuable, are difficult to work. When wet they are very sticky, and during the dry season they become hard and develop deep cracks. Mollisols are temperate grassland soils common in Central Asia, and North and South America. They generally have a dark topsoil, owing to the accumulation of organic matter. Because they have been little affected by leaching they are rich in nutrients and therefore quite fertile.

The Degradation of Soils

Soil degradation occurs in four main ways: water erosion; wind erosion and compaction; and waterlogging, salinization and alkalinization. Normally, vegetation protects soil from being washed away by rains, and also from "splash erosion" caused by the direct impact of raindrops, the effect of which can be quite considerable in drylands receiving infrequent rainfall. The raindrops first disturb soil particles and then pack them together on the surface, sealing pores, decreasing infiltration (sometimes causing plants to die for lack of water) and increasing runoff. A more serious form of water erosion is "sheet erosion", in which fine layers of topsoil are washed away, removing soil nutrients and leading to declining yields unless the nutrients are replenished artificially. Rapid runoff also depletes soil moisture and further increases its vulnerability to erosion. Soil carried away by water may be deposited in irrigation canals and reservoirs, reducing their capacity and leading to floods which damage crops and settlements. If erosion is allowed to develop, then water flows concentrate in small channels called "rills", and these can become more and more pronounced until they develop into recognizable gullies. Small gullies, which often form along cattle tracks that create smooth channels for water runoff, can become as deep as the height of a man. In extreme situations they can grow into canyons as tall as a house. Gullies can be reclaimed but once they get too deep the land is lost for ever.

Wind erosion blows away the finer components of soil, such as silt, clay and organic matter (which contain most of the soil nutrients), leaving behind the less fertile sand, gravel and other coarser particles. In some areas, sands start to drift and sand dunes are mobilized, sometimes overwhelming nearby cropland and villages. Although sand dunes are often popularly associated with desertification they account for only a small proportion of all land affected by wind erosion. Strong winds can blow away detached soil particles, carrying them in dust storms that damage and sometimes kill crops by shredding foliage, and finally depositing the soil as sediment in rivers, lakes and irrigation channels. The *Harmattan*, a strong northeasterly dry-season wind, lifts dust from the Sahara into clouds up to 6,000 metres high (Morales 1977). More than 100 million tonnes of dust are blown westward into the Atlantic each summer from West Africa (Rapp 1986; Junge 1979) and some travels as far as the West Indies and South America. In August 1987, dust believed to have originated in the Sahara was deposited over a wide area of England from Southampton in the south to Darlington in the north (Anon. 1987c).

In the less serious form of soil compaction, called "surface crusting", high-speed mechanical cultivation or cultivation in the dry season turns crumbs of soil particles into a thin powder which, under the pressure of raindrops, is packed into a smooth hard surface crust. Complete compaction down to a greater depth occurs when soil with poor structure is compacted either by the wheels of heavy machinery or by the hooves of large herds of animals. Crusting and compaction make soil hard and less permeable. Runoff increases, leading to erosion, less water entering the soil for use by plants, and a less pervious soil in which it is difficult for plants to germinate and establish roots.

The salinization, alkalinization and waterlogging of soil result from poor management of irrigated cropping and of water supplies in general. Irrigating lands without paying proper attention to drainage, or applying too much water to crops, makes the soil waterlogged, and as excess water evaporates from the surface of the soil the salts dissolved in the water are left behind, either near to or actually on the surface. Soils with a hardpan near the surface are especially vulnerable to these forms of degradation because the hardpan impedes drainage. Saline and alkaline soils often occur in the same area, the formation of one or the other depending upon the mineral composition of the soil and the state of the ground water. Saline soils are often slightly alkaline but are distinguished from alkaline soils in having a high concentration of neutral chlorides instead of alkaline sodium carbonate; a low concentration of exchangeable sodium; a pH usually less than 8.5 (the pH of alkaline soils exceeds 8.5 and often is higher than 10); and reasonably good structure that makes leaching and reclamation of saline lands easier than for alkaline lands. Saline soils with a higher degree of alkalinity (called saline-alkali or saline-sodic soils) are also found. Salinization, alkalinization and waterlogging severely restrict the growth of plants. Yields drop, and more salt-tolerant crops like barley have to be grown, until eventually the land becomes unproductive and in extreme cases white "saline deserts" form. Because of the overlap between salinization and alkalinization it will be generally assumed in this book, for the sake of brevity, that where mismanagement of irrigated cropland leads to salinization, alkalinization can also occur, unless otherwise stated.

THE DIRECT AND INDIRECT CAUSES OF DESERTIFICATION

What causes soil and vegetation in the drylands to become degraded? UNCOD called desertification "an aspect of the widespread deterioration of ecosystems under the combined pressure of adverse and fluctuating climate and excessive exploitation" (UN 1977). It therefore acknowledged the contribution of both human impact and climate but regarded human impact (i.e. poor land use) as the more important of the two factors.

Experts still disagree on the role of climate, but it is convenient to think of it as being similar to a catalyst in a chemical reaction. The reaction would occur even without the catalyst but moves at a much faster rate when it is present. In the same way, drought creates the conditions whereby human impact on the land increases and the capacity of the land to tolerate it decreases. Crops and natural vegetation grow poorly, forcing people to crop and graze land more intensively to compensate for falling yields so as to produce enough food for their subsistence. This in turn depletes soil fertility and organic matter, and reduces the land's protective vegetative cover, already depleted by the effects of the drought on soil moisture. The result is increased degradation of soil and vegetation.

Seen in this way, drought is an "indirect" cause of desertification because it exacerbates poor land use, which is the "direct" cause of desertification, since it leads to the degradation of soil and vegetation. Drought is not the only indirect cause, for bad land use can also be the result of poverty, ignorance, greed, social and economic changes or misguided government policies. Distinguishing between direct and indirect causes in this way does not imply that the latter are any less important. On the contrary, it shows that programmes which are intended to control desertification may be unsuccessful if they fail to take the indirect causes into account. This was certainly the case with some of the projects described in Chapters 5–9. From another point of view, poor land use, and the desertification which results from it, may be thought of as symptoms of a more fundamental problem caused by drought and widespread poverty. Overuse of the land may well be inevitable when the poorest classes of people are relegated to very marginal areas; it becomes most pronounced, and its effects most apparent, in years of low rainfall when famine and malnutrition, two other symptoms of the deeper problem, may also occur.

UNCOD identified four main types of poor land use as direct causes of desertification: overcultivation, overgrazing, deforestation and mismanagement of irrigated cropland and water resources. Each of these degrades soil and vegetation in different ways. Ideally, pastoralism is the best and most sustainable use of the sparse vegetative cover of low-rainfall areas because livestock harvest the vegetation periodically and at relatively low intensity. If too many animals are concentrated in one area, however, either throughout the year (on pastures surrounding a village) or on a seasonal basis (around a borehole on a main trek route for nomadic herds) valuable perennial grasses are depleted and replaced by less nutritious annual plants, the density of vegetation is reduced, soil compaction occurs because of trampling by livestock herds, and soil erosion is encouraged.

Cropping has a far more intense impact on the soil because it requires complete clearance of vegetation, cultivation of the soil, the growth of crops, and often grazing of the stubble that remains after crops are harvested. The soil is therefore exposed to the elements for long periods each year. Cropping can be sustained in sub-humid areas and the wetter parts of semi-arid areas, but when it spreads to drier, more marginal areas extensive soil erosion can result. Traditional rainfed cropping systems often include extended fallow periods, during which vegetation can regenerate and soil fertility can be replenished. Overcultivation, either by reducing the fallow period or increasing the number of crops planted each year, reduces the potential for replenishing fertility and depletes soil organic matter. This causes a decline in the fertility, structure, permeability and water-holding capacity of the soil, and increases its vulnerability to erosion by wind and water. Organic matter also declines when crop residues are cut for animal feed instead of being ploughed into the soil, and when manure which should be spread on the fields is burnt as fuel. Inappropriate cropping can lead to soil compaction, in the form of either surface crusting, which occurs when high-speed mechanical cultivation powders the soil which is later compacted by rain, or the complete compaction of soil under the weight of tractors and other heavy machinery. The effects which the mismanagement of irrigated cropland has on the soil have been described on p. 32.

Deforestation both degrades the vegetation cover and makes the soil more vulnerable to erosion by subsequent over-cultivation or overgrazing. Trees play a crucial protective role in the drylands because they prevent the soil from being blown away by wind, and their roots lend cohesion to the soil and protect it from erosion by water. Numerous scientific studies have shown how devegetation can lead to increased soil erosion. Drylands covered by woodlands or thickets experience negligible erosion (Staples 1939) but when woodlands are cleared for cultivation soil erosion increases tremendously. One study in Tanzania, for example, found up to a hundredfold increase in runoff and soil loss after deforestation (Christiansson 1981).

DESERTIFICATION AND CIVILIZATION

Desertification is not a new phenomenon but has been occurring throughout human history. One of the most graphic examples of the effects of poor irrigation management is in the Tigris and Euphrates river basin (in present-day Iraq), where the Sumerian, Mesopotamian, Assyrian and Babylonian civilizations successively rose and fell over a period of more than 4,000 years. The Sumerian civilization, the first recorded in the world, developed in the southern part of the basin before 4000 BC. Reeds and other vegetation were cleared and a canal system built to drain the waterlogged lands. The canals later served to irrigate the crops grown on the alluvial soils, but were not sufficient to prevent the land becoming increasingly water-logged and saline over the next 2,000 years. The gradual rise in salinity is evident in the remains of crops found by archaeologists. In 3500 BC, equal areas of wheat and barley were cultivated but a thousand years later the saline-tolerant barley covered more than 80% of the farmland, and by 1700 BC no wheat was being grown in southern Iraq. Overall grain yields fell from an average of 2 tonnes/ha in 2400 BC to 0.7 tonnes/ha by 1700 BC. As Sumeria degenerated, splendid cities like Uruk and Ur, which had seen the invention of writing and mathematics, shrank to mere villages and were later abandoned to become ruins. The spirit of Sumeria survives in the Epic of Gilgamesh, perhaps the oldest story in the world, dating from before 2000 BC. It tells of a great flood that destroyed mankind except for a favoured family which survived by building an ark and gave rise to a new race when the flood subsided. The epic symbolizes the conquest of the wetlands and the threat which floods posed to the Sumerian irrigation system.

Upper Mesopotamia also had an irrigated system of agriculture but was not as prone to waterlogging and salinization. The Sassanid Persians conquered Mesopotamia in the third century AD and built a complex irrigation system that allowed almost all of the region to be cultivated. Populations continued to grow and a powerful centralized state was necessary to maintain the irrigation system and prevent it from becoming clogged with sediment. The system started to break down owing to lack of attention in the middle of the eighth century, before the Islamic invasion. Despite a recovery immediately afterwards, degeneration continued and by the twelfth century the irrigation system was in a state of complete disrepair, and populations had plummeted. Today, much of the land that once supported these great civilizations remains abandoned, the rest is so poor that crop yields are among the lowest in the world, and some 20–30% of all land with potential for irrigation in Iraq is unusable.

Further west, the hills around the Mediterranean were once covered with rich evergreen forests from sea to mountain top, but the depredations of thousands of years of civilization have left behind barren hills and a degraded drought-resistant scrubby vegetation called *maquis* (Eyre 1968). Mount Lebanon was described in the Epic of Gilgamesh as a vast green mountain with tall cedars. Felling of the cedars of Lebanon began as early as 3000 BC, after which they formed the cornerstone of the Phoenicians' international trade. The Egyptian Pharaohs, for example, imported forty ships full of cedar in 2600 BC. Fifteen hundred years later the temples and palaces of Assyria, and subsequently Babylon, were built of cedar extorted as tribute from a subjugated Lebanon. King Solomon used the wood for his temple in Jerusalem, and Alexander the Great made his Euphrates fleet out of cedar in the fourth century BC. Not until the Roman emperor Hadrian in the second century BC was any attempt made to protect the remaining trees. Greece also cleared large areas of hill forests for timber for ship building, fuel or grazing land. The early deforestation of Attica was commented on by Plato in the fourth century BC: "Our land, compared with what it was, is like a skeleton of a body wasted by disease." Athens was forced to build a vast commercial fleet to trade wine and olive oil for wheat and other items of food that its eroded soils could not supply.

DROUGHT AND DESERTIFICATION

Our present understanding of the role of drought in desertification is still inadequate, having been considerably determined by recent history. Because concern about desertification was mainly a response to the first phase of the Sahel drought in the early 1970s, it is quite understandable that many people confused the two phenomena. When rainfall appeared to return to long-term average levels in the Sahel in the mid-1970s, some governments assumed that desertification would also automatically diminish and that programmes to control it were no longer of great importance. The upturn in rainfall also meant that experts advising UNCOD were unable to produce any evidence for a long-term downward trend in rainfall, and consequently UNCOD focused on the human causes of desertification, largely consigning climate to a secondary role. Droughts were seen as limited, albeit frequent, periods of below-average rainfall which promoted poor land use and therefore catalysed the long-term process of desertification, as already discussed.

After UNCOD, rainfall in the Sahel and its extension into northeast Africa declined again, and at the end of 1988 the drought was regarded as still continuing. The prolonged drought has led to speculation as to whether the region is indeed experiencing a long-lasting change in climate, and to some dissatisfaction with the view of the role of drought in desertification adopted at UNCOD. This section looks at droughts and the reasons for their occurrence in dry areas, gives examples of recent droughts and their impact on drylands in the Sahel and other dryland areas, and summarizes some of the most promising explanations as to why the Sahel drought is still continuing. It shows that, except in the Sahel, droughts have generally remained short-term phenomena in recent years. Although there is still no conclusive evidence of a long-term change in climate in the Sahel, it is possible that desertification could be prolonging the drought, and that feedback linkages are drawing the two phenomena ever closer together. There are also suspicions that the prolonged drought could be covered by the "greenhouse effect", which is expected to lead to a long-term change in global climate.

Aridity, Seasonality and Drought in the Drylands

Aridity is the condition of areas that on average receive low amounts of annual rainfall and also usually experience high temperatures. The different categories of aridity have been described earlier in the chapter. One of the characteristics of the arid zones is the great variability of rainfall, and this is apparent in two ways. First, most rainfall occurs in one or two limited seasons of the year, and even within these seasons its exact occurrence is not predictable. Second, the level of annual rainfall is highly variable from year to year, and it is quite usual for several "wet" years, with above-average rainfall, to be followed by several "dry" years, when rainfall is below average. A sequence of dry years is called a drought.

Most rainfall in the drylands occurs in quite short rainy seasons separated by dry seasons in which there is little if any rain. In the Sahel, for example, 80% of annual rainfall is typically received between July and September, compared with less than 3% during the dry season which lasts from November until April. Ethiopia has an additional minor rainy season between late March and May (Lamb 1985a). The alternating rainy and dry seasons dominate dryland farming systems, and crops are planted to coincide with the onset of the rainy season. Michael Glantz, of the US National Center for Atmospheric Research in Boulder, Colorado, thinks that the importance of the seasons in dryland areas has been very much underestimated, especially with regard to food shortages: "Preharvest shortages are a common occurrence in many African societies. Such shortages have been referred to as periods of seasonal hunger ... [and] paradoxically occur during the wet season and become acute just prior to harvest" (Glantz 1987a).

In contrast, droughts are extended periods of below-average annual rainfall. They are not unique to the drylands, although in these areas a drought can have devastating social, economic and environmental consequences. Nevertheless, the occurrence of drought is a normal part of life in dry areas, and farmers have traditionally planned their operations accordingly. When a drought becomes longer or more severe than usual, problems arise, and humans and animals may die from lack of food or water. This was the case, for example, in the years between 1968 and 1972 when the amount of rainfall received in the Sahel was, on average, only half of the long-term mean annual rainfall for the period of 1908 to 1956 (Lamb 1979).

Droughts are an aspect of the extreme variability of annual rainfall in the drylands. Generally, the lower the annual rainfall the greater the variability, and quite often a sequence of "wet" years is followed by a sequence of "dry" years. As an extreme example of variability, the city of Cairo has an average rainfall of 28 mm but only received rain in thirteen of the thirty years from 1890 to 1919, in which year it had 43 mm in one day (Gautier 1970). One way to compare the variability in different areas is by using the coefficient of variation, calculated by dividing the standard deviation of rainfall about the mean for a given period by the mean rainfall. A high value of the coefficient indicates a high variability. For Western Europe and tropical rain forest areas like the Amazon where rainfall is fairly regular the value of the coefficient is normally less than 15%, but for most drylands the value is over 25%, and in the driest (hyper-arid) areas it exceeds 40% (Trewartha 1968).

There is no universally applicable definition of drought (Lockwood 1985), although at least 150 definitions have been attempted (Barry and Chorley 1987). The deficiency in rainfall compared with the long-term average, the length of time without rainfall, and the length of time with below-average rainfall are naturally among the factors to be considered. However, the difficulties of definition arise because whether or not a lack of rainfall constitutes a drought depends very much on local conditions and the particular needs for water which prevail. In some areas in Western Europe, for example, even a few weeks without rain at certain times of the year may constitute a drought, while a dry season typically lasting for five or six months is quite normal in West African countries. In India a reduction in rainfall of 25% is required before a drought can be declared (Indian Planning Commission 1973), while other countries apply different standards. Finally, there will always be debate between experts about which period should be chosen for the calculation of the long-term mean rainfall.

Types of Drought

As a way of overcoming some of these difficulties of definition, three main types of drought are recognized: meteorological drought, in which annual rainfall is below average for a year or more; agricultural drought, in which the rains fail during the crop-growing season; and hydrological drought, in which the flow of rivers is too low to supply the needs of crops grown on adjacent irrigated lands or people's drinking water needs. The distinction between meteorological drought and agricultural drought is an important one, because an area may receive close to its average annual rainfall yet crop production can still be severely affected if the rain is not received during the growing season and at critical times in each crop's growth cycle (Glantz 1987a). In the Indian state of Karnataka, for example, the average sorghum yield in 1966, when rainfall occurred at a time unfavourable for the crop, was only about half of the yield in 1969 when a similar total annual rainfall was received but at a more favourable time (Parry and Carter 1988).

The severity of a drought will depend upon its spatial extent. A country should be able to cope with droughts that are localized to certain districts, but droughts that affect one or more provinces are likely to give rise to national emergencies because of their impact on food production and living conditions. Regional droughts that affect more than one country are even more serious. A study of three districts in central Kenya, for example, showed that, on average, local droughts occur every year, droughts affecting a few districts occur once every three to five years, and national droughts affecting more than one province occur once every ten years (Downing et al. 1988).

Some Recent Droughts

Historically, droughts have occurred frequently in the drylands but have usually only lasted for one or a few years before the rains have returned. The Sahel drought has continued now for twenty years. The crucial question as far as desertification is concerned is whether other regions of the world have been affected in a similar way.

Drought has certainly been a prominent feature of the 1980s, much as in the previous decade, with major droughts in southern and eastern Africa, northern India, northeast Brazil, the USA, Australia, and even places like the island of Borneo which normally receive large amounts of rain all the year round. A drought in the USA in 1983 reduced corn yields by 29% compared with the previous year. Total production actually fell by 50% because the government had previously ordered a cut in planting (Parry *et al.* 1985). In 1988, a combination of low rainfall and high temperatures caused the worst drought in the USA and Canada since the Dust Bowl era of the 1930s. US corn production was forecast to fall by 30 to 50% and spring wheat was also seriously affected. Total grain production in Canada was expected to decline by a third to 30 million tonnes, with a 41% fall in wheat production (Dunne *et al.* 1988; Owen 1988).

Northeast Brazil accounts for 18% of the country's land area and contains 29% of its population. Ironically, though immediately to the east of Amazonia, more than half of the region is semi-arid. There have been at least ten extreme droughts and thirty-three partial droughts in the last 400 years. Extreme droughts cut crop yields by about 60% and appear to be occurring with increasing frequency. There have been five such droughts this century, the most recent in 1983 when rainfall was only 43% of the long-term average and the lowest for twenty-five years. This was the fifth successive year of below-average rainfall, and followed partial droughts in 1979, 1980 and 1981. In 1979 and 1980, when rainfall was only 69% of normal, production of beans fell by 72%, corn by 82%, rice by 52% and cotton by 70% in comparison with 1978, when rains were about average (Magalhães and Rebouças 1988).

India has suffered a number of droughts in recent years. For example, there were eight droughts in the Jodphur area in the western state of Rajasthan between 1941 and 1970. However, no long-term decline in rainfall has been detected, either in the country as a whole or in individual regions (Mooley and Parthasarathy 1984). Over a period of sixty years, drought affects 25% of India once in every two or three years, another 37% once in every four years, and a further 31% once in every five years (Gadgil *et al.* 1988). India's most recent drought was in

1987, but even though this was the worst drought in 125 years, it was only the second extreme drought to hit the country since it became independent in 1947. Although the failure of rains affected at least half the country's cultivated lands, grain production was still only 11% lower than the record 152 million tonnes produced in 1983–4 (Anon. 1987b; Anon. 1988a). Fortunately, rainfall was plentiful in 1988, and a new record grain production of 172 million tonnes was in 1988–9 (Sharma 1989).

Australia, like other countries, suffered a major drought in 1982. This cut national wheat production by 37% compared with the average for the previous five years. (Wheat accounts for 60% of the total cropped area in Australia and is typically grown in areas receiving between 250 and 635 mm of rainfall.) The fall in wheat production would have been even larger if Western Australia had been affected in the same way as states like Victoria, where production was only 16% of the average, and New South Wales, where the proportion was just 29%. Between 1864 and 1984 Australia experienced at least ten major droughts, which affected more than 20% of the country, and another eight less severe (but widespread) droughts. Six of the major droughts have occurred since 1950 – in 1951, 1957, 1961, 1965–7, 1972 and 1982–3. Some states are more subject to droughts than others, e.g. Western Australia expects droughts in fifty-four out of every 100 years, while Tasmania might have only twenty-six droughts in the same period (Hobbs 1988; Hobbs *et al.* 1988).

Outside Africa, therefore, droughts continue to occur frequently but they only last for one or at most a few years, maintaining the general historical pattern. In some regions it appears that the frequency of drought is increasing but, owing to the general lack of reliable weather records in previous centuries (and the first half of this century too), it would be unwise to attach too much importance to such comparisons.

The Sahel Drought

During this century, droughts have also occurred frequently in the Sahel region of West Africa and its extension into northeast Africa. But in 1968 a drought began in these two regions which, twenty years later, still shows no definite signs of ending, despite better rainfall in 1988. (In Ethiopia the drought was interrupted during the second half of the 1970s but has since returned.) The Sahel drought plays a central role in any discussion of desertification. It was this which originally prompted the United Nations Conference on Desertification in 1977 and, together with the similar pattern of drought in northeast Africa, over the last twenty years it has had a massive impact on the African drylands and the people who live there.

The previous drought in the Sahel occurred during the 1940s, when there were six dry years, three of them "moderately severe", according to Peter Lamb, of the Illinois State Water Survey. Meteorological records before 1940 are fewer and less reliable but available data on rainfall and the levels of rivers and lakes suggest that the years 1919, 1921 and 1926 were dry, as were the early 1910s, with 1913 probably having a rainfall deficiency similar to 1972. There is also evidence of severe droughts in the periods 1820 to 1840 and 1736 to 1758 (Lamb 1985a). An analysis of the rates of flow of the rivers Senegal and Niger, by Anders Rapp of the University of Lund in Sweden, showed that the levels of both rivers fell sharply between 1910 and 1914, and also in 1940 and 1944 in the case of the Senegal River, confirming the occurrence of two other major droughts this century (Rapp 1974; Rapp et al. 1976).

In 1968 the Sahel rains were early and heavy, but then stopped at the beginning of May. Seedlings died before the rains returned in June. By the end of the dry season in early 1969 animals were beginning to die of hunger. The rains failed again in 1970, with the farmers and herdsmen who had moved furthest north during the period of good rains being the most affected. When the small 1970 harvest was exhausted, an estimated three million people in the six nations of the Sahel needed emergency food aid. The 1971, 1972 and 1973 rains were also below average, with 1972 thought at the time to be the driest year this century, although it was not until 1973 that the true extent of the disaster became widely recognized. Between 100,000 and 250,000 people died in the region as a result of the drought, according to an UNCOD report. Another estimate, of 50,000–100,000, was made by the Club du Sahel, which co-ordinates the work of donor nations in the region, and CILSS (the Permanent Interstate Committee for Drought Control in the Sahel), which represents the needs of Sahelian countries to the donors (see p. 302). Animals died in their millions, with cattle and sheep suffering the most. According to

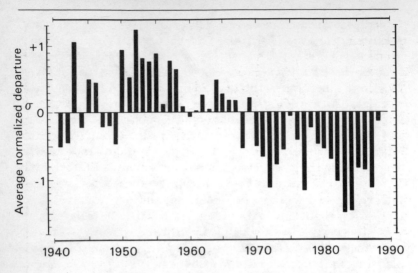

Figure 1.7: Trends in Annual Rainfall in the Sahel Region, 1941–88

Note: based on observations at twenty rainfall monitoring stations in the Sahel region and expressed as the yearly average of the normalized April–October rainfall departures from the long-term mean rainfall.

Source: Lamb (1987), updated by the original author

FAO, an estimated 3.5 million head of cattle, 25% of the total, died in the Sahel in 1972–3 alone.

Although rainfall improved between 1973 and 1975, it has since remained below the average for this century, and the average annual rainfall between 1977 and 1981 was no higher than that for the period from 1968 to 1973. The rainfall then dropped sharply in the early 1980s, and 1983 and 1984 were even drier than 1972 (Fig. 1.7) (Lamb 1982, 1985b, 1986, 1987). The drought became evident in the flow of rivers in West Africa: the Senegal, Niger and Chari rivers each experienced a severe fall in runoff and the level of Lake Chad showed a systematic decrease after 1963 (Sircoulon 1983). Rainfall trends in Sudan have followed a similar but not identical pattern to that in the Sahel since the late 1960s. Rainfall there has remained below average, except in 1978, and the lowest rainfall this century was received in 1984 (Fig. 1.8), (Hulme 1989a). Ethiopia

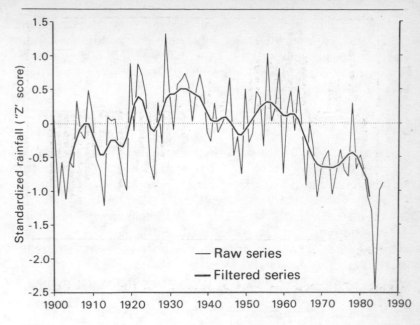

Figure 1.8: Trends in Annual Rainfall in Central Sudan, 1900-1986

Note: based on observations at twenty-six rainfall monitoring stations and expressed relative to the long-term (1921–80) mean rainfall.

Source: Hulme (1989a)

was similarly affected by drought in the early 1970s, but this was followed by a period of good rainfall between 1974 and 1978. Drought returned in 1981, however, and was still continuing at the end of 1988.

The Sahel drought continued during the second half of the 1980s. In 1988 there were three months of regular and heavy rainfall in the summer and excellent crop yields were obtained, although some parts of Burkina Faso, Chad and Mali experienced problems due to flooding (Toulmin 1988; UK Meteorological Office 1988). Sudan also received heavy rains, and these caused massive flooding in Khartoum (Pearce 1988). The Sudan rains were more highly concentrated than those in the Sahel, with some 200 mm falling in only thirteen hours during the night of 4 August (Hulme 1989a). Although it was encouraging to see so much rainfall in the Sahel and Sudan in 1988, the seasonal total was only comparable with the long-term average rainfall for 1951 to 1980, so there is no definite evidence that the drought has ended and a new wet period begun (Hulme 1989a).

The Debate About the Sahel Drought

Since the late 1970s, there has been much debate as to whether or not the Sahel drought has ended, although by now its prolonged nature has been generally recognized by most climatologists. According to Kenneth Hare, of the University of Toronto, Canada, "the sense that the Sahelian drought ended in 1974, and with it the need for political concern, was illusory. Drought has continued through much of the Sahel and has spread at times into parts of East Africa. The past five years have also been dry in many parts of South America, and in some regions of inner Asia" (Hare 1983).

Not all experts agreed with this view, however. Michael Glantz and Richard Katz at the US National Center for Atmospheric Research warned against claims of a "17 year drought" between 1968 and 1985.

Recent weather tends to influence perceptions more heavily than earlier weather, and wet spells more heavily than dry ones. On this basis it has been observed that the inhabitants of the Sahel have adopted practices during the abnormally wet years which end in disaster during the dry years. After a period of abnormally high rainfall in the Sahel which ended in 1965, the inhabitants considered a return to average conditions for rainfall to be a drought. (Glantz and Katz 1985; Glantz 1987b)

Michael Glantz also pointed out that levels of agricultural production achieved in the second half of the 1970s and early 1980s did not support the assumption that conditions remained equally adverse over that period (Glantz 1987b).

There were three reasons why low rainfall in the Sahel in the late 1970s and early 1980s was not translated into reduced agricultural production. First, rainfall varied considerably in different parts of Africa so that not all countries were affected in the same way. Poor rains in Mali in 1983, for example, followed two years of average to above-average harvests, and only affected marginal areas so that national cereal production actually increased slightly (US Department of Agriculture 1984). Second, despite the general shortage of rainfall, this meteorological drought did not have as much impact on crop growth as that of the early 1970s, because it did not translate into agricultural drought, i.e. a shortage of rain specifically in the period when growing crops need it most (Glantz 1987b). Third, agricultural practices seem to have been adapted to handle the 1983–4 drought much better than they did in the very dry years between 1971 and 1973.

The fact that the present drought there has lasted so long clearly sets the Sahel apart from other dry areas, although the pattern of rainfall this century suggests that Sahelian droughts do generally tend to be quite persistent, with both wet and dry periods lasting for between ten and eighteen years (Barry and Chorley 1987; Hulme 1989b). The century began with twenty years of dry weather from 1900 to 1920; this was followed by a wet period in the 1920s and 1930s, a dry period in the 1940s, a return to wetter conditions in the 1950s and early 1960s, and then another dry period which began in the late 1960s and continues to this day (Hulme 1989a). The average length of droughts in different regions can be compared by means of a persistence ratio. This is calculated by dividing the expected frequency of runs of years with either above- or below-average rainfall by the observed frequency of such runs. In areas where both droughts and wet periods last longer than expected there will be, in a given period, fewer runs of dry and wet years and so the value of the persistence ratio will be greater then 1.00. Based on the twentieth-century rainfall record, the Sahel has a persistence ratio of 1.80, compared with values of 0.84 for the northern Kalahari, 1.17 for Brazil and 1.25 for Sudan (Nicholson 1983a; Lamb *et al.* 1986; Hulme 1989b).

The Possibility of a Long-Term Climatic Change in the Sahel

Does the continuing drought in the Sahel mean that a long-term climatic shift is taking place in the African drylands? Experts seem to be divided on the question. During the 1970s some believed that a major climatic change was occurring while others felt that the lower rainfall was simply part of the variation in rainfall common to the drylands. In the absence of hard evidence UNCOD was sceptical about a long-term reduction in rainfall, advised by its chief climatological consultant, Kenneth Hare of the University of Toronto, that: "There is no firm basis for claiming that the extreme weather events of the 1970s are part of a major climatic variation" (Hare 1977). More recently, a study published in 1983 by the US National Research Council concluded that the Sahel had experienced little long-term climatic change in the last 2,500 years and that high variability was a fact of life in that region (NRC 1983).

On the other hand, the work of Peter Lamb at the Illinois State Water Survey shows a clear trend towards drier conditions in the Sahel since 1952 (Lamb 1986). Since that year was the peak of the last wet season, the trend does not begin earlier and so this does not provide convincing proof of a long-term change in climate. In 1983 Kenneth Hare prepared a revised assessment for UNEP in which he reported that the view still dominant among his fellow climatologists was that current climatic events are part of a natural fluctuation due to an oscillation in the general circulation of the atmosphere, and "there is yet no way in which climatologists can decide whether this desiccation will continue." However, his personal view was that: "The possibility of a permanent desiccation of the dry belt climates of Africa cannot ... be ruled out" (Hare 1983, 1984).

The Causes of Drought

Before considering why the Sahel drought has lasted for so long, it is important to take a brief look at the underlying causes of droughts in general, and those which have occurred in recent years in particular. The high degree of variability in rainfall in the drylands results generally from the unreliability of the mechanisms by which they receive rainfall. Because each of the world's main dryland regions relies upon different mechanisms, the immediate reasons for drought vary with the region concerned. In the Sahel, for example, drought is related (among other factors) to a southward expansion of subtropical areas of high atmospheric pressure. This reduces the northward penetration of the rain-bearing monsoon winds into West Africa. Droughts in India are caused by a change in the path of the crucial monsoon winds so that they do not penetrate far inland, and in northeast Brazil to changes in atmospheric circulation resulting from the diversion of ocean currents. The causes of severe and persistent droughts, however, are likely to be more complex (Barry and Chorley 1987).

Despite such differences, atmospheric processes in one region are related to processes in other regions through the global pattern of atmospheric circulation. Inter-regional linkages, called "teleconnections", have been blamed for many, but not all, of the major droughts in recent years. As long ago as 1923 it was discovered that air pressure tends to vary between the eastern and western parts of the Pacific Ocean over a period (Walker 1923). Called the Southern Oscillation, this was found in the late 1960s to be related to variations in rainfall and sea-surface temperature in the eastern Pacific near the Equator, and in particular to higher than average sea-surface temperatures off the coast of Peru, called the "El Niño" ("The Boy") phenomenon because it occurs around the Christmas period (Bjerknes 1966, 1969, 1972). The warming phenomenon occurs each year, disturbing cold ocean currents along the western coast of South America, but once in every 2–10 years it becomes more extensive and has a major impact on fisheries in the region and also on global atmospheric circulation (Barry and Chorley 1987). A variety of explanations have been proposed for the El Niño, one of the most recent being that anomalous eruptions of lava from undersea volcanoes and cracks in the Pacific Ocean floor warm the ocean and atmosphere in the eastern Pacific region, although whether this is the primary cause of the phenomenon is still not certain (Anon. 1988b).

Statistical analysis of historical data has established significant relationships between El Niño events and droughts in a number of countries (Kousky *et al.* 1984; Mooley and Parthasarathy 1983; Fleer 1981). The major droughts in 1982–3 in Africa, India, northeast Brazil, the USA, Australia and Indonesia, mentioned on pp. 40–42, coincided with the most significant El Niño event so far (Parry and Carter 1988). A smaller El Niño event in 1986–7 was associated with drought in Ethiopia; and it has been suggested that the heavy rains in Sudan in 1988, which caused massive flooding in Khartoum, resulted from the opposite phase of the El Niño cycle (Pearce 1988). The links between drought and El Niño events are not necessarily simple ones, for drought may not occur until one or two years after an El Niño event. It would also be mistaken to ascribe all droughts to El Niño events. For example, although poor rainfall in the northern part of northeast Brazil seems to be related to El Niño events, the link is not as strong in the south (Nobre and Molion 1988), and while most of the Australian droughts since 1950 coincided with El Niño events, that in 1961 did not.

Proposed Explanations for the Sahel Drought

Although Africa is clearly not immune to the impact of El Niño events, some other explanation must be found for the prolonged drought in the Sahel and its extension into East Africa. Looking at the various proposals which have been put forward may also provide some insight into whether the continuing drought represents a long-term climatic shift.

Global linkages

Two hypotheses have been proposed to show how dry conditions in Africa could be part of climatic changes on a global scale. (A third possibility, that the drought is linked to the "greenhouse effect", is discussed in detail later.) The first hypothesis blames a cooling of the land masses of the northern hemisphere by about 0.3°C between 1945 and the early 1970s, caused by an increase in atmospheric dust from air pollution and volcanic eruptions. The middle and high latitudes experienced a greater degree of cooling than elsewhere, and this led to a shift in atmospheric circulation. According to Peter Lamb, this hypothesis suggests that: "[This cooling] forces south other components of large-scale atmospheric circulation and, in effect ... squeezes out the northward moving monsoon that ordinarily brings sub-Saharan rain" (Lamb 1986). Evidence in support of this hypothesis is sketchy, and the hypothesis itself seems to be contradicted by the heavy rains which fell in the Sahel during the 1950s when the northern hemisphere cooled rapidly, and by a marked warming in the northern hemisphere in the early 1980s when the Sahel drought became more intense than ten years previously (Lamb 1986).

The second hypothesis links the Sahel drought with changes in sea temperatures in the tropical Atlantic, the source of the southwest monsoon winds which bring rain to the region.

These changes would tend to reduce the northward penetration of the monsoon into West Africa and cause a decline in rainfall in the Sahel. Some support for the hypothesis is given by the higher sea-surface temperatures of the southern oceans that were observed during a number of major drought years (Lamb 1986). While this effect may explain why droughts occur in any one year, so far it has not been shown how it explains the long run of rainfall-deficient years.

Biogeophysical feedback

Two other proposed mechanisms – biogeophysical feedback and the influence of dust storms – explain the drought in terms of regional phenomena. The first of these, biogeophysical feedback, is based upon the fact that the proportion of solar radiation reflected by an area on the earth's surface increases when its vegetative cover is degraded. For example, the proportional reflectance, called the "albedo", increases from 0.25 for land with a good covering of vegetation to 0.37 for a bright sandy desert (Adefolalu 1983). Dry soil also has a higher albedo than moist soil. According to the biogeophysical feedback hypothesis, when the vegetation cover of land is reduced and the environment is made dustier by the effects of drought, overcultivation, overgrazing and deforestation, the surface properties of the land change, and the consequent reflection of a larger proportion of incoming solar radiation inhibits further rainfall and so could make a drought self-perpetuating.

The first person to advocate such a mechanism was Joseph Otterman. In 1974 he argued that, as more sunlight is reflected, the land surface becomes cooler, air has less tendency to rise, rainfall formation is inhibited, and so the region becomes drier. The process would be self-reinforcing as the lower rainfall led to more devegetation due to overcultivation and overgrazing, and to less growth of vegetation, both of which would tend to increase the albedo further. Otterman substantiated his hypothesis by using satellite images of the Middle East to show that surface temperatures in the western Negev Desert, with 35% vegetation cover, were higher than those in northern Sinai, which had only 10% cover (Otterman 1974, 1975). However, his idea was opposed by scientists who presented contradictory evidence that soils in the Sonoran Desert in Mexico and the USA were always warmer than soils covered by vegetation, so that the climatic effects of devegetation could be exactly opposite to those predicted by Otterman (Jackson and Idso 1975).

Later in 1974, Jules Charney of the Massachusetts Institute of Technology proposed another way by which biogeophysical feedback could perpetuate drought. He claimed that the increase in energy reflected by the earth's surface, due to the higher albedo, would change the energy balance in the atmosphere. To maintain thermal equilibrium, air would have to flow down from higher layers to warm lower layers of the atmosphere (by means of the heat given out when it was compressed at lower altitude). This downward flow would reduce the amount of air rising from the ground and therefore inhibit the formation of rainfall (Charney 1974, 1975; Charney et al. 1975).

Charney's hypothesis has been tested by a variety of computer-based models of the atmosphere. Using a fairly simple model, Charney and his co-workers showed that a 15–30% increase in albedo in North Africa would be followed by a decrease in rainfall (Charney et al. 1977). Another team, using a more sophisticated general circulation model, later showed that an increase in albedo would be followed by a decline in rainfall in the Sahel, and by a lesser decline in northeast Brazil and the Thar Desert. On the other hand, when the US Great Plains were tested in this way, no change in rainfall resulted (Sud and Fennessy 1982). Models cannot provide definitive proof but, after reviewing the results of these and other modelling experiments, K. Laval, a scientist at the Laboratoire de Météorologie Dynamique in Paris, stated that: "All these results prove that the variation of albedo, due to a change of vegetation or soil moisture, cannot be ignored in the analysis of the causes of drought" (Laval 1986).

Measurements using satellite data have not yet found the kind of substantial and sustained increase in albedo in the Sahel that would be required to support the Charney hypothesis (Hulme 1989b). Only a few studies of long-term trends in albedo have been carried out, and most of these have only covered relatively small areas. The two most comprehensive studies found that albedo did increase from about 20% after 1967–8 to reach about 28% in 1973–4, but then declined to about 20% by 1979 (Norton et al. 1979; Courel et al. 1984; Rasool 1984). So far, few data are available on albedo trends in the 1980s.

Biogeophysical feedback does not provide firm insights into what is likely to happen in the Sahel in the future, but it does give some explanation why the present drought has been so prolonged. The possibility that human environmental impact could cause changes in regional climates had been viewed with some suspicion by climatologists for many years, but there is now a much greater openness to this possibility. According to Peter Lamb:

> Most climatologists doubt that desertification triggered the sub-Saharan drought, since neither earlier droughts there nor more recent droughts elsewhere in Africa were preceded by a decrease in vegetation. But desertification may have exacerbated the decline in rainfall and may explain why the drought in 1984 was much more severe than the one in, say, 1968, though sea surface temperatures and other conditions were similar in both years. (Lamb 1986)

Kenneth Hare has said that while "the root causes of drought or excessive rainfall are largely global ... it is now becoming clear that the albedo effect ... is a key feedback mechanism in controlling tropical and sub-tropical climates" (Hare 1983). In the view of Sharon Nicholson, "such feedback may be the only way of explaining the decadal persistence that characterizes Sahelian rainfall" (Nicholson 1983b).

The influence of dust storms
Dust-storm frequencies have increased markedly in sub-Saharan Africa since the Sahel drought began in 1968 (Fig. 1.9). The number of days with dust storms in Nouakchott in Mauritania rose from an average of five in the period 1960–4 to fifty-eight during 1980–4. Soil is more easily blown away in drought years when its surface is dry and vegetative cover is sparse. Such wind erosion is also greatly exacerbated by the spread of desertification caused by human impact. In the 1970s some theories proposed that an increase in dust in the atmosphere, from industrial pollution as well as soil erosion in the drylands, could lead to global cooling, but these were later discounted because the situation was found to be much more complicated than was first thought (Gribbin 1979; Grassl 1979; Bach 1979). On a regional, rather than a global, basis, the

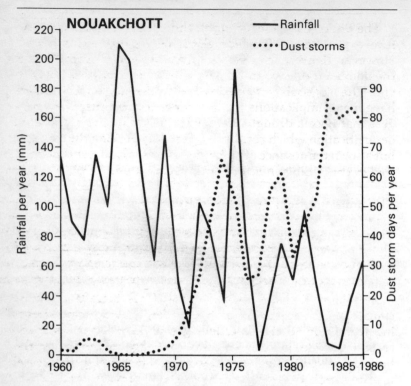

Figure 1.9: Trends in Dust-storm Frequencies and Annual Rainfall at Nouakchott, Mauritania, 1960–86

Source: Middleton (1989)

presence of dust storms could actually warm the atmosphere and inhibit the upward flows of air that are needed for the formation of rain. According to Nick Middleton of the University of Oxford: "The dramatic rise in dust storm activity that has been recorded in the Sahel may, therefore, be acting to prolong the drought in the area" (Middleton 1985, 1987). Whether this is in fact happening has not been proved, for although rainfall has declined as dust-storm frequency has increased, it could be argued that more dust storms are simply a result of lower rainfall, for the reasons outlined above. Another consideration is that our knowledge of the thermal properties of dust particles in the atmosphere is still very limited.

The prolonged Sahel drought, though exceptional, is therefore explicable in terms of global or regional mechanisms. However, there is no evidence to indicate which mechanism is the dominant one, or to suggest when the drought is likely to end. The biogeophysical feedback and dust-storm mechanisms have major implications for desertification, because they suggest that once a drought has begun naturally, the accelerated desertification which results may not only prolong the drought but also, in the absence of an improvement in land-use practices, lead to continuing environmental degradation.

The Impact of the Greenhouse Effect

There is now general agreement among climatologists that global climate is substantially affected as a direct result of worldwide human activities which generate atmospheric pollution. "For the first time we may be on the threshold of man-induced climatic change", says Kenneth Hare (Hare 1983). Most climatologists expect the earth to undergo a long-term warming trend due to the so-called "greenhouse effect" as more carbon dioxide is released into the atmosphere from the burning of fossil fuels, the clearance and burning of tropical forests, and other sources. Carbon dioxide, and other "greenhouse gases" such as methane, nitrous oxide, ozone and chlorofluorocarbon compounds (used as refrigerants and aerosol propellants), are only present in the atmosphere in trace amounts but have a profound effect on the planet's energy balance. They allow incoming solar radiation to enter the atmosphere to fuel photosynthesis and heat the earth, but they absorb heat that would otherwise radiate into space. As the level of greenhouse gases in the atmosphere rises, so the amount of heat trapped in the atmosphere and oceans will increase, and this will lead to a rise in global temperature and to changes in precipitation patterns. One predicted consequence is that in the warmer conditions the polar ice caps will melt, the oceans will expand and sea levels will rise. Could the drought in the Sahel also be due to the greenhouse effect? Might the greenhouse effect have actually caused the region to experience a long-term change in climate?

Initially, forecasts of the impact of the greenhouse effect were confined to projections of the rise in average global

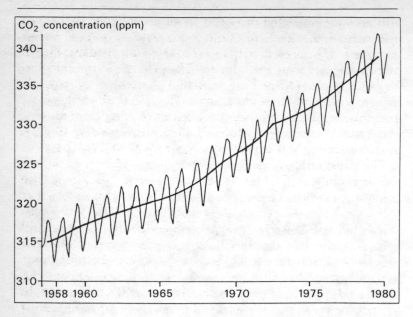

Figure 1.10: Trends in Atmospheric Carbon Dioxide Content, 1958–80

Source: NAS (1983a)

temperature that would result if the carbon dioxide content of the atmosphere were to double. (Since the level of carbon dioxide increased by 10% to 346 parts per million between 1958 and 1985, and by an estimated 25% since the start of the industrial age in the mid-nineteenth century, this was not an academic exercise (Fig. 1.10) (Keeling *et al.* 1976; Neftel *et al.* 1985).) A typical forecast is that global temperature would rise by 1.5–4.5°C (NAS 1983a). However, since it was realized that the actual rise in temperature would lag behind the increase in carbon dioxide content, because of delays in reaching thermal equilibrium between the atmosphere and the oceans, forecasts began to focus more on the "warming commitment" which will have been made by a certain year. This refers to the amount by which the average global temperature would rise by that year assuming, for the sake of simplicity, that the warming is not delayed. The calculation of warming commitment also takes into account projected increases in all greenhouse gases, not just carbon dioxide (although for convenience the overall increase may be presented in terms of the equivalent increase in carbon dioxide content alone). The World Meteorological Organization (WMO) has simulated a number of possible future scenarios, depending upon the rates of emission of greenhouse gases, and these entailed a warming commitment of 0.8 to 4.1° C between 1980 and 2030, although only 40–50% of this warming was expected to occur by 2030 (WMO 1986).

The latest studies, using large general circulation models of the atmosphere (GCMs), have attempted to project future trends on a regional basis. The results suggest that some areas will warm more than others, and that there will also be changes in rainfall patterns, with some regions becoming drier and others wetter than at present. Some studies also conclude that the world's climate will be subject to much more extreme variations, so that droughts could become more frequent. However, our ability to predict future regional changes in temperature and rainfall is very limited, owing to the lack of precision of the present generation of GCMs. A comparison of forecasts of future climatic trends made by three different GCMs, assuming a doubling of atmospheric carbon dioxide content, showed that the forecasts were fairly consistent in predicting an increase in average global temperature of 3.5–4.2°C and an increase in precipitation of 7.1–11.0%. All three models forecast that the warming effect will be lower over the oceans than over land, and will increase from the Equator to the Poles. Temperature increases in the drylands are therefore likely to be smaller than those for areas at medium and high latitudes. There was, however, disagreement about changes in precipitation. All three models predicted that in the months of December, January and February soil moisture would fall in most of Africa, the Arabian Peninsula, Central America, Mexico and the Gulf states, but would rise in Europe, Central Asia, the western and central United States and most of Canada. On the other hand, the forecasts for the months of June, July and August differed markedly. For Africa, one model predicted moister soil everywhere except in North Africa, another predicted drier soil everywhere except in North Africa, and the third predicted drier soil in East Africa and moister soil in West Africa (Schlesinger and Mitchell 1985). Because of such widespread uncertainty about future trends in precipitation, a recent comprehensive assessment of the effects of future climatic change on agricultural production in semi-arid regions, by the International Institute for Applied Systems Analysis (IIASA), decided not to include a GCM-based scenario in its considerations (Parry *et al.* 1988b).

Are the extremes of climate seen in the Sahel in the last twenty years connected in some way with this trend towards warming and these potential changes in rainfall patterns? It is difficult to say, because some GCMs predict that the Sahel will become drier and others that it will become wetter. One study, made by a team whose model had produced projections that indicated the first option, argued that over the last thirty to forty years, there had been a significant decline in rainfall in the northern half of Africa and the Middle East, and an increase in rainfall in mid-latitude regions such as Europe and the USA, and that these trends confirmed their projections (Anon. 1987a). Another leading modeller, James Hansen of the US National Aeronautic and Space Administration's (NASA) Goddard Institute for Space Studies, claimed that the US drought in 1988 was probably due to temperature extremes resulting from the greenhouse effect (Joyce 1988). Until there are improvements in the quality of GCMs, and the projections made with them, it will remain difficult to accept or refute such claims.

The problems are more far-reaching, however, because even with regard to more aggregated global trends it is still not easy to decide whether or not the greenhouse effect is already having an impact on the world's climate. It is commonly argued, on the one hand, that the 0.5°C increase in average global temperature between 1850 and 1980 is sufficient evidence that a global warming trend has been operating since the nineteenth century. The increase was within the range of warming of 0.4–0.8°C predicted for that period by the WMO (1986). However, the greenhouse effect theory does not explain why average global temperature, after rising in the early part of the century, fell between 1940 and the mid-1960s (mainly due to a cooling in medium to high latitudes of the northern hemisphere), before rising rapidly in the last two decades (Fig. 1.11) (Hansen *et al.* 1981). Until it does so, the theory is clearly open to criticism. The years 1987 and 1988 were the warmest since reliable instrumental records began about a hundred years ago,

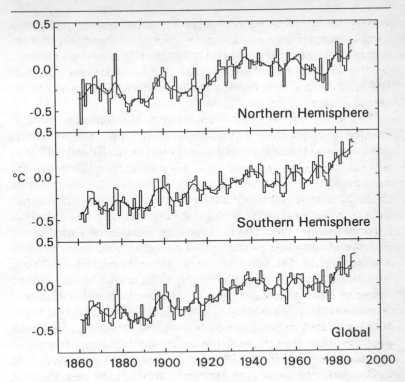

Figure 1.11: Trends in Average Global, Northern Hemisphere and Southern Hemisphere Surface Air Temperatures, 1901–88

Note: the trend in average surface air temperature is expressed relative to the long-term mean.

Source: Jones *et al.* (1988), updated by the original authors

although this could in part be a consequence of a strong El Niño event in 1986–7 (Jones *et al.* 1988; P. D. Jones, personal communication, 1989).

In the midst of such uncertainty, some climatologists, while not denying the potential importance of the greenhouse effect, take a fairly guarded view of its current impact. According to John Lockwood, of the University of Leeds: "At present there is no clear evidence of any climatic changes within the tropics arising as the direct result of increases in atmospheric CO_2 content during this century." In his view, the occurrence of droughts and excessive rainfall in the topics can be explained quite adequately by factors such as the El Niño effect and warmer than average sea-surface temperatures in the southern oceans. In the next century, on the other hand, he feels that the influence of the greenhouse effect could become more important (Lockwood 1988).

It is therefore impossible to say at present whether or not the prolonged Sahel drought is connected in some way with the greenhouse effect. All that can be said is: (a) that the drought is an example of the kind of climatic extremes which we might expect to occur as a consequence of this effect; and (b) that should the Sahel be found to be undergoing a long-term change to a drier climate then this is a consequence consistent with some models of the effect. However, should the greenhouse effect really lead to major shifts in regional climates – and another part of the IIASA study mentioned on p. 58 estimated that an increase of 1°C in average global temperatures could shift the major cereal-growing regions in the northern hemisphere northward by 100 km (Parry *et al.* 1988a) – then it is quite likely that land in some regions will become desertified as a direct result of long-term climatic change. Equally worrying is the possibility that farming would continue on land affected to a lesser extent by the changing climate, even though the area had become marginal for the kinds of land uses previously practised there. This could make the problem of desertification even worse than it is at present, and affect the developed nations just as much as the developing nations.

The Role of Drought

This section has shown that, with the exception of the Sahel and its extension into northeast Africa, droughts in the drylands generally remain short-term phenomena, and so it still seems justifiable to retain the UNCOD understanding of drought as having an indirect, catalytic role in desertification. There is as yet no conclusive evidence of a long-term change in climate in the Sahel; nor is there any firm explanation as to why the drought has continued for so long. One possibility is that desertification is helping to make the drought self-perpetuating, whether through the mechanism of biogeophysical feedback, or by the effects of dust storms, or both. Another possibility is that the drought is one of the first major consequences of the greenhouse effect. In both cases, the proper response is a move towards more sustainable land uses of the kind that would be needed were desertification to be caused by human impact alone. So we certainly shall not misdirect our energies if we continue to focus on reducing the degrading effects of human impact on the drylands.

QUESTIONS REMAIN ABOUT DESERTIFICATION

Although the view of desertification presented in this chapter should be generally acceptable to the majority of those working in this field, it must be apparent that our understanding of the nature and causes of desertification is still quite limited, and so many questions remain to be answered. In such a situation it should come as no surprise that dryland experts differ in their definitions and interpretations of this phenomenon. This section highlights some of the major differences of opinion.

The scientific literature contains over one hundred definitions of desertification (Glantz and Orlovsky 1986), but they leave a lot to be desired in terms of the detail with which they characterize it. Such confusion over meaning tends to inhibit the monitoring of desertification and is one reason for the lack of available data on the extent of the phenomenon. Differences over definitions and lack of data reinforce each other, unfortunately, so that someone who has an entirely erroneous understanding of the meaning of desertification may become quite sceptical about its relevance when there is little hard quantitative evidence to support that point of view. Thus, even in 1989, twelve years after UNCOD, it was being claimed that desertification referred to the expansion of deserts, and because there was no proof that this was actually taking place, then the problem of desertification had been greatly exaggerated (Forse, 1989). The only reliable way to remove such difficulties is to improve the definition and measurement of desertification, a subject discussed in more detail in Chapter 4.

Views also differ concerning the respective roles of human impact and climate. Some scientists, like El Baz, view desertification as the result of a progressive drying of the climate of Africa, a trend that began 5,000 years ago (El Baz 1983); some, like Harold Dregne, consider it to be a "man-induced global phenomenon with no respect for climatic zones" (Dregne 1984a); and others, like Dick Grove, think that there are both human and climatic causes but that it is difficult to distinguish between the two (Grove 1973, 1977).

Adding to the complexity is the fact that poor land use may be influenced by a variety of underlying social and economic factors, such as poverty and population growth. However, it is easy for scientists skilled in the study of the physical environment to neglect this, and focus purely on the degradation of soil and vegetation as an environmental problem which needs to be curtailed. Social scientists object to such an approach, claiming that the real problem is socio-economic, rather than environmental or agricultural, because the farmers and pastoralists who cause desertification in the most marginal areas have such a low economic status that they have no option but to try to gain a living there. Thus, attempts to control desertification by draconian measures which regulate the use of land for cropping or grazing may well exacerbate the poverty and deprivation which are the original causes of the problem. What is needed, according to this view, is more investment in agriculture and a change in government policies in favour of the poorer sections of society. This author acknowledges the need for such a wider view of desertification, and this is reflected in the content of this book, but he would argue that the importance of social and economic considerations does not make the environmental problem any less real or attempts to control it in some way any less important.

Differing views as to the contributions to desertification made by climate and by underlying social and economic factors are accentuated in discussions of famine. The original initiative to convene the UN Conference on Desertification owed more to worldwide horror at the terrible human effects of the Sahel drought and famine in the early 1970s than to concern about long-term degradation of dryland environments. That the latter became the focus of the conference was a tribute to the vision of UN officials and their scientific advisers, who saw that the short-term difficulties of this one region were deeply rooted in a long-term problem common to drylands all over the world. Nevertheless, desertification did come to be somewhat associated with drought and famine, although there is sound evidence to show that famine is not necessarily the result of drought and may be caused by other factors which are more under human control (see Chapter 3). Some development experts therefore object to the whole concept of desertification because they see it as a convenient way of blaming environmental problems for a situation which is largely a result of inadequate economic development and misguided government policies.

The last main source of disagreement concerns the use of the term "desertification". Since the subject first became topical in the early 1970s a large number of terms have been proposed to describe what was taking place in the drylands, including "aridization", "aridification", "xerotization", "desert encroachment" and "desert expansion". These have been reviewed elsewhere by Michel Verstraete (Verstraete 1986). Each term reflected the originator's views of the major cause of the problem, whether climate or human impact, and was also, in the mind of its originator, more appropriate than the others. There are sound reasons for objecting to the term "desertification", since it can give the impression that the phenomenon refers to the spread of deserts or the development of desert-like conditions, whereas the actual degradation is much more subtle and diffuse in character. Some experts question whether we need a special term at all, since soil erosion and vegetation degradation are widespread in humid tropical and montane areas and are not confined to the drylands. They have been answered eloquently by Michel Verstraete:

> Desertification is neither drought, nor soil erosion, nor the destruction of the vegetation cover, nor the cutting of trees, nor even the degradation of the living conditions alone. It is all that and much more. I believe we do need a more encompassing word to refer to the whole process. Whether such a general word should be precisely desertification or something else is another question, but that word certainly has been much more used than any of the other contenders. (Verstraete 1986)

The fact that the term "desertification" is now used all over the world provides a strong justification for its continued use, and instead of changing the word itself it would be far more advantageous to secure wider agreement on what that term does (or should) describe. This brief discussion has shown how easily differences of opinion can arise; the only reliable way to resolve these differences is to undertake scientific field studies and large-scale monitoring of desertification. These would enable us to learn more about the phenomenon and would provide a firmer quantitative basis upon which to form a more objective judgement of its significance.

A COMPLEX PHENOMENON

This chapter has looked mainly at the physical aspects of the process of desertification which is currently taking place throughout the world's drylands, and particularly in Africa and Asia. Our understanding of desertification is still quite limited, although until now it has been thought of as largely the result of human misuse of the land with overcultivation, overgrazing, deforestation and mismanagement of irrigated cropland leading to the degradation of soil and vegetation. The role of climate has been thought of as mainly catalytic, so that desertification accelerates as drought causes people to overfarm the land to compensate for falling yields. However, the boundary between the roles of climatic variation and human impact is now becoming increasingly blurred. This is especially the case in the Sahel, where it has been suggested that the continuing drought is linked with environmental degradation caused by bad land use. Whether this theory is true or not, there seems to be every likelihood that desertification will be influenced in the future by global climatic changes resulting from the activities of the entire human population. Because of the various lags in the world climatic system there is very little we can do to prevent such changes occurring over at least the next twenty to thirty years, and probably longer. The best way to cope with climatic variability and control desertification in the drylands is therefore to make land uses more sustainable and to alleviate the poverty which is another important underlying cause of the problem.

2. The Causes of Desertification

Desertification is caused directly by four main types of poor land use: overcultivation, overgrazing, deforestation and poor irrigation practices. This chapter looks at each type, gives examples from different dryland areas, suggests the reasons for its occurrence, and describes the degradation which results. None of these types should be thought of as the dominant cause of desertification, since they are all interrelated, and a change in the location and intensity of one land use can affect other land uses some distance away. Moreover, as pointed out in the last chapter, poor land use is accelerated by drought, and is also greatly influenced by various underlying social, economic and political factors which, although mentioned in this chapter, will be discussed in more detail in Chapter 3.

OVERCULTIVATION

Overcultivation occurs when farmers try to crop the land more intensively than permitted by its natural fertility, and fail to compensate for the export of nutrients in the crop by using artificial fertilizers or fallowing the land so that its fertility can regenerate naturally. Overcultivation therefore reduces the fertility of the soil, damages its structure, and exposes it to erosion. In dryland areas it is often caused by the breakdown of traditional rainfed cropping systems, either by outside pressures or because they are unable to produce enough food to sustain high population densities. A common form of outside pressure is the expanding cultivation of cash crops or irrigated food crops like rice and wheat. These displace subsistence rainfed cropping to marginal, low-rainfall areas where it is difficult for it to be sustainable. Cash cropping itself causes soil degradation if practised on marginal lands, as does the introduction from temperate countries of mechanical farming practices unsuited to fragile dryland soils. The problems arising from the poor management of irrigated cropping are discussed later in the chapter.

Traditional Farming Systems

The cultivation of arid and semi-arid lands is helped by the fact that the soils are usually rich in nutrients since, unlike wetter areas, the limited rainfall has historically washed only a small proportion of nutrients from the topsoil to lower soil levels. There are also long hours of sunshine throughout the year to provide the raw energy for the growth of plants. On the other hand, the meagre rainfall is not evenly distributed throughout the year, and in many areas comes in a single short irregular rainy season which allows only one major crop per year (some areas have two rainy seasons). Drought periods lasting one or more years occur frequently, leading to crop failures or poor yields. Although soils may be rich in nutrients, they only have a sparse covering of vegetation and a low soil organic matter (humus) content which makes them highly susceptible to erosion.

Traditional systems of rainfed cropping were developed over thousands of years to make the best use of limited rainfall and to sustain food production even in the most adverse climatic conditions. The system generally used in the drylands of West Africa had a number of stabilizing characteristics. Cropping took place in the Sudan savanna zone (400–1,000 mm of rainfall per annum), where rain was most likely to fall in reasonable quantity and with regularity. The two northerly zones – the Sahelian zone (200–400 mm per annum) and the Saharan zone (less than 200 mm per annum) – were generally reserved for the raising of livestock, since the seasonal movement (transhumance) or irregular movement (nomadism) of livestock could best take advantage of pasture grasses, where and when they happened to grow.

Farmers aimed to reduce the risks of a complete crop failure by planting a variety of crops. Because each crop has different water requirements there was a good chance that one of them would survive even if the rains were late or of limited quantity. Essential components of this variety were drought-resistant food crops such as sorghum and millet. More profitable crops, like groundnuts and cotton, that were more demanding in terms of growing conditions, could also be planted to give a second, cash crop in the years when rains were plentiful.

Long fallows were used so that fertility could regenerate. After four or five years of continuous cropping on one plot farmers would move on to another, leaving the first plot idle or to be used as pasture for up to five years or even longer. In some areas, *Acacia senegal* trees were allowed to invade the fallow plot, and after five years they were tapped for gum arabic, a highly profitable commodity, for about seven years. During this fallow period the soil was protected from erosion by tree cover, falling leaves built up a litter layer so that vital nutrients and humus accumulated in the topsoil, and fertility was also enriched by nitrogen-fixing bacteria in the tree roots. Finally, the trees were felled and burned, and the land was cultivated for food crops once again.

Pastoralists and sedentary farmers had a symbiotic relationship, bartering cereals for meat. The cash economy is a recent innovation in this region. Since each group produced something which the other wanted, each benefited from the exchange. Pastoralists also grazed their livestock on village fallow lands in the dry season when more northerly pastures had been exhausted, and the dung which the cattle deposited on the fallow lands contributed to the regeneration of soil fertility.

The Impact of Overcultivation

Traditional rainfed cropping systems are now breaking down as rising demand for food to feed growing populations forces farmers to increase production, either by reducing fallow periods or by increasing the area cultivated. Shorter fallows deplete soil nutrients and organic matter because there is insufficient time for the fertility to be restored before the next crop is planted. The soil becomes more susceptible to erosion because of the lower organic matter content and the fact that the soil is left without vegetative cover for longer periods each year. The exposed topsoil is crusted by the combined impact of rain and sun, blown away by wind, or washed away by water. There is also the possibility that gullies will be formed by the increased amount of water running along the surface. The encroachment of sand dunes on to arable land is only found in some areas but the destruction of crops by strong dust-bearing winds is a common phenomenon.

When declining soil fertility leads to lower yields farmers may either increase the intensity of cropping yet again, which makes matters even worse, or expand the area under cultivation. Much of the new cropland is taken from marginal rangelands unsuited to sustained cultivation; consequently, cropping may only be possible for a few years before the land becomes degraded and has to be abandoned. Farmers practising rainfed cropping are also forced to move to these more marginal lands by the expansion of cash cropping and irrigated cropping, and this leads to similar problems.

The Effect of Growing Populations

Traditional rainfed cropping systems have long satisfied people's basic subsistence needs in the drylands, but there were limits to the population densities which such conservative farming practices could support. All over the drylands, rising populations are now causing those limits to be breached, thereby threatening the sustainability of food production. The *Acacia* fallow system described above was once widely practised in the west Kordofan region of Sudan but is now collapsing as the population grows. "The ecologically balanced cycle of gum gardens, fire, grain crops and fallow is now breaking down", wrote Jon Tinker in 1976:

> Under pressure of a growing population, the cultivation period is extended by several years and the soil becomes too impoverished to recover. Overgrazing in the fallow period prevents the establishment of seedlings. Gum trees are lopped for firewood. More and more widely, *Acacia senegal* no longer returns after the fallow but is replaced by non-gum producing scrub. Without the gum to harvest for cash, the farmers must repeatedly replant their subsistence crops until their land becomes useless sand.

Overcultivation is also widespread in India, a country which has the second highest population in the world after China and has to sustain two-thirds as many people as China on only one-third of the land area. Almost two thirds of India's 320,000 sq km of drylands are in the state of Rajasthan, where they account for 60% of the total area of the state. Only 20% of Rajasthan's drylands are considered by FAO to be suitable for rainfed cropping, but the area under cultivation virtually doubled from 30% of the total dryland area in 1951 to nearly 60% in 1971, mainly at the expense of grazing lands and long fallows. Overall production and average yield per hectare both fell between 1954 and 1970 for three out of the four major crops – jowar, sesame and kharif pulses – despite the fact that the area under sesame and pulses increased over the period.

The UN Conference on Desertification (UNCOD) chose a 1,989 sq km area of land in the eastern part of Rajasthan's arid zone, the Luni Development Block, as the subject of a detailed case study. A third of the area was classified as being under "very high" desertification hazard in 1976 and the remainder was subject to "high" desertification hazard. The factors contributing to such a high incidence of desertification were: 88% of the area was under cultivation by 1972; the human population had trebled since the start of the century so that there were now 48 people per sq km (very high for this type of area); natural vegetation only survived on 13% of the area occupied by common grazing lands; tree and shrub cover was sparse; a third of the area was covered by a sand sheet 0.2–2 metres thick compared with only a 25% coverage in 1958; and existing sand dunes had grown in height by up to 5 metres at their crests. Wells were yielding less water and a growing number were becoming saline (Prakash 1977).

Land Tenure Restrictions

When areas become overcrowded some of the people normally migrate elsewhere, but this is difficult when land tenure regulations restrict people to certain areas, often of low fertility. Cultivation therefore has to become more intensive and the result is overcultivation and desertification. Zimbabwe, only recently independent, inherited from its colonial past a population distribution problem of potentially disastrous proportions. The country has a large area of good agricultural land receiving a reasonable rainfall but there is also a significant area of sub-humid and semi-arid land in the south of the country which was severely affected by drought in the early 1970s and early 1980s.

In colonial times Africans were essentially confined to Reserves and Tribal Trust Lands in these marginal areas while the better lands were reserved for large-scale commercial agriculture. Today 3.7 million people still live and grow food on 16.2 million hectares (ha) of marginal land in what are now called the Communal Areas; another 1.5 million ha of land in the so-called Purchase Areas are farmed by Africans as tenants; and large-scale commercial agriculture continues on what are now designated the Commercial Areas.

Only 2.2 million ha of the Communal Areas are cultivated and the rest is used for grazing, but much of that 2.2 million ha should not be cultivated at all because it is full of steep slopes and the soil has low fertility. Between 1961-2 and 1976-7 the number of cultivators in the Communal Areas rose by 88%, the cultivated area by 91% and the number of cattle by 70% (Stubbs 1977). Land degradation is now widespread on the communal farms and every year an average of 50 tonnes of soil is lost per hectare (Elwell 1979).

Michael Darkoh of Kenyatta University College in Nairobi has described a visit to part of the Communal Areas:

> Big patches of denuded and overgrazed land, bare surfaces of sheet eroded land intersected by inselberg landscapes of kopjes, bare rocky outcrops and dongas as well as sand and silt-choked streams were the commonest features that caught the eye in most of the places visited. The widespread destruction of the vegetation, stream-bank cultivation and subsequent siltation and drying up of streams so common on the communal areas of this catchment stood in sharp contrast to the lush green landscape of the adjoining Purchase Areas and Commercial Areas. (Darkoh 1986)

Lesotho, a small mountainous country completely surrounded by South Africa, also suffers from problems stemming from colonial land division, although strictly speaking the country was only a British protectorate, not a colony. Most of the country is either sub-humid or semi-arid. From the start of British protection in 1868, the indigenous population was confined to an area in the lowland part of the country. As the population rose, overcultivation and overgrazing became widespread and people migrated to the highlands. Ever steeper slopes were cultivated and this has resulted in one of the worst cases of soil erosion in the whole of southern Africa. Erosion occurs on almost 50% of the cultivated land; 12% of the land is severely eroded. Every hectare of arable land loses on average 70 tonnes of topsoil a year due to sheet erosion. Already 4% of the country's cultivated land has been lost to gully formation; another 1% is lost every four years as new gullies develop or old ones increase in size. This is equivalent to an annual loss of about 1,000 ha of productive land (Speece and Wilkinson 1982).

Social Changes After Independence

Overcultivation caused by restrictions on settlement is not only a heritage of colonialism. In most tropical countries which have gained independence within the last thirty years, land distribution is now much fairer than it was in colonial times, but in some countries governments have imposed new patterns of settlement and social grouping which have had detrimental effects. In Tanzania the government has embraced a vigorous policy of "villagization" which brings together into villages farming families who previously lived in scattered homesteads. Concentrating people in this way has led to severe problems of overgrazing and soil erosion around the villages. Traditional farming systems have also been breaking down because of social changes unconnected with villagization. Shifting cultivation and animal husbandry based on the seasonal transhumant movement of herders and livestock between settlement areas and dry-season pastures, the common land-use practices in semi-arid areas of Tanzania, depend for their success on a low density of population. But as populations have risen mobility has become limited, and available land is now cultivated more intensively. New fields are cleared on slopes having highly erodible soils with the result that the topsoil is washed away and rills and gullies form (Christiansson 1981).

The government of Burkina Faso has adopted an entirely different resettlement policy, moving farmers from large collective farms to smaller individual units, but this has also caused problems. Resettlement, in conjunction with rapid population growth, resulted between 1970 and 1976 in continuous cropping becoming almost the norm at the expense of traditional bush fallow techniques. This caused rapid soil deterioration, which in turn reduced cereal production and grazing area and led to an increase in the food deficit and in the number of people migrating from rural to urban areas (Marchal 1983).

Mechanized Farming

Mechanized farming has been the basis of increased farm productivity throughout the temperate world but is unsuited to fragile drylands in both developing and developed countries because it can easily cause soil erosion. The 200 sq km region of Oglat Merteba in southern Tunisia on the fringes of the Sahara is experiencing problems caused by a change to more intensive mechanized cultivation and an increased concentration of population due to the settling of nomadic herders. Only 100–200 mm of rain falls in an average year, and transhumant grazing of sheep and goats on communal rangelands has long been the dominant form of land use. Cultivated land more than tripled between 1948 and 1975, and the increasing use of tractors and disc ploughs caused wind erosion of the dry, sandy soils. Livestock numbers did not drop despite the decrease in the grazing area. The resettling of nomads into villages (sedentarization) concentrated herds around settlements and water-holes, increasing the pressures on both land and water, and leading to desertification (Floret et al. 1977).

Similar problems are afflicting another part of Tunisia. Since 1960 the use of tractors and disc ploughs has become common in the Jeffara area in the south of the country, which comprises the undulating plains between the coast near the island of Djerba and the Matmata mountains. Farmers use money sent home by relatives working in the Libyan oil industry to buy tractors and other equipment, and Libya also provides a market for the grain and meat grown in the area. According to Anders Rapp of the University of Lund: "After 2–3 years of barley harvesting followed by grazing of the stubble, the entire topsoil is often blown off by the wind down to the plough soil at about 20 cm depth. It appears as a denuded crust, compacted by the weight of tractor and plough and marked by parallel straight scratches of the ploughing." Tractors allow farmers to cultivate a much larger area than they would if using animal-drawn ploughs, and this leads to more extensive soil erosion (Rapp 1982; Helldén and Stern 1980; Novikoff and Skouri 1981).

Overcultivation occurs in developed nations as well. Because of government pressure to increase crop production in the USA, marginal lands that were formerly not cultivated have been brought into production again, and more than 40% of US cropland is eroding at rates in excess of tolerable levels. One-third of the cropland risks losing its long-term productivity and every year 3 billion tonnes of soil is lost by water and wind erosion. Soybeans and cotton are now grown on land in the hilly regions of Tennessee which were previously used only for cattle grazing. This is resulting in the annual loss by erosion of 50 tonnes of soil per hectare, causing clogged streams, flooding and siltation of the Mississippi River into which streams from Tennessee drain. Soil erosion is also running at high levels in the US Great Plains. A Department of Agriculture report quoted average reductions of 6% in corn yield for every inch of topsoil lost. One estimate (possibly overstated) predicted that at current erosion rates, yields could fall by 30% over the next fifty years (Risser 1985).

The Expansion of Cash Cropping

The groundnut connection
A large part of agricultural production in tropical countries is for subsistence purposes only, although some cash crops are grown for sale in local, national and international markets. Because most developing countries are poorly industrialized, they still depend on the export of agricultural products and other primary commodities for the bulk of their foreign-exchange earnings; therefore cash crops make an important contribution to national development. Many of the major crops in international trade, like rubber, cocoa and coffee, can only be grown in relatively humid areas but some, like groundnuts and cotton, do not require as much rainfall and can be grown in or adjacent to semi-arid areas. A common feature of all leading cash crops, however, is that they are extremely demanding in their nutrient requirements and in the overall site conditions needed for optimal production. Trying to cultivate these crops on inappropriate sites or without proper management can therefore degrade the land. The expansion of cash-crop

cultivation also indirectly causes desertification by displacing subsistence cropping and pastoralism on to more marginal lands which later become degraded.

Cultivation of groundnuts (also known as peanuts or money-nuts) expanded dramatically in Niger and other Sahelian countries in the 1950s and 1960s, partly in response to the wetter conditions and to price incentives. According to one theory, this may have contributed significantly to the Sahel disaster of the early 1970s, because it caused a large reduction in the amount of fallow land in the agricultural zone where nomadic pastoralists had traditionally grazed their animals in the dry season. Use of grazing lands in the north therefore had to be intensified, causing desertification and increasing the nomads' exposure to the risks of drought (Franke and Chasin 1981).

Groundnuts grow well in areas with an average rainfall of about 500 mm per annum. In the early 1950s the largest groundnut producers were Nigeria and Senegal, two countries which normally receive at least this much rainfall over most of their territory. Senegal now has 1 million ha under groundnuts (half of all the cultivated land in the country), derives about one-third of its foreign currency earnings from nut exports, and is the world's fourth largest producer. In some places where groundnuts have been continuously cultivated for up to ninety years, soil degradation and falling yields have resulted, because farmers under pressure to maintain production have neglected fallows and fertility has declined. At Thyss-Kaymore in Senegal, for example, groundnut yields have fallen from 2.5 tonnes/ha in 1940 to only 1 tonne/ha today.

When a major expansion in groundnut cultivation took place in West Africa in the 1950s and 1960s Niger and Sudan also became prominent producers. Production increased to a lesser extent in Mali, Burkina Faso and Chad. One of the reasons for the expansion was an attempt by France to combat a US campaign to dominate the European vegetable oils market with soya oil. It gave its colonies in the region an incentive to increase groundnut production by guaranteeing an artificially high price that protected producers against shifts in the world market price. In Niger the area under groundnuts rose from 73,000 ha in 1934 to more than 142,000 ha in 1954, and 349,000 ha in 1961. It had reached 432,000 ha in 1968 on the eve of the Sahel drought. In Sudan the area under groundnuts quadrupled between 1960 and 1972 to 810,000 ha. Production in Niger rose from 160,000 tonnes per annum in 1955 to 290,000 tonnes in 1966, and in Sudan from 120,000 to 310,000 tonnes. By 1964, groundnuts or groundnut products accounted for 68% of all of Niger's exports.

Only limited areas of Niger, Sudan and the other three countries receive sufficient. rain for groundnut cultivation. Southern Niger is on the edge of the 500 mm zone and its rainfall is very variable. Most farmers did not wish to become too reliant on groundnuts so they grew them on fallow lands traditionally used by nomadic pastoralists as seasonal grazing. By 1961 half of all the farmers in the Maradi district of Niger depended upon groundnuts for 50 to 60% of their incomes with the other half having a lesser, but still significant, dependence. Despite the guaranteed price, the weight of staple foods which a farmer could exchange for the proceeds from a kilogram of groundnuts declined, so farmers had to grow more groundnuts to maintain their living standards. They did this by using improved seeds which needed shorter growing cycles and less rain. As a result, fallow periods were reduced and cultivation spread north of its previous boundaries into drier areas normally used mainly for grazing. The areas available to pastoralists declined further, and greater pressure was placed upon the more sparsely vegetated lands most vulnerable to desertification.

Cotton cultivation expands

Groundnut cultivation was therefore a significant contributing factor to desertification in Niger and Sudan in the 1960s and 1970s. It was much less important in Mauritania, Mali, Burkina Faso and Chad which did not experience anything like the same expansion in groundnut cultivation, although their lands and peoples were still hit hard by the drought. However, cotton cultivation expanded considerably in these other countries in the 1960s, growing in area by four-fifths in Mali and tripling in Burkina Faso and Chad. In the southern part of Chad, the region's largest cotton producer, a 90,000 ha "showcase" project used irrigation water from Lake Chad and tripled the average yield to 900 kg/ha, but a joint report by the Club du Sahel and CILSS found that the project was implemented partly to the detriment of food-crop production and that the extraordinary growth of cotton production was achieved without any particular concern for the maintenance of soil fertility. Cotton was grown on marginal lands barely able to support crops that were much less demanding of water and nutrients, so soils became exhausted and the water table fell (Adefolalu 1983).

From the early part of this century, peasants in Chad were physically and financially coerced into growing cotton to provide raw materials for French factories (Grove 1978; Stürzinger 1983). They were originally forced to grow cotton to earn cash to pay the poll tax, but since they used part of this income to buy consumer goods they were also drawn into the market economy, even though they would remain marginal to it. In the 1970s the government of the now independent Chad tried to triple the area under cotton, but soils deteriorated despite the use of fertilizers; both cotton and food crops gave poor yields; and peasants were treated as badly by local middlemen as they had been previously by French colonial companies.

There are signs that cash cropping in the Sahel is now on the decline, because peasants are becoming very concerned about maintaining the production of subsistence food crops. In Niger, for example, the area under groundnuts fell from 400,000 ha at the start of the 1970s to 165,000 ha in 1977–8. Some of the land taken out of groundnut cultivation is now under millet and sorghum and the area under niebe, a valuable subsistence bean which enriches the soil with nitrogen, has also expanded.

OVERGRAZING

Nomadic grazers were initially blamed for causing the Sahel disaster. By keeping too many animals, it was said, they spread desertification in fragile lands, although one reason why they were singled out for blame was simply that the impact of their actions was more apparent than that of rainfed cultivators because it occurred on some of the driest lands close to the desert. Overgrazing is indeed a major cause of desertification, and rangelands account for almost 90% of desertified lands (Mabbutt 1984), but the syndrome is not quite as simple as was first thought, since desertification takes place throughout the drylands, not just on the edge of the desert.

Overgrazing results when livestock density becomes excessive and too many animals are grazed on the same area of rangeland, leading to the degradation of vegetation and the compaction and erosion of soil. Livestock density can rise in four main ways. First, herd sizes are allowed to grow too large during wet years to be sustained by the limited pasture growth in dry years. Second, the area available for grazing decreases as nomads are displaced by farmers growing crops either on marginal rangelands previously used for grazing, or on former dry-season pastures in the wetter, mixed agricultural zone. Third, livestock become concentrated around villages by nomad resettlement schemes and along herding routes made popular by the sinking of boreholes. Fourth, traditional controls on the grazing of rangelands break down.

Pastoral Systems

Pastoralism is an excellent way of converting into food the vegetation which grows as a result of the irregular and variable rainfall on large areas of arid and semi-arid rangelands. (Rangelands are wild grasslands – the grasses are not seeded artificially as they are in managed permanent pastures.) The major rangeland areas, accounting for almost half of those found in the world's drylands, are the Sahel and its extension into northeast Africa, North Africa, Western Asia, China, the USSR, South America and Mexico (Mabbutt 1985). Other prominent areas for grazing include the uplands in the Horn of Africa and around the Mediterranean, the Andes Mountains in South America and the Rocky Mountains of the USA; the plains and plateaux in Central Africa, India (the Deccan Plateau), tropical Australia and northeast Brazil; and the temperate grassy steppes of North America, Argentina, Kazakhstan and Mongolia.

Three main types of pastoral systems are practised in these areas: the nomadic/transhumant; the sedentary (or village-based); and ranching. Nomadic pastoralists make the best use of marginal arid lands where rains may or may not come and where the vegetation is low in quantity and sparsely distributed. Herdsmen generally know the best places to take their livestock and leave for the next grazing ground when one has been used up. There are two kinds of nomadism: true nomadism and transhumance. In true nomadism the herdsmen move their livestock more or less continuously, following no set pattern; in transhumance they move livestock along more predetermined routes. Each year they travel from the wet-season grazing lands in the arid zone to fallow lands in the semi-arid agricultural zone which their herds graze in the dry season. True nomadism, like that practised by the Moors in northern Mauritania, usually involves the herding of drought-hardy camels, goats and sheep, with perhaps a few cattle. Transhumance, such as that practised by the Fulani people in Niger, consists chiefly in the herding of cattle over shorter distances (Vermeer 1981).

Pastoralists have over time developed complex methods to ensure their survival in extreme environments. These have been incorporated into their customs and traditions but are constantly being adapted in response to changes in natural environments and social and political circumstances. As one UNEP expert has said:

> Herdsmen have found their ways of coping with the climatic stress that typifies their arid landscapes. Ordinarily they spread their stock thinly over large areas so that grazing pressure is lightened and they can take advantage of the patchwork ecosystems of arid lands where topographical variety usually yields good pasture only here and there. They are mobile, often traversing great distances to reach seasonal pastures. Pastoralists, including nomads, must not be thought of as "wandering". Familiar with the land and what it provides, they have a clear idea of where they are headed. (Anon. 1977)

Donald Vermeer of George Washington University in Washington, DC has described how, in July 1979, within a week of the first rain of the season falling in an area 100 km southeast of Nouakchott in Mauritania, nomads had brought their herds of goats, sheep, cattle and camels to graze the grass that had started to grow there (Vermeer 1981).

Keeping a diverse herd of animals is another way in which nomads insure against drought. Sheep and goats continue to lactate during dry periods and have high reproduction rates. Goats and camels will survive drought better than sheep and cattle. Diverse herds also make the best use of the wide variety of vegetation, since camels and goats, which graze in a more dispersed way and can survive by eating only thorn bush, have a much lower impact on the land than sheep and cattle which feed mainly on grass, stick together in groups, and tend to stay close to sources of water (Swift 1988).

Sedentary livestock raising is practised by farmers who are mainly concerned with rainfed cropping in semi-arid areas, although they also keep some animals, grazing them on fallow lands and communal village grazing lands. Usually only limited areas of pastures are available within easy reach of villages so these are used quite intensively and suffer considerable degradation. Many farmers have long depended on nomadic grazers to supply them with animal products, enjoying a symbiotic relationship in which meat and milk are exchanged for grains and legumes, and fallow lands are grazed by nomadic herds in return for the fertilizing value of the dung, and also for cash or other goods.

Another symbiotic relationship in West Africa is that between the pastoralists in the Sahel and urban populations in the coastal states to the south. Raising livestock in the humid coastal forest belt is virtually impossible because of the debilitating effects on animals of the disease trypanosomiasis carried by the tsetse fly. Countries like Nigeria and the Ivory Coast therefore depend heavily upon meat from animals raised in the dry northern zone (Makinen and Arizo-Niño 1982; Delgado and Staatz 1981; Arizo-Niño *et al*. 1981). To save on transport costs these animals have often been trekked for hundreds of kilometres to the slaughter-houses, losing weight in the process. The development of lucrative markets in countries like Nigeria and the Ivory Coast has encouraged the establishment of cattle ranches in the Sudan Savanna zone and, as in similar ranches in the USA and Australia, overgrazing and consequent desertification are prevalent.

The Impact of Overgrazing

Overgrazing affects the vegetation, the soil, and even the health of the animals themselves. It can lead to a decline in the annual production of rangeland vegetation and to a change in the species composition, with a drop in the proportion of palatable grass species, particularly perennials which are good at holding the soil together. Rangeland degradation in Lesotho has allowed the invasion of drought-resistant shrubs which are of little use either as forage or for preventing soil erosion (Speece and Wilkinson 1982). In Sudan some of the most palatable and nutritive plants have disappeared to be replaced by undesirable species that are either less palatable or totally unpalatable. Siha (*Blepharis* spp.) once dominated the Sudanese semi-desert but has been replaced by nal (*Cymbopogon nervatus*) and gao (*Aristida* spp.). Bogheil (also *Blepharis* spp.) was formerly common in the savanna but is now found only in a few isolated areas (Khogali 1983). Other plants are more ephemeral, springing up with the onset of the rains rather than having a permanent presence, and they decrease the durability of rangelands.

The uncontrolled browsing of trees and shrubs is another aspect of overgrazing and a potent cause of deforestation. Large areas of what are now savanna grasslands in Africa would have a substantial woodland cover if tree regeneration had not been prevented by thousands of years of grazing, together with annual burning to promote the growth of pasture grasses. Overgrazing has contributed to critical levels of deforestation in upland areas such as the Andes and the Ethiopian highlands, leading to flooding and siltation in the adjacent lowlands because rains are no longer held back by the sponge-effect of the trees and so flood down to the lowlands, carrying with them large loads of eroded soil. Steams now tend to dry up in the dry season, severely limiting the amount of water available for livestock (UNEP 1983).

Overgrazing also leads to soil erosion. The degradation of sparse rangeland vegetation by overgrazing exposes the soil to erosion by wind and water. Soil compaction and sealing result from the trampling of livestock near water-holes and the overgrazing of wet-season pastures. Sand dunes which were previously stable can start to move if the vegetation on their crests is removed by grazing and the sand becomes bare. As the quality and quantity of rangelands decrease, the health of livestock declines and production of milk and meat falls.

Growing Herd Sizes

The most common cause of overgrazing is simply an increase in the number of livestock within a given area. The number of cattle in Niger, for example, increased (according to one estimate) by four and a half times between 1938 and 1961, when it reached 3.5 million, and then by a further 29% by 1970. The country had nine million sheep and goats in 1970, a third more than in 1961 and over three times the number in 1938. The number of camels rose by a factor of seven between 1938 and 1966 and the number of donkeys doubled. Similar trends were observed in Sudan. In the fifty years from 1924 to 1974 (when rains returned temporarily after the first phase of the Sahel drought) the number of cattle increased from 1.5 million to 12.9 million, the number of sheep from 1.9 million to 3.1 million, the number of goats from 1.8 million to 11.5 million, and the number of camels from 400,000 to 8.8 million (Khogali 1983).

Drought hit the herds hard in the early 1970s. According to FAO, 39% of Niger's cattle and 10% of its sheep and goats were lost between 1970 and 1974. In the worst affected districts, mortality was much higher: the Agadez department lost 88% of its cattle, 80% of its sheep, 70% of its goats and 45% of its camels between 1968 and 1974. However, no advantage was taken of these losses to reduce livestock numbers to more sustainable levels. Instead, pastoralists restocked their herds: in Sudan, for example, the number of cattle and goats rose 34% and 6% respectively between 1974 and 1979, and the number of sheep increased more than fivefold (Khogali 1983). The camel population, however, declined by 70% to only 2.6 million. Cattle numbers in the whole Sudano-Sahelian region in 1984 were 161% greater than in 1969, and there were 31% more sheep and 17% more goats (Berry 1984a). Experts were warning of the consequences of this trend long before the latest phase of the drought in 1983–5.

Overgrazing is a problem in other regions too. Large areas of semi-arid grasslands in Ethiopia have been degraded; rangelands in northern Iraq carry one million sheep – four times their estimated sustainable capacity; and Syria's rangelands have three times as many sheep as they can safely carry. In Botswana the number of cattle doubled to 3.5 million between 1966 and 1979 and herds of other small livestock tripled to 1.7 million. According to H. J. Cooke of the University of Botswana, rangeland degradation "can be plainly seen all over eastern

Botswana" (Cooke 1983), and a report prepared for the government of Botswana concluded that: "Virtually all range studies conducted over the past two decades point unequivocally to spreading conditions of overstocking and degradation of the vegetation" (Carl Bro International 1982).

Experts disagree on how to estimate the optimum herd sizes required to avoid overgrazing and range degradation. The concept of "carrying capacity" is not precise, and an "expert" may be no better at calculating it than a herdsman. It also pays insufficient regard to the need for herd sizes to keep in step with the cyclical nature of pasture growth, so a herd size which is ideal from the point of view of avoiding overgrazing in dry years may be far too low to take advantage of the much better forage resources that become available when the rains return. Views differ on the extent to which, on the one hand, rangelands are vulnerable to overgrazing and, on the other hand, inherently resilient to it. There is evidence that rangeland degradation may not be permanent, so that even land near water-holes which has been denuded of all its grass by heavy grazing and trampling can recover when rains return, often fertilized by the dung deposited there by animals. Fencing off small areas of degraded rangeland from livestock can also produce dramatic recoveries in pasture growth in a matter of months, although if such areas are left for too long they can be invaded by impenetrable bush which makes them useless for grazing.

Why Do Herd Sizes Increase?

What has caused such dramatic increases in herd sizes? First, the growing human population means that there are simply more mouths to feed. Of the six Sahelian countries (Senegal, Mauritania, Mali, Burkina Faso, Niger and Chad), four have national population growth rates of 2.5% per annum, or more. The current annual increase in population in the region is around 1.5 million, which is at least six times the death toll during the first phase of the drought in the early 1970s, estimated variously at between 50,000 and 250,000 people.

Second, changes in the economic circumstances of nomads lead to greater emphasis on the role of livestock. In the Sahel, and probably in other areas also, nomads have become increasingly impoverished during the course of this century as their traditional trading role in salt, slaves, arms and gold has become irrelevant. Richard Franke and Barbara Chasin wrote that in northern Niger at the beginning of the twentieth century a camel laden with salt "could be converted into 15–20 loads of millet. By 1945–50, the same salt bought only 6–10 loads of millet. By 1974 the salt was worth but two loads of millet in a good year, but often would bring only an equal weight. Overall this amounts to a 95% drop in terms of trade for the nomads." When their trading role declined the nomads lost their main means of acquiring valuables, such as arms, gold and jewels, which could be stored against hard times such as droughts. Increasing the sizes of their herds was therefore the only way in which nomads could save for the future (Franke and Chasin 1981).

Third, the number of herds has increased, and with it the overall size of the national herd. One factor contributing to this trend is the liberation of the slave and vassal groups who used to serve the nomadic herdsmen. When these groups are liberated because of legal requirements or the reduced economic circumstances of the nomads, many want their own herds or croplands or both, and so the number of herds increases.

Fourth, market forces at home and overseas have caused livestock numbers to rise. When European Economic Community (EEC) subsidies raised the price of Botswana beef to 60% above the world market price in 1971, many owners of large herds took the opportunity to make money and promptly increased the size of their herds. Total revenue from beef sales almost quadrupled between 1970 and 1977 while the number of cattle slaughtered only increased by a half. According to H.L. Cooke: "The rapid increase in cattle numbers has been primarily due to an even further westward penetration ... into the fragile ecosystem of the Kalahari sand-veld ... together with increasing pressure on the already crowded communal grazing areas of the east" (Cooke 1983). Previously the Kalahari had been little grazed because of the lack of permanent water supplies, but now boreholes have been drilled. There have been few studies of the environmental impact of this trend on the Kalahari, but one report has concluded that because of the large numbers of cattle and poor range management there is a serious risk of possibly irreversible land degradation (Skarpe 1981).

Fifth, the introduction of better veterinary care for livestock has decreased mortality rates considerably. The lower the mortality rate the higher the net animal population growth rate. Government departments concerned with livestock production have until recently tended to concentrate on vaccinating animals against rinderpest and other infections, rather than on breeding strains suitable for dryland conditions. The health of livestock has also benefited from the better water supplies resulting from the larger number of available water-holes.

To the outsider it may seem strange that herdsmen do not take advantage of improved health care to graze a smaller number of better-quality animals, and even stranger that they do not reduce the sizes of their herds when they know that a drought is expected or has actually arrived. Yet these actions are not illogical to herdsmen, who regard the livestock, rather than the land and its vegetation, as their resource base. The number of animals is therefore their prime consideration, and overgrazing the land is of less concern than the possibility that they might have insufficient animals to take advantage of good pasture growth in wet years and to maintain viable herds in the aftermath of a drought. Problems arise because they are reluctant to reduce stock numbers when the first signs of drought appear, and so the drought-stressed rangelands become overgrazed and degraded.

Shrinking Rangelands

The density of livestock and the propensity for overgrazing are also determined by the area of rangeland. Livestock density may rise when this area is reduced by the encroachment of another land use, such as rainfed cropping whose expansion has meant the ploughing of marginal lands previously used for livestock grazing. In Rajasthan, India, the area of arid land used for rainfed cropping almost doubled from 30% to 60% between 1951 and 1971 at the expense of grazing lands. Despite the decrease in the area of rangeland, animal numbers actually increased and livestock density rose by 75%. In Niger, expanding arable populations have forced cattle herders into increasingly marginal regions and many of them have become the permanent debtors of wealthy entrepreneurs (Blench 1985).

In Algeria the annual grazing cycle of the high plateau has been eroded and the carrying capacity of rangelands impaired by the expansion of arable farming. When the human population was small, and there were fewer than 2.5 million head of sheep and goats grazing the area, nomadic grazing was complementary to the more settled pastoral farming of the lowlands. By 1980 the plateau's population had risen to 4 million and grazing land came under pressure when more than 1,000 ha of the best-watered lands were ploughed up for irrigated agriculture. These changes have impoverished rather than enriched rural populations, nomads have become settled farmers, and there is severe pressure on grazing lands, especially those near water sources (Boukhobza 1982). Nomads generally do not take kindly to the loss of their traditional grazing lands to peasant farmers. The two goups came into conflict in Mali when peasants started to crop traditional rangelands in the Niger delta following the most recent acute phase of the Sahelian drought in 1980–4 and the lower Niger flood.

Deforestation also causes a decline in the fodder content of rangelands. Although we might think of shrinking rangelands merely in terms of a reduction in the area of grass available for grazing, as far as livestock are concerned rangelands are three-dimensional assemblies of fodder, in which grasses, shrubs and small trees are all valuable food sources. Yet in many dryland areas fuelwood shortages are so dire that whole trees are ripped up for fuelwood. This, together with the less extreme hacking of branches, destroys a vital source of food for livestock. It also reduces the soil's protective cover of vegetation, thereby rendering it more prone to erosion by both wind and water.

In some countries the actual area of rangeland used for grazing may be artificially restricted by custom, land tenure or even armed conflict. Hugh Lamprey and Hussein Yussuf of the UNESCO Integrated Project for Arid Lands (IPAL) in Nairobi have shown that in the Mount Kulal area of northern Kenya 40% of the land is totally unused because pastoralists are afraid of raids from other tribes. As a consequence of this restriction 20% of the area is being overgrazed by sheep, goats and camels and is acting as a nucleus from which land degradation and desertification are spreading outwards (Lamprey and Yussuf 1981).

The Sedentarization of Nomads

Livestock density also increases when nomads and their herds are settled in villages or the passage of nomadic herds is concentrated along certain routes by the building of chains of boreholes. Many governments are now "encouraging" nomads to settle and give up their itinerant lives, a process called "sedentarization". There are good administrative reasons for this. Nomads are difficult to tax and control, they cross frontiers and carry arms, and their constant movement hampers the provision of education, health care and other social services. Moreover, the expansion of rainfed cropping is causing conflicts between farmers and herdsmen which governments are anxious to avoid. Unfortunately, the settlement of nomads, whether voluntary or enforced, breaks up large herds into smaller units which become concentrated around villages and cause degradation of vegetation and soil erosion. Traditional nomadic groups in Burkina Faso have increasingly become involved in cattle breeding, and the resulting sedentarization over the last twenty-five years has led to a 60% increase in the area of cropland under millet, and to serious overgrazing and desertification (Krings 1980).

Overgrazing around settlements is likely to become a much more serious problem in the future as the inevitable trend towards nomad sedentarization continues. In some countries like Algeria the transition from nomadic herdsmen to settled pastoralists has been taking place gradually for some time. Herdsmen in the central steppe area of that country traditionally moved north and south with the seasons, also fulfilling an important trading function. However, the growth of settled agriculture and the disruption of trade routes undermined the economic and social basis of nomadic life (Rezig 1982). As settlement increased, the ownership of herds and flocks became more concentrated. Some nomads have retained their traditional lifestyle but they have had to diversify to survive. In other countries nomads have been forced to settle by governments. From 1962 onwards nomads in Iran were prevented from using their traditional seasonal grazing lands by government-enforced land tenure changes which transferred land titles from large landowners to peasants, who then increased both their cropland areas and the size of their flocks and herds. Many nomads subsequently settled, to become pastoralists or farmers or both. Some settled at the edge of their former grazing areas while others chose small pockets of land which had not been taken over by peasants.

Expanding the Number of Boreholes

The routes taken by nomadic herds in their long treks over the vast dry rangelands have always been determined as much by the location of water-holes as by the availability of pastures. In this century numerous attempts have been made to improve the supply of water by sinking boreholes and wells. Boreholes, also called tubewells, are deep wells, often extending down 100 to 200m. They are relatively small in diameter, are constructed using a drilling rig, and require a pump to bring water to the surface. Thousands of new wells and boreholes have been constructed in the Sahel since the end of World War II, and more than 600 were dug in Mauritania, Mali, Burkina Faso and Niger between 1949 and 1954 by one organization alone (Fonds d'Équipement et de Développement Économique et Social). These new water sources, however, act like magnets for livestock. Trekking routes become more concentrated, and when animals reach the water-hole they congregate in the area, overgrazing the vegetation and compacting the soil for distances of up to 20 km from the well itself.

Many of the wells in the region north of Tahoua in Niger were in the 1970s attracting up to four times the number of animals for which they had been designed. A study of the Agadez and Azawak regions in Niger prepared for UNCOD showed that:

> Strategies designed to restrict grazing around these points did not work. Grazing loads were 2–3 times those envisaged. It proved impossible to refuse water to herdsmen in an effort to restrict grazing pressure or to reserve certain pastures. The cumulative effects of trampling and of grazing during the wet season resulted in devastated areas extending 10–12 km around watering points. It was the failure of pastures not of waters, that led to deaths of stock. (Bernus 1977)

The Séno Mango, a semi-desert duneland in Burkina Faso near

the Mali border, could revert entirely to desert if the government carries out its plan to set up a ranching project there, for this would require the reopening of a borehole which produced water in the severe drought years of 1972 and 1973. The new borehole would probably attract countless nomadic groups and their herds to the region, with disastrous results (Benoit 1984).

Social Control Breaks Down

Changes in livestock density that are detrimental to the rangelands are often a reflection of a breakdown in traditional grazing controls. Extensive grazing in the Sahel has been possible for many centuries only because of the rigid social control exerted by the nomads themselves over the movements of their animals. These controls have now broken down, more as a result of outside influences than any other cause. The nomadic Tuaregs in the Sahel, for example, have become impoverished as their commercial and trading functions have been sharply reduced, and their access to pastures seriously affected by the expansion of cash-crop cultivation and the introduction of the cash economy.

The digging of wells has caused a distintegration of the previous sharp divisions between the territories which could be grazed by different peoples. In the absence of such strictly enforced controls arguments can easily ensue. A well-digging programme in Senegal, for example, reduced the diversity of herding routes and caused violent conflicts between herdsmen and farmers, both of whom were attracted by the new water supplies. Richard Franke and Barbara Chasin described the repercussions: "The nomads, who were being driven by the expanding peanut cultivation, were gathering in greater numbers along the trek routes, where the wells were providing water. But this very increase in herd density led to a destruction of vegetation along the trek routes and has thus contributed to soil erosion" (Franke and Chasin, 1981).

The loss of social control on grazing in Somalia has been attributed by Jeremy Swift, of the University of Sussex, to the growing commercialization of the livestock sector in that country from the mid-1950s onwards. This was ostensibly caused by the rise in livestock exports to the nearby oil-producing countries of the Middle East. This led to the growth of a new social class of livestock merchants and began to break down the traditional kinship links between nomadic pastoralists that for generations had formed the basis of social control over the pastoral economy. As a result, there have been private appropriations of what had previously been communal resources, including the fencing of some rangelands, increased private cultivation of fodder, and a growth in the number of private boreholes from which water is sold to pastoralists. Herds have increased in size because water is easier to obtain now that it does not have to be pumped by hand. New pastoral groups have developed whose only qualification for access to pasture is their ability to pay for water, and this has further undermined the power of groups with traditional water rights in that area (Swift 1977).

Cattle Ranching

Large ranches, the typical kind of pastoral system in the dry areas of developed nations, can also suffer from desertification. One cause of overgrazing is the coincidence of wet periods with low market prices: this encourages owners to keep cattle until prices improve and therefore to overstock their ranches. In the USA, more than half of the privately owned rangelands are producing forage at half of their potential or less because of overgrazing and consequent soil erosion (Kates et al. 1977). A survey of a 64,000 sq km area in the Gascoyne Basin of northwest Western Australia (Williams et al. 1977) in 1969–70 showed that, after sixty years of heavy sheep grazing, 15% of the area was so badly degraded and eroded that continued grazing was expected to result in irreversible deterioration. More than half of the area was degraded to some extent and in need of proper management. Too little advantage had been taken of good years to keep stock numbers down and regenerate rangelands. Many of the less productive properties in the Gascoyne Basin are likely to be abandoned during the 1980s because their owners cannot afford to run them at sustainable carrying capacities.

Western-style ranches are springing up in Africa (see also Chapter 6). Cattle ranches owned by members of the urban élite have displaced traditional pastoral farms in parts of the semi-humid grasslands of Gongola state in Nigeria's so-called

"Middle Belt" (Blench 1985). In Niger, between 1964 and 1968, 110,000 ha of land at Ekrafane, 300 km north of Niamey, were closed to traditional transhumant herders so that a cattle ranch producing beef for export could be established. However, ranches in Africa have not been very successful, and according to the Club du Sahel: "the investments called for were too heavy; the farms, placed in regions poor in resources, resulted in mediocre productivity; and finally, marketing of production was hampered because the ranches were too far from large commercial centres" (Club du Sahel 1980). Another reason for the failure of most commercial ranches in Africa has been their orientation towards meat production, whereas local people have traditionally used livestock mainly to produce milk. Commercial ranches have rarely even attempted to integrate their work with the pastoralists already there.

POOR IRRIGATION MANAGEMENT

An increase in irrigated cropping might seem to be the logical way to solve the food problems of dryland areas. Using water from rivers flowing through the drylands or from underground aquifers, irrigated cropping is supposedly independent of the cyclical variations in rainfall, and the threat of crop failure during droughts is therefore removed. Watering crops can indeed increase the yield of cereals sixfold and the yield of root crops fivefold. By producing large yields on relatively small areas of cropland it can in principle allow agricultural production to keep pace with the needs of rapidly growing populations, and help to stem the vicious circle of desertification in which the area of (often marginal) land under cultivation has to be continually increased to compensate for falling yields caused by poor farming practices.

For these reasons irrigation schemes in dryland areas have been promoted by both large international agencies and smaller non-governmental organizations active in supporting development. In 1980 the World Bank and its soft-loan affiliate, the International Development Association, lent $1.2 billion (over a third of all bank lending for agricultural development) to developing countries for irrigation schemes with budgets totalling $2.6 billion. Some 250 million ha of irrigated farmlands now exist in the world, accounting for about 13% of all land under cultivation. Despite the theoretical advantages of irrigation, the reality is very different and the poor management of many irrigation projects usually causes productivity to fall after a few years of operation. If allowed to continue this can lead to soil salinization, alkalinization (see p. 32) and waterlogging which will eventually make the land unproductive. Irrigation is, paradoxically, therefore a cause of desertification as well as being a potential cure.

Problems with Irrigation

If not properly designed and managed, irrigated cropping can turn land into desert. The key to good irrigation is good drainage. If irrigated land is not drained properly, the soil becomes first waterlogged, and then salinized or alkalinized. A salt crust may form on the surface. Too often attention focuses on the building of dams and irrigation channels, which brings large profits for construction companies, while the need for adequate drainage is neglected. If irrigation water cannot drain away, the soil becomes waterlogged and difficult to cultivate. When the dry season comes, high temperatures increase the evaporation of water from the soil. This produces a "pumping" action which brings more water from lower soil levels up to the surface. As the water evaporates, soil water becomes saltier and salts are deposited as a white crust on the surface. Waterlogging and salinization lead to lower crop yields; if the situation is not corrected the land eventually has to be abandoned.

Throughout the drylands, an estimated 40 million ha of irrigated cropland is affected by salinity, alkalinity or waterlogging. Another 500,000 ha, or about one eighth of the area brought under irrigation each year, becomes desertified annually (Dregne 1983a; Mabbutt 1984; Winrock International 1985). Half of all desertified irrigated cropland is in South Asia, with Pakistan and India being the most seriously affected countries. Pakistan is very dependent upon irrigated cultivation, which accounts for three quarters of all cropland. Yet, according to various estimates, 30 to 60% of its 15 million ha of irrigated cropland suffers from salinity and/or waterlogging. After China, India has the second largest area of irrigated cropland in the world – 40 million ha – but about 7 million ha is affected by salinity and waterlogging. A particularly bad case is the Ramganga irrigation project in Uttar Pradesh where in the early 1980s 195,000 ha were waterlogged and 352,420 ha salinized; the affected areas were increasing at annual rates of about 50,000 ha and 28,000 ha respectively (Joshi and Agnihotri 1984).

Egypt is almost entirely within the hyper-arid zone and has long depended on irrigated agriculture which uses flood waters from the River Nile. In the 1950s, faced by a rapidly growing population, the government embarked on the massive Aswan High Dam project. Completed in 1960, this extended the country's irrigated lands by 552,000 ha, with each field being cropped two or even three times a year. Overall food production has increased by a modest amount but has failed to keep pace with population growth. Even that gain is under threat: waterlogging and salinization now affect 28% of all Egypt's farmland and average yields in affected areas have fallen by about 30% (Speece and Wilkinson 1982).

Land use on the Golodnaya Steppe in Soviet Central Asia was once dominated by nomadic cattle grazing, as in the Sahel. Irrigation of the fertile loess-type (fine, dusty) soil began in 1902, but within seven years the water table had risen from 10–20 metres to 1.2–1.6 metres below the surface. This made 44% of the irrigated land saline: 6% was irretrievably lost to productive use with the remaining 38% severely saline. The construction of drains to cure salinization began in the 1920s (when more than 20,000 ha had been irrigated) but was not completed until the 1950s. By 1953–4, 92% of the irrigated land was saline. During the 1960s up to 20,000 ha of new land were irrigated every year; between 6,000 and 10,000 new settlers arrived annually, and energetic attempts were made to drain the soil. Despite this effort, 86% of the more than 250,000 ha of land which had been irrigated by 1970 was saline (USSR 1977).

Groundwater is likely to be saltier than surface water and can become even more so if it is continually recharged with salty water that has been used for irrigation. In Jordan and the United Arab Emirates, the excessive use of well water has lowered the water table, made wells saline, and thereby salinized the soil. Even the United States is not immune to such problems, for salinity affects over 12 million people and 1 million irrigated acres in the Colorado River basin (Gardner 1984). In the San Joaquin Valley of California, the major area for fruit and vegetable production in the USA, the evaporation of irrigation water has made the groundwater inflow increasingly saline and in some areas the land is visibly caked with white salt. Fruit and vegetables with a low saline tolerance cannot be grown there any more (Risser 1985).

Other Causes of Salinization

Soils can be made saline in two other ways that are not directly connected with irrigation. First, in coastal or estuarine areas, seawater may encroach into the subsoil when excessive use of wells has lowered the (fresh) water table. Oman and Yemen are suffering from this kind of problem and, according to Mark Speece of the University of Arizona and Justin Wilkinson of the University of Chicago: "The aquifer becomes polluted with salt water, soils become salinized, the land loses its capacity for agricultural production, and the desert moves in. Along the Barinah coast in Oman ... date palm groves are dying because of salinized water, even though the date palm is a relatively salt tolerant tree" (Speece and Wilkinson 1982).

The second type of problem occurs where the deforestation of watersheds leads to saline seepage. Replacing trees in upland areas with crops like wheat (which have a lower evapotranspiration rate) reduces the amount of water transferred from the ground into the atmosphere, and causes water to build up in the soil. Some of this water will run off underground, but in doing so it dissolves salts (such as sodium chloride) out of the soil and becomes saline. Lowland areas now receive more water than previously, and without adequate drainage they can become increasingly waterlogged and salinized. The rivers into which the saline water eventually flows also become saltier, and croplands irrigated by water from these rivers are therefore at risk of salinization. Saline seepage is a major problem in Australia, increasing the salinity of rivers and reservoirs and causing waterlogging and salinity in lowland agricultural areas. The salinity of the Murray River rose by 84% between 1938 and 1981, while in the country as a whole about 8% of all irrigated lands (123,000 ha) have been salinized and about four times this area suffers from shallow groundwater tables (less than 2 metres below the surface).

Irrigation in Iraq

Farmers in the basin of the rivers Tigris and Euphrates, in present-day Iraq, have practised irrigated agriculture for 4,000 years, but whenever their management of irrigation has lapsed the land has become waterlogged and saline (see Chapter 1). In 1953 Iraq started yet another irrigation project, the Greater Mussayeb Irrigation Project. At the outset 60% of the land was saline; irrigated land was poorly surveyed and levelled; there were no field trials to determine the water requirements of different crops; managerial staff were inadequate; irrigation canals silted up; and drains were badly maintained. By 1969 waterlogging was common and two-thirds of the soil was saline. Yields fell and barley was substituted widely for wheat. Many farmers reverted to the traditional "nirin" cropping system which involves cultivating cereals under flood irrigation until the land becomes saline, then leaving it fallow so that the rain can leach salts out of the soil.

In 1965 a $10 million scheme began to rehabilitate the Greater Mussayeb Irrigation Project, receiving support from FAO and the UN Development Programme (UNDP) from 1970. The aim was to increase the number of technical staff, install tile drains to prevent waterlogging, repair canals, and reclaim saline lands. By 1974, one-third of the saline land had been reclaimed, but agricultural extension services were inadequate to give proper technical support to farmers; within five years from the start of the scheme they began to revert to the "nirin" system. The canals still silted up (2,700 tonnes of silt were deposited each day), drainage ditches filled up with windblown silt, and few tile drains were installed. UNCOD was told that in 1976 the project remained a failure: 64,000 ha supporting only 32,000 people. Farm income represented less than a 5% return on capital invested, and most farmers grew subsistence rather than cash crops (Dougrameji and Clor 1977).

Many people dream of using irrigation to turn the deserts green, but examples such as Iraq are sobering reminders of the technical and organizational limitations which make the reality of irrigated cropping very different from the ideal. In trying to solve one problem we create another. The installation of many slow-moving irrigation canals, for example, may spread debilitating diseases by encouraging the breeding of malarial mosquitoes or the snails which act as hosts for the bilharzia fluke. The costs of failing to design and manage irrigation projects properly are huge, both in human terms and in the effects on the land and environment in general. In Algeria the first stage of a rehabilitation project involving some 18,500 families in the Cheliff Basin is costing $13.3 million. Between 1978 and 1983 Pakistan was spending 40% of all funds for water development on the reclamation of waterlogged and salinized land while new irrigation schemes received only 25% (Speece and Wilkinson 1982). It is likely that in the remaining years of this century many other countries will also have to devote increasing funds to the costly process of rehabilitating existing irrigation schemes and place less emphasis on new schemes. Standing still is better than falling behind in the race to feed growing populations. Designing irrigation projects properly in the first place would be even better.

DEFORESTATION

The open woodlands that still cover large areas of the drylands consist of trees and shrubs of modest height that are widely scattered across grasslands. Their name derives from the fact that, in contrast to closed forests like the tropical rain forests, the forest canopy is more open and covers a lower proportion of the ground surface. Open woodlands may look less impressive than the dense forests of the humid tropics, but they have a very important role in the local economy as sources of building poles, fuelwood and fodder. Fuelwood was previously gathered mainly from dead wood, but now large-scale fuelwood cutting is a major cause of deforestation. Tree leaves, seeds and pods are significant sources of fodder, accounting for about a third of the total annual food supply for livestock in the Sahel. Their contribution is crucial towards the end of the dry season when other vegetation has withered away. They are the source of up to 45% of all fodder eaten. Local people also derive a wide variety of other products from woodlands, including fruits, gums, honey and medicines.

Marilyn Hoskins of FAO has given an excellent description of the importance of woodlands to people in Burkina Faso:

Local women ... collected shea nuts (*Vitellaria paradoxa*) from which they made cooking oil, they gathered leaves and seeds essential for the nutritional sauces they put over their starchy staple grains, they searched for grasses and bark for weaving and dyeing mats and elaborate baskets, concocted home remedies from leaves, pods and roots, and let their goats browse on the shrubs and bushes in this unused looking area. Women also piled their heads high with dead branches and sticks to carry home for cooking fuel. Their children ate the nutritious monkey bread or baobab (*Adansonia digitata*) fruit or hunted small animals in the undergrowth. Their husbands cut chew sticks (the local substitute for tooth brushes) and twisted bark into ropes. The whole family picked and ate "desert raisins" (*Cissus* fruits) and other fruits and nuts, and various family members earned small sums selling firewood or other surplus items which the bushland provided. (Hoskins 1982)

There are about 734 million ha of open woodlands, two-thirds of which (486 million ha) are in Africa, and they are being lost at a rate of about 3.8 million ha per year. As might be expected, Africa is the worst-affected continent, losing 2.3 million ha every year (Table 2.1), (Lanly 1982). These figures are necessarily very inaccurate and refer only to clearance. Woodlands are also being degraded in other ways but it is difficult to estimate the scale of the problem. The main causes of deforestation and degradation are agricultural clearance, fuelwood cutting, livestock browsing and burning due to bush fires. While local people are usually selective in culling branches for fuelwood, commercial fuelwood cutters clear substantial

---

Table 2.1: Deforestation Rates for Tropical Open Woodlands, 1976–80 (million hectares per annum)

Africa	2.3
Asia	0.2
Latin America	1.3
Total	3.8

Source: Lanly (1982)

areas at a time. The fuelwood problem is discussed in detail in the next section.

Agricultural Causes of Deforestation

The outright clearance of woodlands to provide extra cropland and rangeland is an important cause of deforestation in the drylands but it has tended to be overshadowed by deforestation for fuelwood cutting (Catterson *et al.* 1985). Agricultural expansion is estimated to result in the clearance of 50,000 ha of woodland in Burkina Faso every year and 60,000 ha in Senegal (Berry 1984a). In Sudan's Kordofan and Darfur provinces about 88,000 ha of woodlands are cleared each year for conversion to mechanized agriculture. An estimated 42,000 ha of this land, after being cropped for sorghum continuously for three to four years, has become degraded and barren, and then been abandoned (World Bank 1986a). In Mozambique, Tanzania and Zimbabwe and other East African countries, large areas of woodlands are cleared for tobacco plantations and to supply the fuelwood needed to cure the tobacco. In Zimbabwe, for example, about 70,000 ha of woodland are cleared each year for tobacco cultivation, and another 75,000–100,000 ha are cleared for fuelwood (Milas and Asrat 1985).

Throughout history, vast areas of open woodlands in dry areas have been cleared for the purposes of livestock raising. Chapter 1 described how this, together with the regular burning of rangelands to promote the growth of edible grasses, has meant that an artificially modified savanna vegetation has become the norm over much of Africa. The main purpose of burning is to suppress and clear shrubs only suitable for grazing by goats and camels, and to provide a fertilizing ash to promote the quick growth of new grass when the rains come. However, the fires prevent the regeneration of trees and shrubs that would otherwise grow to produce a substantial woody cover, and they also spread into denser woodlands where they cause considerable damage.

Those trees that remain are often key sources of animal fodder, which is either cut from them by herders or directly browsed by livestock. Browsing can sometimes even kill a tree, especially when degraded rangelands provide inadequate food for animals. Too much browsing changes the species composition of woodlands so that less palatable species become dominant. In the Phewa Tal catchment area of Nepal a quarter of all fodder comes from forests around the villages, mainly as leafy branches cut by villagers from live trees. According to John Wyatt-Smith, former Forestry Adviser to the UK Overseas Development Administration, the area of forest land needed to supply fodder for the average farm is three to five times that required to provide fuelwood and timber (Wyatt-Smith 1982).

The destructive effect which uncontrolled grazing can have on forests is shown vividly on the Mediterranean island of Cyprus. When the island was placed under British administration in 1878 the forests had been reduced in area by a third in the previous twenty years and were in a critical state because of grazing and the cutting of fuelwood and timber. There were 250,000 goats on the island, two for each person and ninety-six per square kilometre. It took the Forest Department more than seventy years to bring the goat under control by such measures as establishing goat-free villages on the forest margins and giving compensation for loss of grazing rights (Thirgood 1986).

Although deforestation has its specific causes, it is closely linked to overcultivation and overgrazing. According to Seifulaziz Milas and Mesobework Asrat of UNEP: "Increased pressures on arable lands tend to be reflected in encroachment of cultivation on forests and rangelands. This in turn leads to intensified overgrazing and deforestation as grazing areas are diminished, forests are cleared and soil erosion increases" (Milas and Asrat 1985). Tree cover also declines when farmers who have hitherto used tree fallows like those of *Acacia senegal* in Sudan (mentioned on p. 68) reduce the length of the fallows or abandon them altogether in order to intensify cropping.

Fuelwood Cutting

Fuelwood is the predominant forest product used in the drylands. However hot the days may be, the nights are cold, and wood is needed to provide warmth and to cook the evening meal. Half of all the wood used in the world is burned as fuel, mainly in developing nations where it accounts for four-fifths of wood harvested. About 90% of people in developing countries depend upon wood or charcoal as their main source of household fuel. In some countries, such as Burkina Faso, Chad, Ethiopia, Somalia and Tanzania, wood accounts for over 90% of total national energy consumption (Eckholm *et al.* 1984). Traditionally fuelwood has been gathered from dead wood but now the wholesale cutting of trees for fuelwood is an important cause of deforestation.

Of the total volume of wood used as fuel, perhaps half is used for cooking, a third for heating water and providing general warmth, and the remainder for agricultural and other processing purposes (making bricks, baking bread, etc.). The average citizen in a developing country burns as much wood every year (1 cubic metre or 1,000 kg) as a North American consumes in the form of paper. Consumption is above average in forest-rich countries in the humid tropics, but in areas with low forest cover in South Asia, West Asia and West Africa, annual per capita consumption may be only half a cubic metre (500 kg) or even less. Surveys of seventeen villages in a forest-deficient area of the Indian state of Tamil Nadu, for example, showed an average per capita fuelwood consumption of 344–676 kg and in six Sahelian villages in Mali and Niger consumption ranged between 440 and 660 kg per capita (Eckholm *et al.* 1984).

Charcoal is often used in place of fuelwood, especially in cities, where annual per capita consumption typically ranges between 100 and 200 kg per year (Eckholm *et al.* 1984). It is easier to use, contains more heat per unit weight than fuelwood, and is therefore cheaper to transport over long distances to urban areas whose nearby forest resources have been depleted. Consumer preferences vary, however, and while urban dwellers in African countries tend to prefer charcoal those in Latin American countries often have a greater liking for fuelwood. Such preferences are quite fundamental, and it may be as hard to persuade fuelwood consumers to switch to charcoal as it would be to persuade them to switch to kerosene.

Demand for fuelwood in many dryland areas far exceeds the natural growth of woodlands and is a prime cause of deforestation. Rising populations and diminishing forest resources lead in turn to widespread fuelwood shortages. In Burkina Faso, CILSS estimates that demand for wood already exceeds the regeneration capacity of forests and woodlands, and women may have to walk for four to six hours, three times a week, to gather wood to cook the evening meal. In Kenya some women spend up to twenty-four hours a week collecting fuelwood. In Central Tanzania it may take 300 man-days (usually woman-days) per year of work to provide the wood for an average household.

In rural areas people usually gather fuelwood, preferably as dead wood, from trees and shrubs on farms or common lands. Sometimes they are also allowed to gather dead wood from forest reserves, but most wood comes from trees outside defined areas of forests and woodlands. Dead wood is easier to cut and also burns better than green wood. In normal situations such gathering is not very intensive and would not present much of a problem, but rising populations and diminishing forest resources make the impact of fuelwood gathering more significant and, as Erik Eckholm, Gerald Foley, Geoffrey Barnard and Lloyd Timberlake have pointed out: "fuelwood scarcity is as much a consequence as a cause of deforestation" (Eckholm *et al*. 1984). It should not therefore be seen in isolation from agricultural clearance and the overgrazing of forests.

The one exception to this is the cutting of fuelwood for sale in towns and cities by gangs who scour the countryside felling live trees. Even forest reserves are not immune to their depredations. Some of the wood is converted into charcoal in simple earth kilns before being transported to cities by trucks. According to Willem Floor and Jean Gorse of the World Bank it is the rising demand in urban, rather than rural, areas that is responsible for most of the fuelwood scarcity in the drylands. "Urban inhabitants in some cases consume a disproportionate percentage of total fuelwood .... Thus, an inhabitant of Dakar [Senegal] draws, on average, 2–3 times as much from national wood resources as a rural inhabitant. The situation is similar in urban areas in Mauritania, such as Nouakchott" (Floor and Gorse 1987). Ouagadougou's fuelwood consumption, for example, accounts for a startling 95% of total forest production in Burkina Faso. Although urban dwellers usually buy wood or charcoal from suppliers it is still cheaper than competing fuels and so price does not constrain demand very much. In Nairobi, for example, firewood is about one-eighth of the price of kerosene and one-twentieth the price of electricity for comparable amounts of energy, while charcoal is two and a half times the price of wood (O'Keefe *et al*. 1984).

Growing urban demand for fuelwood is strikingly evident in

the "haloes" of denuded land now found around many dryland cities. Dense *Acacia* woodlands were common around the Sudanese capital Khartoum in 1955, but now only isolated woodlots survive within 100 km of the city, a mere 10 million ha of woodland survive within a radius of 500 km, and even this will disappear if management is not introduced quickly (World Bank 1986a). Almost all trees within 40 km of Ouagadougou have been felled and most farmland within a similar distance of the city of Kano in northern Nigeria has been stripped of trees. As a result of such shortages, fuelwood and charcoal have to be shipped to cities from ever increasing distances, up to 500 km in the case of some of the charcoal brought to Khartoum and Dakar. Erik Eckholm has described the racketeering that is inevitable in such situations:

> Well organised syndicates bring fuel by truck, camel and donkey into cities like Ouagadougou in Upper Volta [Burkina Faso] and Niamey in Niger, damaging the landscape in a widening circle. In Sudan, forest rangers have accosted armed crews as they fill trucks with illegally cut wood that will be converted into charcoal for sale in the cities. Shoot-outs reminiscent of the American Wild West have occurred. Rising fuelwood prices can tempt desperate individuals as well as greedy big-time entrepreneurs into cutting live trees. Near Bhopal, the capital of India's Madhya Pradesh State, the forest department has granted people the right to collect headloads of dead wood in the forest reserves for personal use. Yet throughout these reserves are signs that dead wood is being actively manufactured: trees with their bark girdled and trees axed outright. (Eckholm 1982)

Not all the fuelwood cutting is for domestic purposes, for wood is also an important industrial fuel source. In Tanzania, for example, 1 cubic metre of wood is needed to dry 150 kg of tea and brew 400 litres of beer, while making the 25,000 bricks needed for an average family house requires 35 cubic metres of wood, and curing the tobacco grown on a hectare of plantation requires the felling of a hectare of woodland (Eckholm *et al.* 1984).

Most countries in the Sudano-Sahelian region have or are

expected to have fuelwood deficits. Guinea-Bissau, southern Senegal, southern Sudan and Gambia are in fact the only areas where fuelwood resources now exceed present and foreseeable needs (Berry 1984a). Gambia has the highest national average fuelwood consumption rate of any Sahelian country (1.8 cubic metres per capita per year) but the government has heeded the potential threat to future supplies by banning charcoal production, since this tends to lead to large-scale destructive fuelwood cutting of the kind mentioned above (Weber 1982). Although forest resources are still plentiful in southern Sudan long distances and poor communications make the flow of fuelwood to the north of the country virtually non-existent (World Bank 1986a).

Fuelwood shortages are not confined to Africa, but are also common in India, Pakistan, Bangladesh, Central America and arid parts of South America such as northeast Brazil. An FAO survey found that 100 million people in twenty-six countries already experience acute fuelwood scarcity, defined as "a very negative balance where the fuelwood supply level is so notoriously inadequate that even overcutting of the resources does not provide the people with a sufficient supply and fuelwood consumption is, therefore, clearly below minimum requirements." Areas experiencing acute scarcity include: the Sahel; East and Southeast Africa; mountain areas and islands of Africa; the Himalayas; the Andes; and densely populated areas of Central America and the Caribbean. A further 1.3 billion people, nearly two-fifths of the population of developing nations, live in deficit situations in which they are only able to meet their minimum fuelwood needs by overcutting existing forest resources and jeopardizing future supplies. The fuelwood crisis is expected to get even worse by the end of the century, when acute scarcity will affect 150 million people and 1.8 billion will be in deficit situations (FAO 1981).

The total fuelwood deficit in the drylands in 1980 was estimated, on the basis of FAO data, at 207 million cubic metres, of which 90 million cubic metres were in Asia and 72 million cubic metres in Africa. By the year 2000 the total deficit could more than double to 473 million cubic metres. To put these figures in perspective, the current deficit of fuelwood supplies in the drylands is equivalent to the sustained production of almost 26 million ha of intensively managed, fast-growing fuelwood plantations. By the end of the century, this could rise to nearly 62 million ha (Grainger 1988). The US Agency for International Development (USAID) has estimated that if the present rate of overcutting in the Sahel persists until the year 2000, native woodlands will then supply only 20% of regional fuelwood demand. Just to prevent the forecast shortfall in fuelwood production in sub-Saharan Africa by the year 2000 would require the establishment of plantations which, if collected together, would form a forest belt 34 km deep and 6,000 km long across the Sahel from Senegal to Ethiopia.

Action to address the fuelwood problem is being hindered by a general lack of awareness by urban dwellers of any long-term gap between supply and demand. People in rural areas are much more aware because they have to walk for longer distances to gather wood, although to the casual observer, says Peter Freeman, "the presence of trees, bushes and brushland is deceptive. Some trees such as baobab are not good for fuel, wild fruit trees are also spared if at all possible, and fruit trees around compounds, such as mangos, aren't used for fuel" (Freeman 1986). Town dwellers, however, may not realize that the situation is bad, for while firewood prices in urban areas have risen sharply because of higher transport distances and fuel costs (prices doubled between 1975 and 1985 in Ouagadougou), they have mostly stayed constant in real terms. Some low-income families may spend as much as a fifth of their income on buying fuelwood for charcoal, but studies in the Sahel have shown that for higher-income families the corresponding share is as low as only 5-8% (Floor and Gorse 1987). A survey of women in Niamey, Niger, in 1984 showed that 80% did not consider fuelwood expenditures excessive; they were far more worried about the prices of food and clothing.

Deforestation and Desertification

Deforestation is the first step along the road to desertification. In dry areas where vegetation is relatively sparse, trees and open woodlands play a vital role in stabilizing soil and water and giving shade to people and animals. When the trees are removed croplands and rangelands become more exposed to the elements, unprotected soil is baked by the sun and eroded by the wind and rain, the whole area becomes more arid, and towns and villages are exposed to frequent dust storms. It is for these reasons that Jack Mabbutt has called deforestation the most dynamic indicator of desertification, cutting across all the main land uses (Mabbutt 1985). In the words of Leonard Berry: "this visible and pervasive problem is an indicator of the presence of significant desertification. It reflects long run trends, and is presently signalling trouble" (Berry 1984a).

Deforestation around Niamey, the capital of Niger, has set in train a whole set of land degradation processes. Soil erosion by wind has increased, and sand dunes have been reactivated, continuing human and animal pressure prevents the regeneration of vegetation, and severe desertification is taking place within a 40–50 km radius of the city (NRC 1983). Nouakchott, the capital of Mauritania, is suffering similar problems and is threatened by sand dune encroachment and frequent dust storms (Berry 1984a).

Deforestation in the Ethiopian highlands is causing "massive soil erosion [that] is inexorably converting large areas ... into stone deserts" (UNEP 1983). Increasing populations and desertification in lowland areas have forced farmers to deforest and cultivate steep mountain slopes in regions such as Wollo and Tigrai and as much as 40,000 sq km of these highlands have been almost irreversibly degraded. Ethiopia is said to be losing up to 1,000 million tonnes of topsoil every year and national forest cover has fallen from 16% a few decades ago to just 3% (Milas and Asrat 1985). Overgrazing of mountain pastures is also widespread, as almost a third of Ethiopia's 77 million livestock, the largest number of any African country, live in these highlands. Once the trees are gone the soil is washed away by rains, and in one area the average annual soil loss is 20 tonnes/ha. Rains flood down into the valleys instead of being released gradually by the watershed forest cover, and streams which once flowed all the time now dry up soon after the rains end. Wollo and Tigrai have been heavily affected by drought and famine since the early 1970s and tens of thousands of people have died.

The effects of deforestation are often experienced far from where it actually occurs. Egyptian agriculture has depended for thousands of years on the fertilizing qualities of the annual floods of the River Nile, rich in silt eroded from the Ethiopian highlands, but the river's sediment load has now become so great as to pose a serious threat to irrigated agriculture in both

Egypt and Sudan, reducing the operational life of reservoirs and inhibiting the flow of water in irrigation canals. The Khashm el Birba reservoir on the Atbara River in Sudan, completed in 1976, lost half its capacity in just six years. The Roseires Dam on the Blue Nile had lost a third of its capacity by 1981 and its hydroelectric output has also been greatly reduced, by as much as 80% during peak flooding periods; this seriously affects power supplies to the greater Khartoum area (World Bank 1986a). Silt deposition in all of Sudan's rivers has caused primary water channels to divert from their courses, away from established agricultural areas (Freeman 1986).

Farmers in dry lowland areas of India and Pakistan are suffering from the flooding and sedimentation of rivers rising far away in the Himalayas , where deforestation has reduced forest cover to only 25%, too little to hold back the monsoon rains and release them in an orderly flow over a period of months. Seventeen of India's major reservoirs are silting up at three times the expected rate, and thousands of people and livestock have died in India in the last decade because of the effects of floods. Direct expenditure in India to offset flood damage averaged $250 million a year between 1953 and 1978.

Deforestation also causes soil erosion and desertification in an indirect way. When wood becomes very scarce people burn animal dung instead; because this would otherwise be used to fertilize croplands and rangelands, burning it depletes soil fertility and exacerbates desertification. In India, 60–80 million tonnes of dried dung are burned as fuel each year, providing 10% of national energy requirements. The practice is becoming more common in other arid countries, even those like Morocco and Turkey which are considered to be relatively prosperous. Michael Arnold, of the University of Oxford, estimates that the loss of grain production in Asia, Africa and the Near East resulting from the diversion of dung from the soil to the fire could be as much as 20 million tonnes per year: enough to feed 100 million people, one-third as much again as the annual population increase in Africa, Asia and Latin America (Arnold and Jongma 1978).

CAUSES AND SOLUTIONS

Desertification has four main direct causes, overcultivation, overgrazing, deforestation and mismanagement of irrigated cropland. These do not arise by accident, but are greatly influenced by the effects of growing populations, economic development and conscious policy decisions by governments and aid agencies (see the next chapter). The displacement of one land use by another is also an important factor. Thus an expansion of cash-crop cultivation to increase foreign-exchange earnings can displace rainfed cropping on to former rangelands; this pushes herds of livestock on to ever more marginal land, which becomes desertified as a result of overgrazing. The expansion of cropping and grazing also causes deforestation, which in turn makes those two land uses even less stable by removing soil protection and depleting fodder resources. Ironically, irrigated cropping, which could make drylands more productive and less susceptible to drought, is itself a major cause of desertification when poorly managed. How one land use impinges upon another is often not predictable, and so the only practical approach to controlling desertification is to ensure that the way in which an area of land is used does not exceed its capabilities, and that rainfed cropping, irrigated cropping, livestock raising and wood production are made as productive and as resilient to climatic variation as possible.

3. Desertification, People and Policy

The last chapter explained the growth of desertification in terms of the four main direct causes – overcultivation, overgrazing, deforestation and the mismanagement of irrigated cropland. It also indicated that these causes were greatly influenced by a variety of underlying social, economic and political factors, and were not just a consequence of faults in a particular land use, although such faults undoubtedly exist. This chapter looks in more detail at these underlying causes of desertification. Population growth and economic development are the two main driving forces which lead to an expansion of agriculture and to changes in the types of agriculture practised. Economic development is also generally accompanied by a growth in urban populations, and serious degradation of natural resources is common around towns and cities. The benefits of economic development are not shared equally among the inhabitants of a country, and it is often the poorer people, forced to live on the worst lands, who are the most directly involved in causing desertification, the most seriously affected by it, and the least able to prevent it from happening. They are also often affected by famine, although that is not an inevitable consequence of drought or desertification, and can occur when government policies constrain food production in particular areas and fail to alleviate poverty. This chapter argues that in order to assess the causes of desertification correctly, and devise programmes to bring it under control, it is necessary to take into account the pattern of land use not only in the desertified areas themselves but in the country as a whole, and the social and economic factors and government policies which determine that pattern. Although it may appear from this analysis that desertification is yet one more regrettable symptom of underdevelopment, at the same time it is also shown to be amenable to control by better land-use planning and improved government policies.

POPULATION GROWTH

Populations are growing rapidly in most developing countries and the drylands are no exception. Since the mid-1970s, population growth rates have exceeded 2% per annum in four-fifths of the sixty-six countries containing appreciable areas of arid or semi-arid lands, and exceeded 3% per annum in a third of those countries. They have also been greater than 3% per annum in over half of the thirty-four predominantly dry countries (those in which more than three-quarters of the total area is arid – see Chapter 1), and in the Sudano-Sahelian region, where in the seven years from 1977 to 1984 the population increased by almost a quarter (Berry 1984a).

Population growth is a major cause of land-use change, as farmers try to grow more food to feed rising numbers of people by increasing either the yield per hectare or the area under cultivation. Food production has more than kept pace with population growth in Latin America and Asia, but not in Africa. Within Asia there is a great difference between the performance of agricultural sectors in the countries of the Far East (South and Southeast Asia) which have adopted the intensive Green Revolution approach so enthusiastically, and the predominantly dry countries in the Middle East (West Asia) which have lagged behind. Between 1961-5 and 1985-6, food production per capita fell by 13% in Africa, but rose by 17% in the Far East, by 8% in the Middle East and by 5% in Latin America (FAO 1975, 1987). Such comparisons are inevitably biased by the effects of drought on food production in Africa especially, but do illustrate general trends.

Population growth does not necessarily result in desertification (Blaikie 1985), but can help to induce it if the only way to increase food production is to increase the area being cropped. Yields per hectare can be increased by more intensive cropping, but these can only be sustained through investment in fertilizers and more productive cropping systems. Many dryland countries have insufficient capital to invest in agriculture and, without fertilizers to replenish nutrients taken from the soil, more intensive cropping leads to declining fertility and falling yields. To maintain production a larger area of land has to be cropped, and since this is often marginal land unsuitable for cropping, the result is further soil erosion. Agricultural productivity in the largely dry areas of Africa and the Middle East (West Asia) has grown much more slowly in recent years than that in Latin America and the Far East (South and Southeast Asia). Between 1961–5 and 1985–6, for example, cereal yield per hectare rose by 39% in Africa and 35% in the Middle East, but by 68% in the Far East and 58% in Latin America. In the Sahel the rise was only 8%((FAO 1975, 1987).

Often the reason for overcropping, and the overuse of natural resources generally, is not so much that the population in the country as a whole has risen, but that there has been a growth in population density in certain areas. Many farmers are unable to move elsewhere when their areas become overcrowded, and are too poor to invest in more productive types of farming (Blaikie 1985). So overcrowding causes farmers to crop their lands more intensively, e.g. by reducing fallow periods or cultivating a particular patch of land for a longer time. Traditional cropping systems (such as the *Acacia* bush fallow system in Sudan, described in Chapter 2) which have proved sustainable for hundreds of years because of their long fallow periods, break down when the fallow periods are reduced. Yields decline, and the land becomes degraded.

Population density can also rise because of land tenure constraints and political reasons. Chapter 2 included examples of land degradation in Lesotho and Zimbabwe, where large numbers of peasants are confined by long-standing land tenure rules to relatively small areas of unproductive land. As their population rises, so does population density, and degradation occurs. The government of Tanzania has for some time been resettling people into new villages and the resulting concentration of population is leading to desertification because of overgrazing and overcultivation in the vicinity. Desertification due to overgrazing also occurs around villages in which nomadic herdsmen have been resettled (or "sedentarized"). Rings of rangeland degradation and deforestation are found around settlements, whether small towns or large cities, as their inhabitants graze livestock or gather fuelwood.

Migration is one way to alleviate overcrowding, although in Kenya the continuing movement of people from the highlands to arid and semi-arid lands is causing great concern as to whether these lands are able to support an ever increasing population. Despite government efforts to mitigate the situation, life in the Kenyan drylands is becoming ever more hazardous (Bernard 1985). Many recent migrants in the African drylands are refugees fleeing from drought, famine and war, and anxious to re-establish their livelihoods elsewhere. But in the process they can create tremendous social and environmental problems. When circumstances are very dire migration even transcends national boundaries; for example, the Ivory Coast has received a large number of migrants from Burkina Faso and Mali.

ECONOMIC DEVELOPMENT

One of the most worrying kinds of migration for government planners is the constant stream of people travelling from rural to urban areas in search of jobs in commerce and industry as the market economy grows. Urban populations in dryland areas have been growing at 5–6% per annum over the last twenty years, much faster than rural populations (MacGregor and Valverde 1975; Berry 1984a), and a significant proportion of this growth (50% in the case of the Sahel) is accounted for by migration from rural areas (Berry 1984a). Rural–urban migration is typical of countries undergoing economic development, but places a tremendous burden on governments which have to provide costly urban services. Agricultural development also suffers, since the cities consume a growing proportion of the national budget, the rural labour force declines, and farm prices have to be kept down by governments to satisfy urban populations and maintain political stability. Rises in food prices can easily lead to riots which may provide the excuse for a *coup d'etat* by ambitious army personnel.

The growth of urban centres represents a serious environmental problem for many developing countries because they lack the sophisticated production, technology and supply networks found in western cities. Therefore local forest and pasture resources usually have to be exploited for fuelwood and fodder (Kates *et al.* 1977). Other key resources, such as groundwater, also become depleted. The Sahel is one of the least urbanized regions in the developing world, with less than

20% of its total population living in towns and cities compared with an average of 29% for Africa as a whole and 31% for all developing countries. Because there are so few large towns and cities, urban growth in the region has particularly harmful environmental effects because it is concentrated in the capital cities. The capitals of Chad, Mali, Niger and Burkina Faso contain more than half of all the four nations' urban populations (that is, people living in towns with more than 10,000 inhabitants). The population of Ouagadougou, the capital of Burkina Faso, nearly tripled between 1961 and 1979 while that of Khartoum (Sudan) increased even faster.

Another consequence of economic development is that the general pattern of land use undergoes changes. Some of these may be beneficial, for there is an increased ability to invest in more productive cropping techniques, improved seeds, fertilizers, pesticides, better crop storage, etc. Other changes lead to mixed benefits. A developing country will usually seek to improve its international trading position by growing more cash crops for export, and while this improves its foreign currency earnings the expansion of cash-crop cultivation may displace traditional land uses. Farmers practising subsistence rainfed cropping then have to move to land previously considered too marginal for this use, which in turn shifts pastoralism to even more marginal areas.

The growth of the market economy also leads to a breakdown in traditional forms of social control over land use, for example the way in which nomadic herdsmen used to control the grazing of rangelands (see Chapter 2). The importance of such traditional mechanisms may not be realized until they have largely been dismantled, by which time it is usually too late to restore them. Numerous attempts have been made to introduce new forms of rangeland management in Africa, but all have failed.

The adoption of technological innovations characteristic of economic development in the industrialized countries can often take a long time to occur in developing countries. In the case of energy, for example, many urban dwellers remain committed to the use of fuelwood or charcoal for cooking and heating, and this leads to tremendous pressure on local wood resources and therefore to deforestation. On the other hand, technological change overemphasized at the expense of traditional methods can also have detrimental effects. The last chapter showed how huge funds have been poured into expanding irrigated cropping while subsistence rainfed cropping has been neglected. Much of this money has in effect been wasted, because the poor design and management of irrigated cultivation schemes have resulted in salinization and waterlogging. The introduction of mechanized farming techniques from developed nations can also cause problems, for example in Tunisia where the disc ploughing of fragile soils has led to widespread soil erosion. The immunization of animals against debilitating diseases should, by reducing animal mortality, lead to an increase in the productivity of livestock raising, lessening pressure on rangelands. But what often happens is that pastoralists continue to keep their herds as large as possible, to ensure that some animals survive the frequent droughts. The result is overgrazing. Well-meaning attempts to improve livestock health and production by sinking more boreholes also lead to desertification because they tend to channel nomadic herds along the small number of routes which link the boreholes. Pastures along these routes become overgrazed and the soil becomes compacted after being trampled on by thousands of hooves.

DESERTIFICATION AND UNDERDEVELOPMENT

Although population growth and economic development are the two basic driving forces which lead to changes in land use, the nature of these changes depends very much on a particular country's economic status. Poorly developed countries usually lack the capital to invest in making agriculture more productive and sustainable, many people live in conditions of abject poverty, and the skilled personnel needed to plan the wise use of land (and of natural resources generally) are in short supply. Most of the countries which are experiencing serious desertification are economically underdeveloped, and therefore lack the resources to respond to change in a sustainable way (Caldwell 1984). Piers Blaikie of the University of East Anglia described the dilemma: "environmental degradation is seen as a result of underdevelopment (of poverty, inequality and exploitation), a symptom of underdevelopment, and a cause of underdevelopment (contributing to a failure to produce, invest and improve productivity)" (Blaikie 1985).

Some evidence for a link between desertification and underdevelopment is apparent in the results of the UNEP survey of desertification, completed in 1984 and described in the next chapter. It showed that desertification affected 70% of drylands in developing nations but only 40% or less of drylands in developed nations (Mabbutt 1984). The wider Sudano-Sahelian region (embracing the dry zones of both West and East Africa) contains some of the poorest countries in the world. This may explain why desertification and related problems have affected its land and people so strongly in comparison with other regions. Mali, Burkina Faso, Niger and Chad, together with Sudan, Ethiopia and Somalia, each had a GNP per capita of less than $400 in 1984; for Mauritania and Senegal it was under $500. In contrast, the USA had a GNP per capita of $14,080.

One of the most serious human consequences of economic underdevelopment is that large numbers of people live in conditions of extreme poverty and deprivation. When populations rise, many small farmers simply lack the means to increase food production without degrading land, for the only land to which they have access is of the poorest quality and therefore vulnerable to desertification. Governments, forced to make a choice between food security for the poor and the expansion of cash-crop cultivation to improve foreign currency earnings, often choose the latter. Seen in this way, "the coincidence between eroded areas and politically marginalized and powerless people", in the words of Randall Baker of the University of Indiana, is no accident (Baker 1984). Desertification does not therefore just affect the least developed countries more than the industralized, it selectively hurts the poorest sections of the population, the poorest of the world's poor (Kates et al. 1977).

This social and economic dimension to desertification indicates the need for the utmost care in formulating strategies to bring it under control. For, as Piers Blaikie has put it, "conservation is as much about social processes as physical ones, and ... the major constraints are not technical (in the agricultural engineering sense), but social" (Blaikie 1985). The reality of such constraints will become apparent in the remaining chapters of this book; they can lead to pessimism about the feasibility of bringing desertification under control, if this appears also to require fundamental changes in the structure of society (Blaikie 1985).

The need for care is perhaps most evident in the very delicate subject of family planning. Might not promoting this more vigorously, and thereby reducing the rate of population growth, have some impact on desertification? To some extent the likely future trends in desertification can be gauged from the predicted balance between population and resources. According to John Caldwell, of the Australian National University in Canberra, the problem may not be too serious in parts of the Asian drylands, since the availability of water and generally good grassland soils will allow them to absorb a greater fraction of their future population growth than is likely for the region's non-arid lands. In Latin America the proportion of the population located in dry areas could fall as young people migrate to the cities. However, he warns that in Africa most rural populations will not have access to irrigated land or to rich urban areas, so the majority of population growth will occur in rural areas: "Africa is the one continent where the early stage of demographic transition [the relationship between falling mortality and rising/falling birth rates] and low levels of urbanization mean that a huge growth in rural population is still to take place" (Caldwell 1984). Bearing in mind the economic dimensions of the problem of desertification described above, it is very doubtful whether controlling population growth will check desertification. Caldwell observes that family planning campaigns have so far not proved very successful in the African drylands, and he argues that the only way to reduce birth and death rates and the likelihood of death during drought and famine is by continued economic development.

THE ROLE OF GOVERNMENT

If desertification is partly a consequence of underdevelopment, what is the role of government, which is responsible for guiding the course of development? Discussions of land degradation often assume that the government's role is essentially neutral, but this is mistaken (Blaikie 1985). Government policies may actually promote desertification, albeit indirectly and unwittingly.

To promote national development, governments in most developing countries encourage the growth of industry and urban centres, and the building of roads, electric power generators and distribution lines, and other infrastructure and facilities characteristic of a developed society. However, urban growth reduces the agricultural labour force and causes environmental degradation in the hinterlands. Fulfilling the aspirations of urban dwellers means keeping food prices low, but this makes agriculture unprofitable for many farmers, who therefore have little incentive to improve productivity.

Foreign currency is an important prerequisite for development, because it enables a country to import the vast range of goods needed to build and maintain the fabric of a modern society. Governments therefore promote the cultivation of cash crops and the exploitation of minerals and other natural resources which may be exported to earn this foreign currency (Schmidt-Wulffen 1985). Ironically, in doing so they continue the export practices of their former colonial rulers (Bothomani 1984). Expanding cash-crop cultivation displaces subsistence cropping and in turn livestock grazing, so that land uses become inappropriate to the land on which they are practised, and degradation results. Too great an emphasis on cash crops can also result in plummeting food-crop production; and even more cash-crop exports are then needed to pay for imports of food. When food deficits first developed in Ghana and Nigeria the governments increased imports to resolve the situation, but local production of staple crops fell, so even more cash crops had to be exported to earn sufficient foreign currency to pay for imported food (Anson-Meyer 1983). The import of wheat flour and rice increases local demand for these grains, and if such a trend continues it can bias a country's whole agricultural development effort towards production of exotic grains, diverting substantial funds towards establishing expensive irrigated cropping projects.

Another development goal for many governments is social change. Thus, nomadic pastoralists are settled in villages so that they may be more easily regulated, taxed, educated and given proper health care. But such policies have led to widespread overgrazing around the new settlements. Many aspects of traditional society, such as subsistence cropping and tribal practices, are regarded by governments as old-fashioned, and while they may not be deliberately dismantled or discriminated against, their further development is certainly not encouraged. Traditional customs and practices are in any case highly vulnerable to the encroachment of the market economy and may quickly become redundant, even without government intervention. In some countries, like Zimbabwe and Lesotho, people are still constrained by land tenure regulations imposed during colonial times, but governments find it difficult to replace them even though they have resulted in high population densities and land degradation.

Governments of a Marxist persuasion, such as those of Ethiopia and Tanzania, often attempt to introduce central planning and communal forms of social organization. Tanzania's villagization programme aimed to resettle scattered farm populations in villages but, as with the sedentarization of nomads, this has had a detrimental effect on the local environment. In other respects, for example in promoting cash-crop cultivation, socialist countries are not very different from countries adhering to a capitalist philosophy. By favouring state-controlled cash-crop cultivation at the expense of domestic food production, and concentrating agricultural production in a small number of regions, the Ethiopian government increased the likelihood that the 1984–5 drought would lead to a major famine (Kelemen 1985; Griffin and Hay 1985).

There are a number of possible reasons why desertification is ignored in government policies. First, despite UNCOD and the activities which flowed from it, there is still much confusion, misunderstanding and ignorance about the meaning and causes of desertification. Second, desertification, and environmental conservation in general, do not fit easily within the conventional "development model", which only considers the environment and natural resources as sources of income to be exploited, rather than elements vital to ensure the sustainability of the whole economic development process. The great achievement of the World Commission on Environment and Development (the "Brundtland Commission") was to establish a new frame of reference for development models, and show that development and environmental conservation are self-reinforcing, not contradictory (World Commission on Environment and Development 1987). It will, however, take time for this concept to permeate through to planners and policy-

makers. The third and least charitable reason is that governments do not care much about desertification unless it directly
affects the interests or the power of the ruling political groups
and their friends (Blaikie 1985).

FAMINE, POVERTY AND DESERTIFICATION

Underdevelopment and misguided government policies can
combine with drought and desertification to cause hunger or
even famine. The United Nations Conference on Desertification owed its genesis more to worldwide horror at the terrible
effects of famine during the first period of drought in the Sahel
in the early 1970s than to concern about dryland environments
alone. Famine generally occurs when poor crop yields lead to
severe shortages of food in a particular area, and market
mechanisms or government interventions are insufficient to
balance these shortages with supplies from elsewhere in the
country or from overseas, but it does not have to be a
consequence of desertification and drought.

The causes of famine have been debated at length. A period of
prolonged, extended drought is obviously a contributing factor:
Michael Glantz has referred to it as "setting the stage" for
famine (Glantz 1987a). Amartya Sen, on the other hand, has
asked: "Why is it that producers of food are the first and most
seriously affected by drought and famine and why do so few
town dwellers die from hunger while rural areas are decimated
by starvation and death?" (Sen 1981). The implication is that
famine is selective and the causes of famine are heavily
influenced by social and economic factors. Famine can arise, for
example, from misguided agricultural policies or the inadequacy
of transport to ship food from one part of a country to another.
Thus, a country may have an overall food surplus yet some of
its people can still starve. According to a report by the
Independent Commission on International Humanitarian
Issues, famine is really a "man-made disaster" and over-
intensive agriculture, loss of topsoil and desertification help to
establish the preconditions for it long before drought delivers
the "*coup de grâce*" (Brown 1985).

Numerous studies have linked famine with government
policies that encourage the cultivation of cash crops at the
expense of subsistence food crops, not necessarily with the

laudable aim of preventing hunger. Government price policies which forced small farmers to produce export crops at expense of food crops were blamed for the famine in Mali from 1969 to 1973 (Schmidt-Wulffen 1985). These exports were needed to earn foreign currency to finance imports, create jobs in industry, and increase the size of the government bureaucracy, but only farmers in favoured areas could buy enough food for their subsistence from the profits of cash-crop production. In northern Mali neither cash-crop profits nor food supplies were adequate. No exchange of food was possible between different regions because the south of the country did not have a food surplus to meet the needs of the north. A similar criticism has been expressed about policies in Kenya (Hegmar *et al.* 1982).

As Ghana and Nigeria have experienced, too great an emphasis on cash crops can result in a vicious circle in which domestic food-crop production falls and an ever greater volume of cash crops has to be exported to pay for imports of food which could easily be grown in the country itself (Anson-Meyer 1983). Rising food imports, whether through normal trade or as "food aid" from developed nations concerned about famine, can change consumers' tastes away from locally grown staples and so depress local food production. Per capita consumption of wheat and rice in the Sahel, for example, grew by 4.6% and 2.0% respectively between 1966–70 and 1976–80 while that of maize, millet and sorghum fell (Delgado and Miller 1985). Continuing demand for such "exotic" imported grains can bias a country's whole agricultural development effort towards producing them, often by expensive irrigated cropping, and this seriously hampers the rehabilitation of the subsistence farming sector. Rice production in the Sahel and other parts of West Africa has been expanding in recent years, but not as rapidly as consumption. Only limited areas are suitable for wheat growing in the region. It is likely, therefore, that food imports will continue to rise and there is a danger that increased emphasis will be placed on growing cash crops to pay for these imports. Furthermore, in times of drought migrants are drawn to the cities in search of food and employment, depriving rural areas of the farm workers needed to grow crops and further reducing agricultural production (Glantz 1987b).

Famine may also occur as a consequence of serious political conflict inside a country. Michael Glantz has pointed out that:

Although there were major droughts and climate-induced food shortages around the globe [in 1982–4] – in various parts of Africa, India, China, Indonesia, Brazil – famines occurred only in Africa. Even a cursory review of the historical record of any drought-plagued region shows that famine does not necessarily follow drought. Most recently, for example, 31 countries in sub-Saharan Africa were declared to have drought-related food shortages. Only five of them, however, were plagued by famine. Each of those five (Mozambique, Angola, Chad, Ethiopia and the Sudan) were also subjected to internal wars. (Glantz 1987a).

The misery of the peoples of Ethiopia and Sudan provides a good illustration of how, even in countries affected by drought, civil war can make the difference between food shortages and famine. In Ethiopia, many hundreds of thousands of people died from starvation in 1984–5 (one estimate put the number as high as a million) because the government did not alert international agencies early enough to the food shortfalls likely to result from the drought (MacKenzie 1987). Although in the late 1980s the supply of food aid is now planned more in advance, with better co-ordination between the Ethiopian government and international agencies, the delivery of food supplies to badly affected areas in Eritrea and Tigray has been impeded by the civil wars raging there. Even the supply of food aid can be seen as a political weapon – the Eritrean rebels have organized their own relief supplies through Sudan but have fired on UN food convoys. In one incident, twenty-three trucks were destroyed. Malnutrition is therefore widespread and the threat of death from starvation ever present (Forse 1987; Buckoke 1988; Anon. 1988c).

The civil war in southern Sudan has led to a famine of massive proportions as millions of people have been driven from their homes and lands by rebel troops or by intertribal conflict encouraged by both government and rebels. It was estimated that during the summer of 1988 hundreds of people in the region died each day from hunger and disease. Large areas of crops have been burned and hundreds of thousands of cattle killed, so that people are unable to feed themselves. In October 1988 the Sudanese government ordered the UN Children's Fund (UNICEF) to stop distributing food in rebel-controlled areas (McElvoy 1988). The rebels laid siege to towns like Torit that were still held by government troops, and prevented food supplies from reaching them (Coughlin 1988). Up to 1 million people from the south fled from the horrors of the war zone to the capital Khartoum or to Ethiopia. Some of them were to lose their new homes in shanty towns or refugee camps when the floods came in August 1988. One camp alone housed 213,000 refugees (Powell 1988; Anon. 1988d). To add to these problems, after heavy rains had raised hopes of high crop yields, Sudan was infested by a plague of locusts that in some areas destroyed 60–80% of all crops. The locust swarms then spread westward and northward, and by the middle of October an estimated 7 million ha had been infested in Sudan, Chad and Niger, with large areas also infested in Mauritania, Morocco, Algeria, Tunisia and Libya (Hooper 1988; Ozanne 1988).

Famine is therefore as much a result of inadequate government policies and underdevelopment as it is of drought. According to Bradford Morse, the former administrator of the UN Development Programme (UNDP), "drought itself is not the fundamental problem in sub-Saharan Africa …. The present drought has, however, intensified the interaction of the factors impeding development in Africa; it has laid bare the African development crisis" (Morse 1987). The danger is that if famine is blamed on drought, then the deeper socio-economic and agricultural changes needed to ensure food security for all will not be made, since these require long-term action and this is not attractive to politicians. Famine certainly exacerbates the land-use pressures which cause desertification, for when food is in short supply people become desperate and overcultivate the land to get what food they can from it, even though they know that in the long term the land will be the poorer for it. Famine most seriously affects the poorest people, and they are often forced onto marginal lands that are the most vulnerable to being desertified. If desertification does occur, then it will reduce the area of farmland available for growing food, total food production will decline, and these people will be even more marginalized than before.

DESERTIFICATION IN DEVELOPED NATIONS

So far in this chapter, desertification has been strongly linked with underdevelopment. This might have given the impression that when dryland countries become more developed economically, desertification will go away. However, desertification is not confined to developing nations, and poor land management can occur anywhere, even in the USA. More than 13 million ha of US farmland is so highly vulnerable to erosion that the annual soil loss can only be kept to tolerable levels by the most restrictive land-use practices. Half of all cropland requires conservation measures to keep soil loss within tolerable limits, 8% is so vulnerable that intensive cultivation leads to unacceptable rates of soil loss, and only one-third is not inherently susceptible to erosion under any type of management (Bills and Heimlich 1984).

The consequences of poor management of drylands in the Great Plains became apparent in the 1930s when huge areas were turned into the infamous Dust Bowl. The situation was saved only by a massive mobilization of manpower and funds by the federal government, which established the Soil Conservation Service and the Civilian Conservation Corps to advise farmers on conservative cropping regimes, plant shelterbelts and do other soil conservation work. The fact that wheat yields in the Great Plains did eventually recover after the Dust Bowl, and even exceeded former levels, gives rise to optimism that desertification can be brought under control.

Soil conservation, however, demands continued vigilance, and the USA dropped its guard. In the early 1970s the Department of Agriculture told farmers to bring back into production 24 million ha of idled cropland to increase American grain exports. According to James Risser of Stanford University:

Soil conservation structures were destroyed so that all land could be put into row crops. Terraces and rows of trees were removed because they kept valuable land idle and because they got in the way of the big new farm machinery that was being developed. And hilly, erosive, marginally usable land was ploughed and planted – some of it for the first time. Crop rotations were abandoned and farmers adopted monocultural cropping of corn, wheat or soybeans. (Risser 1985)

Just what effect this had on soil erosion rates is not clear, although Risser claims that total US soil erosion is now greater than in the worst years of the Dust Bowl. Fortunately, at the same time as some of the windbreaks were removed between 1970 and 1975, new windbreaks were established elsewhere, giving an apparent net gain of 1,098 km of windbreaks; and in 1985 the Department of Agriculture decided to introduce a "set-aside" scheme which would aim to take more than 18 million ha of cropland out of production in order to reduce food surpluses. By 1988 about 11 million ha had already been set aside, although this is less than half of the area brought back into production a decade earlier (Garnett 1988).

The USA was not alone in slackening its grip on soil conservation. The expansion of croplands into marginal areas has also increased the problem of wind erosion in Australia and the Soviet Union. In addition, Australia, Canada and the USA are also experiencing worsening trends in the desertification of rainfed croplands in areas where saline groundwater is seeping to the surface (Dregne 1985). In 1988 a massive salinization problem of a different kind in the Soviet Union was publicized for the first time. Since 1956, water has been taken by canal from the Amu-Dar'ya River in Soviet Central Asia to irrigate millions of hectares of land under cotton cultivation in the Kara-Kum Desert several hundred miles away. The Amu-Dar'ya is one of two rivers that are the main sources of water for the Aral Sea, previously the fourth largest lake in the world. The diversion of water has reduced the area of the sea by two-thirds since 1960 and the exposed lake bed has been turned into a salt desert said to be thousands of square kilometres in extent. Surface salt has been swept up by winds and transported for hundreds of miles, causing considerable damage to farmland at its destination (Miller 1988; Lindley 1988). The fact that such desertification problems can occur in developed countries offers a warning that the future of the drylands may not be secure even if developing nations do succeed in dramatically improving their economic status.

A BROADER PERSPECTIVE

It is clear, even from this brief discussion, that assessing the causes of desertification in a particular area, and devising possible solutions to it, requires much more than simply focusing on the type of land use which appears to be the main cause of the problem. A broader perspective is needed which takes into account social and economic factors, government policies which influence the way in which land is used (both locally and nationally), and the pattern of land use over a much wider area than that which is actually suffering from desertification. Such an approach is very different from the classic "colonial model" in which soil erosion was seen as a strictly environmental problem, caused by the people actually using the desertified land, doing so too intensively (because of overpopulation), and employing primitive types of agriculture. Their social difficulties were ignored, and the general view was that the problem should be prevented by force (Blaikie 1985).

This chapter has shown how population growth and economic development are the major driving forces that lead to changes in land use in dryland areas, that the changes which occur in a particular country are greatly influenced by its level of economic development and by government policies on agriculture, trade and development, and that underdevelopment and misguided policies can easily result in land degradation. This analysis has tended to look, for convenience, at the effects of each of these factors individually; in the real world, however, they are interrelated. This can be illustrated by three mechanisms which combine socio-economic factors, policy choices and land use, and lead to desertification: confinement, displacement, and the tragedy of the commons.

Confinement occurs when a land use has to become more intensive than it has been previously, but without the technical inputs (such as fertilizers) or changes in cropping techniques that would allow it to be sustainable at the new intensity. Confinement can result when a population grows within a limited area; or when an extensive land use, like nomadic grazing, becomes more intensive because part of the area available to it is taken over for another land use. Examples of confinement given in this chapter include the resettlement of pastoralists and farmers in villages and the degradation which results when their grazing and cropping become too intensive. Such resettlement usually takes place as a matter of express government rural policy to concentrate previously dispersed populations, but other forms of confinement are the result of more general development policies, for example the promotion of urbanization, which leads to deforestation and land degradation in the vicinity of towns and cities; and the growth of population density in an area because land tenure regulations prevent migration from it when the population rises.

Displacement occurs when a land use which has been fairly stable in one area is forced to move to a less suitable area when it is replaced by another type of land use. The expansion of cash-crop cultivation in a semi-arid area, for example, can force rainfed cropping on to lands that are marginal for that purpose; this in turn displaces the nomadic herds which used to graze there on to even more fragile arid lands. Consequently, overcultivation and overgrazing may occur, leading to desertification. When a land use cannot be moved elsewhere, it is made more intensive in the area that remains to it after the expansion of the other land use, and the result is confinement (see above). Thus grazing often becomes more intensive when the rangeland area is reduced by the expansion of cropping. The two key implications of the displacement mechanism are that: (a) contrary to the colonial model of soil erosion, the real cause of desertification in a particular area may not be local, but some distance away, where the first displacement occurred; and (b) the real cause may be an "advanced" (rather than a "primitive") form of agriculture which is directly promoted by government policy. The poor farmers and grazers who are forced on to marginal lands which then become desertified should therefore be considered the victims of desertification rather than its perpetrators.

The "tragedy of the commons" is the overexploitation of communal land to which all have free and unregulated access, often as a result of the breakdown of traditional social controls over land use (Hardin 1972). It is found in the overgrazing of arid rangelands and village lands, and in deforestation generally. Common lands have not always been as "common" as their name implies, and the extensive grazing of arid areas such as the Sahel, for example, was quite rigorously controlled by nomadic pastoralists until their societies began to disintegrate, for reasons discussed in Chapter 2. Bad though the degradation may be, the real tragedy of the commons is the extreme difficulty of persuading those who believe they have some rights to these lands to agree on ways to restore communal management so that productivity can be improved for mutual benefit. Thus attempts to introduce new forms of pasture management to make livestock raising more sustainable, and to afforest degraded village lands in community forestry schemes, to give just two examples, often encounter insuperable obstacles.

The three mechanisms described here were arbitrarily chosen to show the interrelationships between the various direct and indirect causes of desertification. Other mechanisms undoubtedly exist, and better ways of classifying them could be found, but the main point was simply to justify the need for taking a broader perspective of the causes of desertification. Two reservations should be stated. First, such an approach inevitably oversimplifies what happens in the real world, and the various relationships and mechanisms described here (and in this chapter generally) will not necessarily operate under all circumstances. Second, a possible contradiction arises between this approach and the general strategy for controlling desertification advocated in this book, which is to make land uses more productive and sustainable on the lands best suited to them. That strategy suggests that an expansion of cash cropping is beneficial if it helps to improve the productivity of agriculture on the more fertile lands, so that marginal lands vulnerable to desertification can be taken out of food production. On the other hand, according to our broader perspective an increase in cash-crop cultivation is apparently detrimental because it displaces subsistence food cropping on to marginal lands, and this can cause desertification. The solution to the contradiction, however, is better planning of the expansion of cash-crop cultivation. Displacement is then minimized. Where it is inevitable, more consideration must be shown for the needs of subsistence rainfed croppers. If the country as a whole is to gain a net advantage from an increase in cash cropping, and not enter into a vicious spiral of having to continue to increase cash-crop exports to pay for more food imports, then appropriate support will have to be given to farmers growing food crops, so that they have an incentive to make their activities as productive and sustainable as possible.

This broader perspective on desertification therefore suggests that programmes intended to bring it under control should not only tackle flawed land uses but take into account the social and economic factors which contribute to them. The limitations of a purely technical approach to desertification control will be seen in a number of the projects described in Chapters 5–9 in which social and economic aspects have been given a low priority. Of course, individual agricultural and forestry development projects by themselves cannot really aim to do more than improve the living standards of people in a relatively small area, so they need to be complemented by government policies which create better economic conditions in the country as a whole, in the hope that these will encourage all farmers to make their land-use practices more productive and sustainable in the areas to which they are best suited.

This chapter has also shown that government policies are not neutral with respect to desertification, and that by failing to alleviate the poverty that accompanies underdevelopment or by promoting land uses that carry a high risk of degradation, they represent a major (indirect) cause of the problem. Governments concerned with controlling desertification would therefore be well advised to reassess all their development policies, not only those specifically concerned with rural areas, to reduce any risks of this kind. The general subject of reforming development policy is outside the scope of this book, but some specific policy recommendations for controlling desertification will be made in Chapter 10.

4. The Scale of Desertification

Desertification is a global phenomenon which afflicts drylands in more than 100 countries, of which about sixty with substantial areas of drylands are the most critically affected. The first part of this chapter gives a broad overview of where desertification occurs and the relative significance in each region of the four major direct causes: overcultivation, overgrazing, poor irrigation and deforestation. Numerous detailed examples of each cause were given in Chapter 2. The second part summarizes the most recent estimates of the actual extent and rate of increase of desertification (made in 1983 and 1984), but also shows how inaccurate these estimates are. Few large-scale measurements have been made, desertification is still quite loosely defined, and we are much too reliant on the subjective assessments of experts for these estimates. The shortage of reliable data on the extent of desertification, and on its social and economic impact, leads to scepticism among the scientific community about the reality and relevance of the phenomenon, and is a major obstacle to securing the political and financial support of governments and international agencies for programmes which could bring it under control. Improving the monitoring of desertification is therefore of the utmost importance. Consequently, the final part of the chapter discusses possible ways to do this by means of remote-sensing techniques, and the concomitant need for measurable "indicators" which will allow the degree of desertification in an area to be estimated with far more accuracy than in the past.

THE REGIONAL DISTRIBUTION OF DESERTIFICATION

Desertification affects drylands all over the world but tends to be concentrated in Asia and Africa, each of which accounts for

37% of all desertified land. The general review of desertification in this section is mainly qualitative in order to highlight its regional distribution and to avoid many of the problems arising from inadequate quantitative estimates, such as those contained in the global assessments of Mabbutt (1984) and Dregne (1983a) described in detail in the next section. This review does, however, make use of some statistics from those assessments, and where the two authors differ the estimates given by Mabbutt have been adopted for convenience.

It is usually difficult to pinpoint a dominant cause of desertification in a particular region since all four main direct causes are active throughout the drylands. However, there are regional distinctions and these will be indicated where appropriate. Overgrazed rangelands account for more than 90% of the total desertified area, largely because grazing is the only feasible use for the majority of arid and semi-arid lands, although this should not obscure the importance of the other causes. Overcultivation and poor irrigation management affect only a small percentage of the total desertified area but have a considerable social and economic impact because the vast majority of the rural population in the drylands lives on farms. Moreover, overgrazing is often an indirect consequence of the expansion of subsistence and cash-crop cultivation which forces herds from traditional pastures on to drier and more marginal lands. The impact of deforestation is probably much higher than estimates of annual deforestation rates (although these are quite substantial) would indicate, because deforestation not only degrades the vegetative cover but also makes drylands more vulnerable to desertification caused by overgrazing, overcultivation and poor irrigation management. Those estimates also understate the true magnitude of the reduction in tree cover as they ignore the removal of scattered trees, which have an important role in protecting soil from erosion.

Africa

More than 80% of Africa's drylands are moderately or severely desertified. Africa accounts for over a third of all desertified land in the world, and for a similar proportion of all desertified rainfed croplands and rangelands. The region most affected by desertification is the Sudano-Sahelian region (as defined by the

United Nations; see Fig. 1.3, p.16). It consists of four main dryland zones, most of which are arid or semi-arid: the Sahel zone on the southern fringe of the Sahara together with Mauritania and the Cape Verde Islands; the more humid Sudan savanna zone to the south of the Sahel; northeast Africa (defined in Chapter 1 as Sudan and Ethiopia); and the East African countries of Somalia, Djibouti, Kenya and Uganda. To the south of the Sudano-Sahelian region are Tanzania, Zimbabwe, Mozambique and other countries in Central and East Africa; and southern Africa, including Lesotho, Botswana and Namibia. The more developed North Africa region consists of Morocco, Algeria, Tunisia, Libya and Egypt.

Overgrazing and overcultivation are widespread both in the Sahel region and on the northern fringe of the Sahara in Morocco, Algeria, Tunisia and Libya. Degraded lands in these two regions are subject to serious wind erosion and in extreme cases are overrun by sand dunes and drifting sand. Sand-dune encroachment on rainfed croplands is also a critical problem in coastal areas of Senegal, Somalia and other countries. Overgrazing and overcultivation in Ethiopia, Somalia, Sudan and northern Kenya, and in Botswana, Namibia and Zimbabwe, lead to problems similar to those in the Sahel, although deforested highlands in Ethiopia and Lesotho also suffer from serious water erosion. Deforestation exposes soils to erosion by wind and water, and is occurring rapidly all over the continent. Grey haloes of deforested and degraded rangelands spread outwards from cities for distances of up to 100 km or more as trees are browsed by animals or cut down to provide fuelwood or charcoal for use in the cities. Irrigated cultivation is not as widespread in Africa as it is in Asia, but the salinization and waterlogging of irrigated lands is prevalent in Egypt, Algeria, Tunisia and parts of Sudan, in Ethiopia and Somalia, and to a lesser extent in Niger, Chad and Gambia. In Tunisia and other North African countries the introduction of intensive mechanized cultivation techniques such as disc ploughing has led to soil erosion on fragile lands.

Asia

An estimated 36% of Asia's drylands are desertified and they account for more than a third of all desertified lands in the world. Asia has similar areas of desertified rangelands and rainfed croplands to those of Africa but contains 70% of all irrigated croplands affected by salinization and waterlogging. Dry areas are mainly found in the western part of the continent in West Asia, Central Asia and South Asia. West Asia may be divided for convenience into two areas. The first includes Iraq, Syria, Jordan and other countries with semi-arid and sub-humid climates located close to the Mediterranean or in the Fertile Crescent of the rivers Tigris and Euphrates, where substantial areas of irrigated croplands are affected by salinization and waterlogging. The second area is Arabia, where the extensive hyper-arid and arid lands are mainly used for grazing, and rangeland degradation is the predominant form of desertification. Overgrazing is important in the extensive Central Asian drylands of China, Mongolia and the USSR, although salinization and waterlogging of irrigated croplands are found there too. In South Asia, overgrazing is common on the large areas of rangelands in Iran, Afghanistan, and the Indian state of Rajasthan, while salinization and waterlogging are widespread in the vast irrigated croplands of the Indo-Gangetic Plain in Pakistan and India, which together account for up to 40% of all desertified irrigated croplands in the world. Water erosion of rainfed cropland occurs all over Asia but is a particular problem in northern Pakistan, parts of western India, and on the loess plateau of China where the Yellow River derives its name from the colour imparted to it by the huge loads of eroded soil it has carried for thousands of years.

Mediterranean Europe

Europe's semi-arid and sub-humid drylands are located around the Mediterranean in Spain, Portugal, Italy, Greece, Turkey, Cyprus and Malta. The region was the centre of western civilization for thousands of years, but large-scale deforestation, overgrazing and overcultivation during and since that time left the vegetation heavily degraded and the soils in hilly and mountainous areas severely eroded. More recently the problem has been somewhat ameliorated by the afforestation of uplands and the control of grazing and cultivation. A fifth of the region's drylands are thought to be desertified but they account for a mere 1% of all desertified land in the world. Salinization and waterlogging of irrigated cropland are serious problems in Spain, particularly in the Guadalquivir Valley in the south of the country and on the Ebro River watershed in the northeast, and also in Greece and Portugal.

Australia

Australia accounts for about a tenth of all desertified land in the world. Most of the arid interior and the semi-arid or sub-humid lands in the north of the continent are only used as rangeland. Irrigated and rainfed cropping are concentrated in moister coastal areas, although rainfed cropping also extends to semi-arid lands in the interior. Grazing does not date back more than 125 years and as a result of this (and a progressive improvement in range management), as little as a quarter of all Australian rangelands have so far been desertified. Overgrazing has degraded the vegetative cover by replacing perennial shrubs and grasses with annuals, and it exposes the soil to wind and water erosion which in the most extreme cases leads to gullying. About 30% of rainfed croplands are affected by soil erosion due to overcultivation, but another serious problem affecting rainfed croplands in parts of West Australia and South Australia is the saline seepage which results indirectly from the deforestation of watersheds. When tree cover is removed, the amount of water lost by evapotranspiration into the atmosphere falls, groundwater levels rise, and more groundwater flows down into the river basin, becoming saline as it dissolves salts from lower soil levels. Some of the water flows into rivers and makes them more saline, while the remainder builds up in neighbouring lowland agricultural areas which then become waterlogged and saline. Irrigated croplands in the Murray River system of Victoria and New South Wales have also been waterlogged and salinized by the more usual cause of poor irrigation management.

North America and Mexico

North America and Mexico account for about a tenth of all desertified land worldwide. The arid zone includes northern Mexico and parts of California, New Mexico, Arizona, Utah and Nevada in the United States, but most of the key grain-growing areas in the Great Plains and Prairie regions of the USA and Canada are semi-arid, and overcultivation there has caused severe wind erosion of the soil. Although large-scale soil conservation schemes were introduced in North America after the notorious Dust Bowl episode of the 1930s, soil erosion still continues and saline seepage is also a problem. Overgrazing has led to widespread degradation of vegetation and soil on the dry rangelands of the USA and Mexico. Salinization and waterlogging affect irrigated croplands in a number of valleys along the Colorado and Gila rivers, including the San Joaquin Valley in California, the Imperial and Mexicali valleys, and the lower valley of the Rio Grande.

South America

Dry areas in South America are essentially confined to northeast Brazil; Argentina; areas on the Caribbean coast of Colombia and Venezuela; the Andes mountains; and the narrow western coastal strip in Chile, Peru and Ecuador centred on the Atacama Desert, one of the driest regions in the world, which receives less than 10 mm of annual rainfall. South America accounts for about a tenth of all desertified land. Deforestation and overcultivation of rainfed crops are causing serious soil erosion in the densely populated Andes and northeast Brazil. Overgrazing has degraded vegetation and caused wind and water erosion of soil on the extensive rangelands in Argentina, and in the Andes, northeast Brazil and other northern coastal areas. Irrigated croplands in western Argentina, northeast Brazil and along the coast of Peru are affected by salinization and waterlogging.

ESTIMATES OF THE EXTENT AND RATE OF DESERTIFICATION

Estimating the actual extent of desertified land is much more difficult than producing general assessments, such as the one given in the previous section. The ideal procedure for making estimates would involve: first, defining a set of categories that would distinguish between different degrees of degradation; second, measuring the area and degree of desertification throughout each country by a combination of ground tests at

sample plots and large area surveys by remote-sensing techniques such as aerial photography and satellite imaging; and
third, classifying the resulting data to arrive at estimates of
desertified areas disaggregated by land use and degree of
desertification. Global estimates of desertification could be
made either in a decentralized way, with government agencies
in each country collecting data and forwarding them to an
international agency like UNEP for collation; or in a centralized
manner, with the international agency making measurements
itself, using its permanent staff, together with consultants and
local personnel in the countries being monitored. In practice,
however, estimates of the extent of desertification have largely
depended upon the subjective assessments of consultants
appointed by UNEP. These assessments have been based upon
a limited amount of local data, little of which has been measured
and categorized in a rigorous manner. This section discusses the
results of these assessments and the criteria upon which they
were based.

Criteria for Assessing the Degree of Desertification

The degree of desertification in an area is typically categorized
as slight, moderate, severe or very severe, but the criteria
adopted for placing land in these categories have so far been
fairly general. Slight desertification is normally disregarded
when making estimates of desertified areas, so only land in the
other three categories is included. Thus all estimates of
desertified areas quoted in this chapter imply that the land
suffers from "at least moderate desertification".

The World Map of Desertification, prepared by FAO,
UNESCO and the World Meteorological Organization (WMO),
was one of four maps commissioned by UNCOD (see p. 135).
Although it later became the most widely publicized of the four,
it showed only the degree of desertification hazard, not the
status of desertification. Desertification hazard was assessed as
moderate, high or very high by a subjective evaluation of
climatic conditions, the inherent vulnerability of the land, and
human or animal pressure (FAO/UNESCO/WMO 1977).

The first proper map of the status of desertification, another
of the four UNCOD maps, was prepared by Harold Dregne. He

used separate definitions of desertification status for rangeland, rainfed cropland and irrigated cropland. He defined moderate desertification in terms of a significant increase in undesirable shrubs (for rangeland) and small dunes or gullies (for rainfed cropland), and a reduction in the yields of irrigated crops of up to 50% owing to soil salinity. In his definition of severe desertification, undesirable shrubs dominated the flora; large gullies were present, or sheet erosion by wind and water had largely denuded the land of vegetation; and irrigated crop yields were less than 50% of potential levels owing to salinity. Very severe desertification resulted in large, shifting, barren sand dunes; large, deep and numerous gullies; or the development of salt crusts on almost impermeable irrigated soils (Dregne 1977).

Dregne used a rather more specific set of criteria for his second assessment in 1983. He classed desertification as moderate if 26 to 50% of the plant community consisted of climax species (i.e. those belonging to the climax vegetation – see Chapter 1), 25 to 75% of the original topsoil had been lost, and soil salinity had reduced crop yields by 10 to 50%. Severe desertification required that only 10 to 25% of the plant community consisted of climax species, soil erosion had removed all or practically all of the topsoil, and soil salinity had reduced crop yields by more than 50%. On very severely desertified land, less than 10% of the plant community would consist of climax species, the land would be covered by many sand dunes or deep gullies, and salt crusts would have developed on irrigated cropland (Dregne 1983a).

In his 1984 global assessment of desertification for UNEP, Jack Mabbutt classed desertification of rainfed and irrigated croplands as moderate if there was widespread erosion or salinization and waterlogging, and losses of up to 25% of crop production; severe if production losses were between 25% and 50%; and very severe if they exceeded 50%. Rangelands were moderately desertified if there was a significant reduction in vegetative cover and deterioration in species composition, a significant level of soil erosion, and a 25% decline in livestock carrying capacity. They would be severely and very severely desertified if the proportional reductions in carrying capacity were 25 to 50% and greater than 50% respectively. Very severely desertified rangelands were those which could not be reclaimed economically (Mabbutt 1984).

The UNCOD World Map of Desertification

The World Map of Desertification (Fig. 4.1), (FAO/UNESCO/ WMO 1977), specially commissioned for UNCOD, identified the degree of potential desertification hazard to which each dryland area was exposed. An estimated 37.6 million sq km of drylands, two-thirds of which were in Asia and Africa, were said to be at serious risk of desertification (Table 4.1). Because this was not a map of desertification status, it was not known whether and to what degree desertification was actually occurring within this vulnerable area.

Recent Estimates of the Extent of Desertification by Mabbutt and Dregne

The two most recent sets of estimates of the actual extent of desertification were published in 1983 and 1984 by Harold Dregne and Jack Mabbutt respectively. Harold Dregne estimated that 32 million sq km of the world's drylands were suffering from moderate or severe desertification (i.e. "at least moderate desertification"). This represented almost one-quarter of the earth's land surface, and was close to the total area thought at the time of UNCOD to be at risk of, but not necessarily suffering from, desertification. Of the affected area

Table 4.1: UNCOD Estimates of Areas Subject to Moderate or Severe Hazard of Desertification

	Area (million sq km)	% Total*
Asia	13.7	36
Africa	10.4	28
Australia	5.7	15
North & Central America	4.3	11
South America	3.3	9
Europe	0.2	0.5
Total	37.6	100

*Due to rounding, the sum of these percentages is not 100.
Source: FAO/UNESCO/WMO (1977)

Source: FAO/UNESCO/WMO (1977)

Figure 4.1: World Map of Desertification

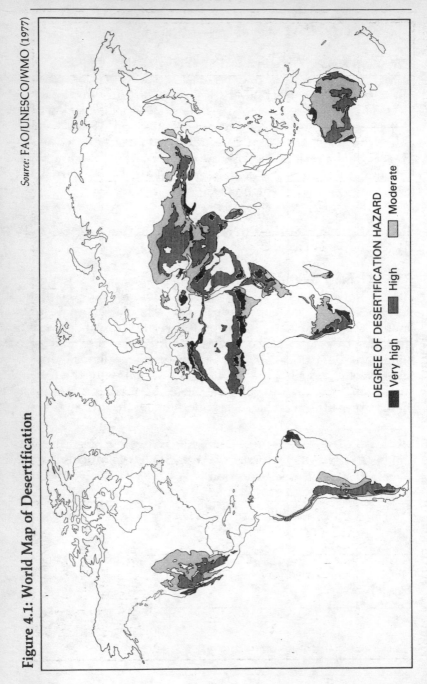

DEGREE OF DESERTIFICATION HAZARD

Very high   High   Moderate

Figure 4.2: World Map of the Status of Desertification

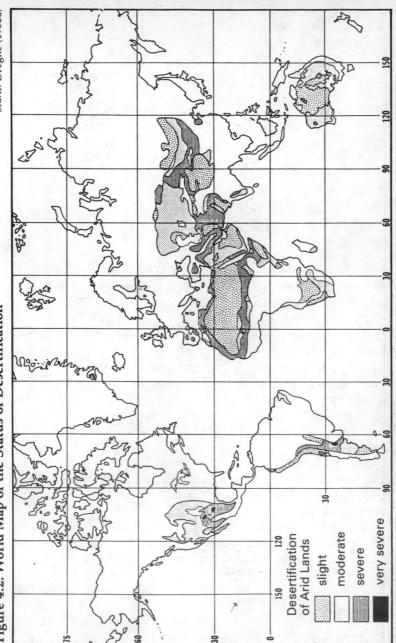

Source: Dregne (1983a)

Desertification
of Arid Lands

slight

moderate

severe

very severe

Table 4.2: Dregne's Assessment of Areas Suffering at Least Moderate Desertification (thousand sq km)

	Rainfed cropland		Irrigated cropland		Rangeland		Total	
	area	%	area	%	area	%	area	%
Africa	396	23	14	5	10,268	34	10,678	33
Asia	912	53	206	76	10,889	36	12,007	37
Australia	15	1	2	1	3,070	10	3,087	10
Europe (Spain)	42	2	9	3	155	1	206	1
N. America & Mexico	247	14	28	10	2,910	10	3,185	10
S. America	119	7	12	5	3,194	10	3,325	10
Total	1,731	100	271	100	30,486	100	32,488	100

Source: Dregne (1983a)

Note: Due to rounding, the sum of percentages may not equal 100.

of 32 million sq km, 30.7 million sq km were rangelands, 1.7 million sq km rainfed croplands, and 0.27 million sq km irrigated croplands (Table 4.2 and Fig. 4.2; Dregne 1983a).

Jack Mabbutt's estimate was much lower. Using data obtained from replies to a special questionnaire circulated to governments of all dryland countries by UNEP, and the findings of regional studies specially commissioned as part of the same programme, he estimated that 20 million sq km of land in the world were suffering from at least moderate desertification, and of this 16.15 million sq km were rangelands, 3.46 million sq km rainfed croplands, and 0.40 million sq km irrigated croplands (Table 4.3; Mabbutt 1984). Mabbutt's current assessment (personal communication, 1988) is that the overall area of desertified land is between 15 and 20 million sq km.

According to Mabbutt's estimate, the world's most desertified areas were the Sudano-Sahelian region, Africa south of the Sudano-Sahelian region, North Africa, and West Asia (the "Middle East"), all of which had more than 80% of their drylands at least moderately desertified. Developing countries, with 60% of their drylands desertified, were more affected than developed countries for which the corresponding proportion was only 40%. Of the 20 million sq km of productive land suffering from at least moderate desertification, 4.2 million sq km (21%) were in the Sudano-Sahelian region and 3.3 million sq km (16%) elsewhere in Africa; 2.5 million sq km (13%) in India

Table 4.3: Mabbutt's Assessment of Areas Suffering at Least Moderate Desertification (thousand sq km)

	Rainfed cropland area	%	Irrigated cropland area	%	Rangeland area	%	Total area	%
Africa	1,290	37	19	5	6,100	38	7,409	37
Sudano-Sahel	720	21	8	2	3,420	21	4,148	21
S. of Sudano-Sahel	420	12	6	2	2,000	12	2,426	12
North Africa	150	4	5	1	680	4	835	4
Asia	1,350	39	280	70	5,850	36	7,480	37
W. Asia	150	4	30	8	980	6	1,160	6
S. Asia	1,050	30	200	50	1,270	8	2,520	13
USSR Asia	120	3	20	5	1,500	9	1,640	8
China & Mongolia	30	1	30	8	2,100	13	2,160	11
Australia	120	3	3	1	1,000	6	1,123	6
Med. Europe	130	4	16	4	150	1	296	1
N. America	240	7	40	10	1,800	11	2,080	10
S. America & Mexico	330	10	40	10	1,250	8	1,620	8
Total	3,460		398		16,150		20,008	
Alternative total*	3,350		400		31,000		34,750	

* After adjustment to include remote, non-productive rangelands
Source: Mabbutt (1984)
Note: All percentages are of the Total in each column. Due to rounding, regional percentages may not equal the sum of sub-regional percentages, and the sum of regional percentages may not equal 100.

and Pakistan; and 5 million sq km (25%) in other parts of Asia. Asia and Africa each accounted for 35–40% of all desertified rainfed croplands and rangelands, but Asia alone had 70% of all irrigated land affected by desertification, of which 57% was in India and Pakistan (Mabbutt 1984, 1985).

Most of the difference between the two estimates of the total desertified area was due to the disparity in their rangeland components: 30.5 million sq km in Dregne's estimate but only 16.2 million sq km in Mabbutt's. On the other hand, it should be noted that Dregne's estimate of the area of desertified rainfed cropland was only half of the figure which Mabbutt gave, and that for irrigated cropland was just two-thirds. Both experts agreed that Asia had the largest area of desertified irrigated cropland of any region, but Dregne considered that it also had by far the largest area of desertified rainfed cropland.

There are two main reasons why Mabbutt and Dregne produced such different estimates: a lack of reliable data (discussed on p.143), and differences in the overall areas of drylands within which desertified lands were identified.

Table 4.4: Distribution of Drylands (Including Sub-Humid Lands) by Region

	Area (million sq km)	$ total
Africa	21.5	35
Asia	20.7	34
Australia	6.5	11
North America and Mexico	6.1	10
South America	4.8	8
Europe	1.2	2
Total	60.8	100

Note: Includes hyper-arid, semi-arid and sub-humid zones
Source: Mabbutt (1983)

Dregne's study covered a smaller area than Mabbutt's, and included hyper-arid, arid and semi-arid lands with an overall area of 47 million sq km (see Table 1.1). On the other hand, Mabbutt's study also covered sub-humid lands, and this increased the overall area covered to 60.8 million sq km (Table 4.4). The smaller areas of desertified rainfed and irrigated croplands in Dregne's estimate are largely a result of the fact that his study did not include the sub-humid zone. The great difference between the two estimates of the area of desertified rangelands is explained in the following way. Both Mabbutt and Dregne excluded about 6 million sq km of slightly desertified areas in natural deserts, such as the Sahara, Arabian and Takla Makan, where the natural biological productivity was too low to support cropping or grazing and so human impact was minimal. However, Mabbutt also left out other large areas of remote, difficult or unwatered terrain, classified for convenience as rangelands but seldom if ever used as such. He assumed that only about 30 million sq km of the total area of 47 million sq km of hyper-arid, arid and semi-arid lands were productive in an agricultural sense (Mabbutt 1984).

To achieve a greater degree of comparability with Dregne's estimate, Mabbutt (at the request of UNEP) made a second estimate which included these remote, unproductive "rangelands'. The area of desertified rangelands increased from 16.2 to 31.0 million sq km, the area of desertified irrigated cropland remained about the same, but for some reason the area of desertified rainfed cropland was slightly lower than in the first estimate. The total area of drylands suffering from at least moderate desertification rose to 34.75 million sq km, which was close to Dregne's estimate (Table 4.3; Mabbutt 1984). On the basis of the larger total Mabbutt estimated that about 80% of all rangelands, 60% of rainfed croplands, and 30% of irrigated lands were already at least moderately desertified. Severe desertification, in which productivity has fallen by more than a half, affected 33% of all rangelands, 30% of all rainfed croplands and 10% of all irrigated lands, a total area of almost 15 million sq km.

The Social Impact of Desertification

Desertification has a tremendous impact on the lives of people in the drylands. Jack Mabbutt estimated that about 280 million rural people are affected by at least moderate desertification, a

Table 4.5: Mabbutt's Assessment of Populations Affected by At Least Moderate Desertification (millions of people)

| | Cropland | | | | | | |
	rainfed	irrigated	Rangeland	All moderate	%	All severe	%
Africa	79.00	3.50	25.50	108.00	38	61.00	43
Sudano-Sahel	36.00	1.50	13.50	51.00	18	27.50	19
S. of Sudano-Sahel	32.00	1.00	8.00	41.00	15	25.00	18
North Africa	11.00	1.00	4.00	16.00	6	8.50	6
Asia	56.00	50.00	17.00	123.00	44	55.50	39
W. Asia	16.00	12.00	4.00	32.00	11	8.50	6
S. Asia	34.50	23.00	9.00	66.50	24	16.00	11
USSR Asia	1.50	4.50	1.00	7.00	2	29.00	20
China & Mongolia	4.00	10.50	3.00	17.50	6	2.00	1
Australia	0.10	0.10	0.03	0.23	0	6.50	5
Med. Europe	13.00	1.50	2.00	16.50	6	0.03	0
N. America	2.00	1.00	1.50	4.50	2	6.00	4
S. America & Mexico	22.50	2.50	4.00	29.00	10	13.50	9
Total	172.60	58.60	50.03	281.23		142.53	

Source: Mabbutt (1984)

Note: All percentages are the total in each column. Due to rounding, regional percentages may not equal sub-regional percentages.

figure which rises to 470 million (10% of the world population) if urban dwellers in these areas are also considered. Most of the rural people affected by desertification live in rainfed cropland areas in the Sudano-Sahelian region, Africa south of the Sudano-Sahelian region, the Andean region of South America, and parts of South Asia (particularly Nepal). Farm-based populations account for 85% of the rural population in the drylands but occupy only 15% of the dryland area, so the social impact of desertification caused by rainfed cropping is likely to be quite serious (Mabbutt 1984). Overall, fewer people are affected by the desertification of irrigated lands (the rainfed and irrigated farming populations are similar in Asia), but this will cause the greatest losses in production and investment, particularly in West Asia and in South Asian countries such as Pakistan (Table 4.5).

Estimates of Rates of Desertification

The only available estimate of the annual rate of desertification is a slightly revised version of one made by Harold Dregne for UNCOD, according to which over 202,000 sq km of agricultural land (an area the size of Senegal) was said to be desertified each year to the point of yielding zero or negative net economic returns. In the revised estimate, this comprises 177,000 sq km of rangeland, 20,000 sq km of rainfed cropland, and 5,460 sq km of irrigated cropland. The potential value of agricultural production lost every year in the late 1970s was estimated at more than $26 billion (Table 4.6; Dregne 1983a). The rate of desertification of 202,460 sq km per annum compares with an

Table 4.6: Estimate of Annual Rates of Desertification

	Area (sq km/yr)	% total
Rangelands	177,000	87
Rainfed Croplands	20,000	10
Irrigated Lands	5,460	3
Total	202,460	100

Source: Dregne (1983a)

estimated rate of clearance of open woodlands in the drylands of 38,000 sq km per annum, of which Africa accounts for 23,000 sq km (Table 2.1), (Lanly 1982).

No new estimate of desertification rates emerged from UNEP's special assessment in 1982–3, which included Mabbutt's 1984 study (mentioned above) and contributions by Leonard Berry (1984a) and Harold Dregne (1983b), although qualitative indications were given of trends in the rates of desertification. Mabbutt commented that desertification was accelerating on rangelands in Africa south of the Sahel and in Andean South America, as well as on rainfed croplands in tropical Africa, South Asia, South America and subtropical Mexico. Desertification of irrigated lands was relatively "static" everywhere. Deforestation rates were either unchanged or accelerating in most drylands, particularly in Africa, Latin America and parts of South Asia. Depletion of groundwater reserves was important in West Asia, Mediterranean Africa and the USA (Mabbutt 1984).

Leonard Berry reported that desertification continued to worsen in the nineteen countries of the Sudano-Sahelian region. He used some data from the responses to the UNEP Desertification Questionnaire, heavily supplemented with data from UN agencies and knowledgeable individuals. Each country was ranked according to whether the rate of desertification since 1977 had remained stable or increased slightly or significantly in five main categories: sand dune encroachment, rangeland deterioration, forest depletion, deterioration of irrigation systems and degradation of rainfed cropland. Berry concluded that there was a "moderately worsening situation in 12 countries and a seriously worsening situation in 7. The picture is a sombre one but realistic according to the judgement of most field observers" (Berry 1984a).

Harold Dregne took a different view, remarking that desertification rates were particularly high in the rangelands and cultivated lands of Morocco, Egypt, Sudan, Ethiopia, Kenya, Somalia, Iraq, Iran, Pakistan and China. Rates of degradation of rangelands in the developing nations as a whole were static, and either static or declining in developed nations. On the other hand, the erosion of soil by water was increasing in Ethiopia and North Africa, and the wind erosion of rainfed croplands had become more serious in the Sahel, Ethiopia, India, and in the

USA, Australia and the USSR. Dregne agreed with Mabbutt that the rate of salinization and waterlogging of irrigated lands was not increasing, although the seepage of saline groundwater to the surface of rainfed croplands was becoming more of a problem in Australia, Canada and the USA (Dregne 1983a, 1985).

The Reliability of Estimates of the Scale of Desertification

How reliable are the estimates of the extent of desertification made by Mabbutt and Dregne? In contrast to the ideal procedure for making assessments outlined at the beginning of this section, in both studies the desertification categories were only loosely defined, very few actual measurements were made to obtain basic data, and the subjective judgement of the two experts was predominant. Although Mabbutt's estimates were prepared for UNEP they were not officially released by it, mainly because of the very poor quality and quantity of the data contained in the returned questionnaires. Thus both estimates should be treated as giving only a very approximate idea of the true scale and intensity of desertification, and there is no reason why more credence should be placed upon Mabbutt's figures than Dregne's. According to Harold Dregne: "Given that both estimates are based on informed opinion, it is impossible to know which set of figures is more nearly correct" (Dregne 1983b).

Estimates of the current rates of desertification are even more tenuous than those of the overall desertified area, and so the figure of 202,000 sq km per annum quoted by Dregne should be regarded with great caution. The significance of any estimate of the rate of increase in a phenomenon is of course limited if the overall scale of that phenomenon is not known with any great accuracy, since the ways in which these two types of estimates are made are closely related. Thus, estimating the rate of desertification requires the use of maps, aerial photographs and satellite images which, when compared, will show changes in the extent of desertification over a period of five, ten or more years. Furthermore, it is clearly quite important that desertification be regularly monitored over time, since estimates of the extent of the problem made in only one year could be very misleading. Satellite monitoring of the

Sahel (described on p. 155), for example, has shown that the apparent southern border of the Sahara Desert can shift either south or north over a period of a few years, as the growth of vegetation changes with the incident rainfall. In the absence of regular monitoring that would show these dynamic aspects, comparing the area apparently desertified in one year with that affected, say, five years previously, could show either an increase or a decrease in the extent of desertification, depending upon the two years chosen for comparison.

There is, therefore, an urgent need to obtain improved estimates of the extent and rate of desertification. The remainder of this chapter discusses the two main requirements for achieving this goal: a better definition of the phenomenon in terms of measurable characteristics; and the collection of more reliable data through comprehensive ground measurements and the use of a variety of remote-sensing techniques, such as aerial photography and satellite imaging, to monitor the problem continuously over large areas.

DEVELOPING A NEW SYSTEM OF DESERTIFICATION INDICATORS

Desertification will remain an ephemeral concept to many people until better estimates of its extent and rate of increase can be made on the basis of actual measurements. However, before a global monitoring system for desertification can become operational there will be a need to specify in greater detail what is actually to be measured. The very general criteria used previously for ranking areas according to their degree of desertification will have to be replaced by a comprehensive and coherent set of measurable characteristics, or "indicators", such as the soil erosion rate and the density of vegetative cover. The word "measurable" is absolutely crucial here, for indicators do not only have to be quantifiable, but also amenable to practical measurement on the ground and by remote-sensing techniques. Selection of the indicators will require a great deal of discernment; also a pragmatic approach so that indicators which may be ideal from a theoretical standpoint are rejected if they cannot be easily measured. This section describes how suitable sets of indicators could be chosen and lists some which have already been proposed.

Criteria for Selecting Indicators

There are four main types of indicators: physical indicators describing ground conditions; indicators of trends in agricultural production; climatic indicators; and socio-economic indicators that show trends in human health, income and welfare. Jack Mabbutt has suggested some criteria which potential indicators should satisfy. They should be as specific as possible to desertification so as to avoid confusion with other phenomena; sensitive enough to show the gradual development of desertification in an area; easily quantified by observation on the ground or by remote-sensing techniques, or (especially in the case of socio-economic indicators) available in the form of published statistics; comprehensive enough to be widely applicable to different types of areas; suitable for repeated scanning by ground observation or remote sensing, or capable of periodic updating if obtained from published statistics; and, particularly in the case of developing countries with limited numbers of skilled personnel, recognizable or usable without specialized training (Mabbutt 1986).

To what extent should climatic, socio-economic and agricultural production indicators be used in monitoring desertification? Even if we define desertification solely in terms of changes in the physical characteristics of the land, we still need to monitor climatic changes in order to distinguish between long-term degradation caused by human action and (a) temporary changes in ground conditions (e.g. poor plant growth) caused by low rainfall; (b) the effects of long-term regional or global climatic change. If data on the physical condition of the land are not readily available, then trends in agricultural production and in the social and economic characteristics of dryland areas could in principle also be useful indicators, since a decline in soil fertility and vegetative cover translates into lower crop and livestock yields, and reduced human and animal health. Some would also argue that if we are to persuade governments and aid agencies to give a higher priority to desertification control, then estimating the magnitude of these trends is just as important as estimating the physical extent of desertification.

However, a problem arises because socio-economic and agricultural production indicators are more ambiguous and unreliable than physical indicators. It is difficult to separate the impact of desertification from that of poverty, misguided government food policies and political instability; and most data on food consumption, malnutrition, famine and human mortality in tropical countries are very inadequate (Meuer 1984). Trends in agricultural production could also be misleading, since data are generally sparse and unreliable; production is heavily affected by annual rainfall; the use of improved varieties and fertilizers can mask the effects of land degradation; and straightforward economic substitution can occur (between alternative products or production in different areas of the same country) which has nothing to do with desertification. Thus, declines in the indices of food production per hectare or per capita, such as those quoted in the previous chapter, may not signify either desertification or even poor farm productivity, but merely the fact that crop yields were low in the latest year in the comparison because of low rainfall. Concerning substitution, sugar cane production in Brazil's dry northeastern region fell in the early 1980s, but this was not necessarily due to desertification, since there was competition from other regions and livestock production actually increased (Sales 1982). Social and economic characteristics cannot therefore at present be used as practical desertification indicators, and trends in agricultural production should be used with great care.

Indicators should be selected with due regard to the instruments which will be used to monitor them, and limited to those which are feasible for scientists and organizations to monitor. A group of sophisticated soil structure indicators, for example, would be of little value since the majority of measurements will have to be made by soil laboratories in tropical countries where only a limited range of equipment is generally available. If remote-sensing satellites are to be used for the large-scale monitoring of some indicators, then the resolution with which satellite sensors can detect features on earth has to be taken into account in defining these indicators. It is also important that indicators be defined, and where necessary, appropriate measurement techniques specified, so that the measurements produced by different scientists will be comparable. Without homogeneous data it will be impossible to make aggregate regional and global assessments, and some results may even be directly contradictory (Mabbutt 1986).

Comprehensive Sets of Indicators

An early comprehensive group of indicators (dating from 1977) included the surface reflectivity of the land (albedo), which is determined by the type and density of vegetative cover; frequency of dust storms; annual rainfall; rate of soil erosion; sedimentation of streams resulting from soil erosion; salinization of irrigated lands; standing biomass; biological productivity; climate; nutrition; human well-being; and human perception. Only a few of these indicators can be easily measured. Although monitoring rainfall is important if all the factors contributing to desertification are to be fully understood, it would not be easy to construct a single indicator to describe all climatic phenomena. The same is true for biological productivity, since in a particular area this is so dependent on rainfall and site conditions. Indicators of human well-being and perception would also be difficult to quantify (Berry and Ford 1977).

A longer list of indicators, prepared at a seminar in Nairobi directly preceding UNCOD, included soil depth, the amount of soil organic matter and the degree of soil crusting; groundwater depth and quality; the extent and persistence of surface water; the general status of rivers and streams, their turbidity and rate of discharge; the extent and composition of vegetation canopy cover, the annual amount of biomass produced by plants and the distribution and frequency of key plant species; animal species distribution and the size, composition, production and yield of domestic livestock herds. The additional soil and water indicators would probably require repeated measurements on the ground but most of the vegetation indicators could be measured by satellites. Animal indicators are not so useful since they are indirect and could be confusing: for example, large herd numbers could indicate either plentiful pastures or overgrazing and the potential for future desertification (Reining 1978).

FAO has produced its own list of indicators describing: the degradation of vegetative cover; soil erosion by wind and water; salinization; soil crusting and compaction; and organic matter reduction (FAO/UNEP, 1984). However, the suitability of these indicators for broad-scale international assessments is open to question. Measuring some of the indicators would require considerable technical skill and a large number of them, particularly the soil indicators, would have to be measured on the ground and so would be difficult to estimate for large areas using remote-sensing techniques (Mabbutt 1986).

The fact that there is still little agreement on the choice of indicators reflects a continuing uncertainty about the nature of desertification which will probably only be removed when large-scale monitoring is introduced and those responsible for it are forced to choose a practical and effective set of indicators. Otherwise, the debate about the virtues of possible sets of indicators will remain very much in the realm of theory. It is likely that shortages of personnel and funding for monitoring programmes will mean that the indicators which will be used most frequently will be restricted to those physical indicators measurable from aerial photographs or satellite images and by limited ground measurements.

USING REMOTE SENSING TO IMPROVE THE MONITORING OF DESERTIFICATION

The lack of reliable data on the extent and rate of increase of desertification is very worrying in view of the scale and seriousness of the problem. The most obvious way to improve the availability and accuracy of data quickly would seem to be to employ remote-sensing techniques, such as aerial photography and satellite imaging, which are designed to survey ground conditions over large areas. A well-designed remote-sensing programme could in theory tell us how large an area was desertified, and by carrying out regular surveys we could detect increases in the intensity and extent of degradation in different areas. So far, however, there have been relatively few attempts to monitor desertification in this way, and although this can be explained in part by a lack of available imagery, another reason is that the spatial resolution, that is the amount of detail which sensors can distinguish on the ground, has been too low to monitor the phenomenon effectively.

Aerial Photography

Aerial photography from light aircraft has been used for decades to map land use and vegetation characteristics such as forest cover, and is capable of high spatial resolution. The camera is normally mounted so that it automatically takes a sequence of vertical photographs of the ground, and the aircraft is flown along routes that are planned to ensure a complete coverage of the area to be surveyed. Despite its accuracy, aerial photography is costly and needs good organization to get the sensor (the camera) and the sensing platform (the aircraft) into operation. Aerial surveys of large areas take a lot of time and produce a huge number of photographs, each of which has to be interpreted separately by a trained operator. For these reasons, aerial photography could be used for a detailed baseline survey of a whole country or part of a country, but would not be suitable for frequent monitoring of large areas on, say, an annual basis. A more appropriate use would be to monitor desertification in relatively localized areas where problems were most acute. An easier but less rigorous survey method is aerial reconnaissance, in which a trained observer flies over areas of interest and takes note of ground conditions, perhaps supplemented by photographs taken obliquely out of the aircraft window. The widely publicized but controversial finding that the Sahara Desert was moving south at a rate of over 5 km per annum (see Chapter 1) was made by comparing aerial reconnaissance observations with an eighteen-year-old map of the southern border of the desert (Lamprey 1975).

Satellite Imaging

Earth resources satellites, such as the American Landsat satellites, often carry video cameras but their most widely used sensors have hitherto been multi-spectral scanners (MSS) which take images of the visible and infra-red radiation emitted from the earth in discrete wavelength bands. The Landsat MSS has four of these bands: green (0.5–0.6 µm), lower red (0.6–0.7 µm), upper red to near infra red (0.7–0.8 µm), and near infra red (0.8–1.1 µm). The smallest unit of radiation collected by the MSS represents the average radiation emitted from an area of 0.62 ha, called a picture element or pixel. In 25 seconds the MSS can survey millions of these pixels and produce an image of an area of 34,225 sq km. The image is actually a set of numbers on a computer tape that correspond to the radiation intensity for every pixel in each of the four bands, but it can easily be converted into a photographic format in which the intensity of radiation emitted from particular areas is represented in terms of shades of grey. The relatively low resolution of satellite images, however, and the restricted wavelength bands in which they collect radiation, mean that they are not as representative as a photograph which collects light over the entire visible spectrum. Before images can be properly interpreted the features which appear on them need to be carefully matched with what is known by experts to exist on the ground. This operation, called "ground truthing", is an absolutely essential part of satellite remote sensing.

Satellites have four main advantages which make them far more appropriate than aerial photography for monitoring large-area conditions on a regular basis. First, once the satellite has been launched it travels automatically over every part of the earth once every one or two weeks, and sometimes even more frequently. This obviates the need for constant launches of the "platform" which carries the sensor (in aerial photography this would be an aircraft). Second, large areas can be surveyed at relatively low cost. A single Landsat MSS image, for example, covering 34,225 sq km, is the equivalent of 5,000 aerial photographs taken at a scale of 1:20,000. A variety of photographic image products can be obtained for a few hundred dollars and digital tapes for $660. Third, a better balance is struck between resolution and the volume of output data so that the actual cost of interpreting an image is between a tenth and a third of that required to analyse an aerial photographic survey of the same area. Fourth, because the image is stored in digital format its interpretation can be automated, thus saving more time (Grainger 1984a).

There have only been a small number of attempts to use satellite imagery to monitor desertification. Not all of them yielded reliable results, and most were essentially preliminary studies designed to evaluate or develop techniques rather than actively to monitor desertification.

The most thorough satellite-based studies of desertification have been made by a team at the University of Lund in Sweden. In an early pilot project, Ulf Helldén and Mikael Stern used computerized analysis of Landsat images to monitor the appearance of new patches of desert in the El Hamma and Medenine areas of southern Tunisia. Some of the desert patches already evident in 1972 had increased in size by 1979, and new ones had also formed by that time (Helldén and Stern 1980). In a later study, Lennart Olsson compared Landsat images collected in 1973, 1977 and 1979 of an area in the province of Northern Kordofan in Sudan. By generating maps of the distribution of vegetation and the magnitude of albedo (the proportion of solar energy reflected by the earth's surface) from each image, he showed how the land recovered in the late

1970s as the rainfall increased following the first phase of the Sahelian drought. Olsson pointed out that most previous reports on desertification had compared an area's environmental situation before a drought with that during or shortly after it. "Not surprisingly", said Olsson, "the environmental conditions degraded from the first to the second situation." He, however, had made the opposite comparison, and although "signs of desertification can mainly be seen during extremely dry periods ... it has not been possible to find a consistent trend of a degrading landscape." Olsson was handicapped by the fact that no Landsat images of the area were available between 1980 and 1984, when he conducted his field studies, and so he was limited to using images collected in the 1970s, even though he could not provide proper ground truth for them. Nevertheless, his work confirmed the findings of previous studies by the Lund team concerning the limitations of Landsat MSS imagery for mapping vegetation and land use in dry areas (Olsson 1985).

Obstacles to the Use of Satellites to Monitor Desertification

Why has so little progress been made in using satellites to monitor desertification? One reason is that the resolution of satellites has been too low to detect the fairly subtle changes in ground conditions which characterize the phenomenon. The resolution of the Landsat MSS, for example, is 79 metres, and so the smallest area which can be identified is 0.62 ha. Foresters all over the world have used Landsat imagery to survey areas of closed forest (large contiguous blocks of closely spaced trees), but this resolution is generally only sufficient to distinguish between forested and non-forested areas, rather than between different types of forest. Using satellite images to detect tree cover in dry areas, where the density of trees is much lower, is more difficult because the reflective character of those areas is dominated by grasses, shrubs and soil. Bearing in mind that the deforestation of open woodland is one of the most visible forms of desertification, the technical obstacles to monitoring the phenomenon with satellite imagery become apparent. The situation is improving, however, for the new generation of satellites is equipped with sensors that have a much higher resolution. Landsats 4 and 5, launched in 1982 and 1984, each carry a Thematic Mapper sensor with a resolution of 30 metres in addition to an MSS. The sensor of the French SPOT satellite, launched in 1986, has an even higher resolution of 20 metres. No appropriate applications of these high-resolution sensors in the monitoring of desertification were identified at the time of writing of this book. Despite the increase in sensor resolution during the 1980s, Paul Tueller of the University of Nevada in Reno has argued that to be more successful in monitoring drylands we still need to learn far more about their spectral characteristics, and in particular the way in which these represent the combined properties of soil and the sparse vegetative cover. Improved knowledge of this crucial topic will greatly enhance our ability to interpret satellite images of drylands and monitor desertification (Tueller 1987).

Another obstacle concerns the availability of images. Although satellites like Landsat normally take (and transmit) images continuously, only a small proportion are recorded by ground receiving stations. Each ground station can receive images transmitted by the satellite within a radius of about 1,500 km (near the Equator) but some areas are not served by such stations. Currently, dryland areas are served by Landsat receiving stations in Hyderabad, India (covering most of south Asia); Cuiaba, Brazil; and Mar Chiquita, Argentina (covering all but the northwest tip of South America). Although the European ground station at Fucino in Italy covers the northern fringe of the Sahara, there was no station in West or Central Africa to cover the Sahel until one recently opened at Maspalomas in the Canary Islands, and its coverage only extends as far east as Niger. Similarly, the ground station in Riyadh, Saudi Arabia, covering a large part of West Asia, only opened a few years ago. In the absence of a ground station the only way in which an image of an area can be obtained is by recording it on video-tape in the satellite for later transmission to central ground stations (in the case of Landsat these are in the USA). Because the satellite's video-tape recorder has a limited capacity only a few images may be recorded on each complete passage around the earth. Until now, areas suffering from desertification have usually not been accorded sufficient priority to be recorded in this way very often, and therefore the number of images available for them has been fairly small. This was certainly the experience of the Lund team in Sudan, referred to above. The dependence on local ground stations and the onboard video-tape recorder was finally removed in February 1989 with the completion of NASA's Tracking Data Relay Satellite System (TDRSS), a network of communications satellites which can relay to a US ground station signals from Landsat (and other) satellites, regardless of their orbital position. Just what effect this will have in practice on the collection of key global environmental data, however, remains to be seen.

Finally, there has been a startling lack of interest by international agencies in developing an effective system for monitoring desertification. The United Nations, for example, has not yet even established a continuous, satellite-based global environmental monitoring system for tropical forests, called for in a resolution of the United Nations Conference on the Human Environment as long ago as 1972 (Grainger 1984a). With the current generation of remote-sensing satellites this would be fairly easy technically, compared with monitoring desertification. UNEP and its sister organization FAO have opted instead for a decentralized approach (see p. 133) to monitoring natural resources in the tropics and elsewhere, inviting government agencies in individual countries to supply data on forest areas etc., and supplementing these with other data from special UNEP or FAO studies and the work of scientists in different parts of the world. This approach assumes, of course, that government agencies have the interest, skills, personnel and funds to carry out the work, which is not generally the case at present. Setting up a centralized monitoring system would obviously be an ambitious task, requiring a sophisticated computer system for image interpretation, and a worldwide network of collaborating scientists to conduct ground-truthing checks. The system would also have to process a large number of satellite images – 150 would be required for the Sahel alone. The question is whether there is any practical alternative if we are to be assured of measuring the extent and rate of increase of desertification during the next ten years. There are two other possible options, indirect monitoring and the use of weather satellites. These are described in the next two sections.

Indirect Monitoring

In view of the limitations of satellites for direct monitoring of the development of desertification it might be more feasible to observe the side-effects, such as dust storms, to which desertified areas are very susceptible (Goudie 1978; McTainsh 1980). Satellites can monitor the paths of dust storms over quite long distances and since they can also show the main sources of dust, the desertified areas most vulnerable to wind erosion can be identified. Geostationary weather satellites are well suited to this job, for despite their low resolution they give daily coverage of an area. In February 1977 a major dust storm developed in New Mexico and for four days travelled right across the continental United States to the Atlantic Ocean. The dust storm was tracked by two US geostationary satellites, the SMS-2 and GOES, and the source of the dust was pinpointed as the Clovis-Portales area of New Mexico (Breed and McCauley 1986).

The satellite monitoring of dust could be supplemented by other data. Grant McTainsh of Griffith University in Brisbane, Australia, has proposed a single index of desertification based on the amount of dust mobilized from desertified areas in Africa and carried out to the Atlantic by the *Harmattan* (see p. 31). With a knowledge of the main dust mobilization and deposition mechanisms, McTainsh claims that it should be possible to develop a low-cost monitoring system by building on the existing network of dust measurement stations. The system would use simple techniques such as dust deposition traps, measurements of atmospheric turbidity by recording the strength of sunshine, and sampling dust in the air by means of filter pumps (McTainsh 1985). Of course such data would be highly aggregated and, while giving some idea of the total annual rate of soil erosion in a region like West Africa, the technique would not identify the places where desertification was actually taking place.

The Contribution of Weather Satellites

One way to overcome the obstacles of poor image availability and the large number of images needed for broad-scale assessments may be to use weather satellites. They take images of each area on the earth more frequently than Landsat, and have a coarser resolution so that each image covers a much larger area. A group of scientists at NASA's Goddard Space Flight

Center at Greenbelt, Maryland, led by Compton J. Tucker, has been conducting experiments with three polar-orbiting weather satellites operated by the US National Oceanographic and Atmospheric Administration (NOAA). Called NOAA 6, 7 and 8, these satellites carry sensors called Advanced Very High Resolution Radiometers (AVHRR) which, despite their name, have a resolution of only 1 km or 4 km (compared with 79 m for the Landsat MSS). Because AVHRR images cover much larger areas, the amount of computer processing required for each image is reduced. The images are also potentially available every one or two days, giving a higher monitoring frequency than Landsat.

An early result of the project was a map of African vegetation (Tucker *et al.* 1985) which resembled a UNESCO vegetation map of Africa that had taken more than a decade to produce by conventional means. It was followed by a study which tracked the production of vegetation in semi-arid areas and the way it varied throughout the year and from one year to another. Then Tucker and his colleague Christopher Justice used the NOAA satellite data to estimate the spatial extent of true deserts by detecting the absence or otherwise of transient vegetation, a tricky operation because a complete development cycle can occur within a mere eighteen days (Tucker and Justice 1986). Most recently, Compton Tucker and Bhaskar Choudhury monitored the boundary between the desert and the Sahelian region between 1981 and 1986, showing that it shifted south in 1981 but then moved north in 1985 (Tucker and Choudhury 1987). This clearly demonstrates the need for continuous long-term monitoring if erroneous estimates of the spread of deserts are to be avoided. (A single survey might show an increase or a decrease in the area of desert, in comparison with one made five or ten years before, depending upon the years chosen for comparison.) Although the NOAA satellites can be used to monitor true deserts, whether natural or expanded by human impact, it has yet to be shown that they are suitable for monitoring the much more dispersed process of land degradation characteristic of desertification, which it is difficult to detect even with Landsat satellites of much higher resolution. However, the NOAA images would allow us to identify areas where the growth of vegetation does not return as expected when rainfall increases, and these areas could then be investigated in more detail by other techniques in order to check for desertification.

THE DATA IMPERATIVE

The shortage of reliable data on the extent and rate of spread of desertification is a major reason why there are so many differing views on the phenomenon and why there has been little effective action to bring it under control. Unless desertification can be accurately defined and measured, the possibilities for scepticism and misunderstanding abound, and policy-makers will be reluctant to allocate funds for programmes to control it. The subjective judgements of a few experts are insufficient evidence for such a major component of global environmental change. If we wish to obtain better estimates of the scale of desertification in the near future, then, in the opinion of this author, the United Nations or another organization must establish a satellite-based global monitoring system with the resources to purchase and process imagery, and the status to give priority to the collection of images of areas subject to or threatened by desertification. A centralized system is necessary because even in the most ideal circumstances it would take a considerable amount of time before government agencies in most dryland countries would have the necessary technical resources and skilled personnel to carry out such monitoring themselves. We cannot afford to wait that long. Satellites alone would not be sufficient to monitor this complex phenomenon, however, and they would need to be supplemented by a worldwide network of scientists to conduct ground checks and aerial surveys. Monitoring desertification in the drylands is much more difficult than monitoring deforestation in the humid tropics, but it should not be beyond the ingenuity of scientists to devise appropriate instruments and procedures. More doubtful is whether the United Nations and the governments of threatened dryland countries have the will to ensure that such a monitoring system is established. If they do not, then our present state of ignorance is likely to remain for many years to come.

5. Controlling Desertification by Improving Crop Production

The direct cause of desertification was shown in previous chapters to be poor land use. If desertification is to be controlled it is therefore imperative to improve the productivity and sustainability of each major type of land use in the areas best suited to it. This should be accompanied by efforts to prevent soil erosion on cropland and rangeland by planting trees or introducing soil conservation techniques, and to reclaim land that has already been significantly desertified. Chapters 5 to 9 of this book describe the four main types of techniques which the UNCOD Plan of Action (Table 5.1) recommended for improving land use and controlling desertification:

1. increasing the productivity of rainfed cropping on good land so that soil erosion and the expansion of cropping on to marginal land are halted; improving the management of irrigated cropping and rehabilitating failed irrigation schemes to reduce salinization and waterlogging (Chapter 5);
2. improving range management and developing new livestock breeds to increase productivity and reduce pressures on the land; intensifying rainfed cropping on better lands and restoring traditional fallows should make more pastures available for the grazing of livestock in the dry season (Chapter 6);
3. restoring tree and woodland cover to stabilize cropping and pastoralism (by reducing soil erosion and providing more supplementary fodder), halt widespread deforestation for fuelwood, and reclaim desertified land so that it can be more productive again (Chapters 7 and 8).
4. introducing soil conservation practices and stabilizing sand dunes (Chapter 9).

Table 5.1: UNCOD Plan of Action: Checklist of Priority Measures to Combat Desertification

A. Land Use and Rehabilitation

1. Introduce methods of planning land use in ecologically sound ways.
2. Improve livestock raising by means of new breeds of livestock and better range management.
3. Improve rainfed cropping by introducing more sustainable techniques.
4. Rehabilitate irrigated cropping schemes that have failed owing to waterlogging, salinization and alkalinization.
5. Manage water resources in environmentally sound ways.
6. Protect existing trees, woodlands, and other vegetative cover and restore tree cover and vegetation to denuded lands.
7. Establish woodlots as sustainable sources of fuelwood and encourage the development of alternative energy sources.
8. Conserve flora and fauna.
9. Ensure the fullest possible public participation in measures to combat desertification.

B. Socio-Economic and Institutional Measures

1. Investigate the social, economic and political factors connected with desertification.
2. Introduce measures to control population growth, as appropriate.
3. Improve health services.
4. Improve scientific capabilities.
5. Expand local awareness of desertification and skills with which to combat it by training and education, both by means of mass media and courses at various educational institutions.
6. Assess the impact of settlements and industries on desertification, and keep desertification in mind when planning or expanding new settlements and industries.

On the other hand, climatic fluctuation, underdevelopment and related social and economic factors, and misguided government policies also play important roles in causing desertification and need to be taken into account when formulating a strategy to bring it under control. Climatic fluctuation – and the uncertainty attached to this – is inevitable in the future and the only effective response is to make land use more productive and sustainable so that it is resilient to these changes. At the time of UNCOD it was felt that we had at our disposal all the physical techniques necessary to control desertification. Recent experience, including that obtained in the course of projects described here, demonstrates that this is not the case and that such techniques alone are not sufficient. Underdevelopment cannot be solved overnight, but this is no reason to ignore the importance of social and economic factors, as many agricultural and forestry projects (and to some extent the Plan of Action) have done, and focus instead on purely physical techniques. As Chapters 5–9 show, projects which do this are likely to fall far short of their objectives. Other projects, which involve local people in their design and implementation, and improve their standard of living without too much risk on their part, stand a much better chance of success. Finally, although well-designed projects which improve land use can overcome some of the constraints of underdevelopment, their long-term success and the widespread adoption of the techniques which they incorporate can be severely limited if misguided government policies, for example emphasizing the expansion of irrigated cropping to the exclusion of rainfed cropping, are not modified.

This discussion of ways to control desertification is structured so that each chapter discusses one of the main types of land use and the techniques which could be used to improve it. This structure is merely a matter of convenience, and does not indicate that the points made in the preceding paragraph, and in Chapter 2, have been ignored. On the contrary, attention is frequently drawn to the effects of the inclusion (or exclusion) of social and economic considerations in project design. Similarly, the subject of policy is referred to as appropriate, mainly in the context of identifiable constraints on project implementation or suggested policy changes which will remove them. General recommendations for policy changes are summarized in Chapter 10.

This book, like the Plan of Action, advocates the improvement of land use throughout the drylands, not just in severely desertified areas or on the fringes of deserts. It is important to emphasize this point because one of the great misconceptions about desertification is that it only occurs in the driest areas on the fringes of deserts, implying that preventive action is not needed anywhere else. This understates the role of cropping and gives the impression that the predominant cause of desertification is overgrazing (because livestock raising is the main land use in the driest areas). The reality, as shown in Chapters 2 and 3, is that desertification is caused by imbalanced land use over a wide range of climatic zones, with rainfed and irrigated cropping in more humid areas having a major role in the process by displacing other land uses on to marginal lands that are highly vulnerable to desertification. Improving the sustainability and productivity of crop growing, which forms the focus of this chapter, is therefore vital if desertification is to be controlled, and needs to be accompanied by better land-use planning so that such displacements are minimized.

IMPROVING RAINFED CROPPING

Rainfed cropping is still the dominant form of crop production in most dryland areas. In the Sahel, for example, it accounts for 95% of cereal production and 96% of the cultivated area. It is unlikely that this situation will change very much in the medium term, so urgent action is needed to improve the productivity and sustainability of rainfed cropping, both to meet food production targets and to bring desertification under control. "Until the turn of the century", states the Club du Sahel, "cereals will still be grown throughout the Sahel mainly using rainfed farming methods. It is probable that to attain food self-sufficiency the share of development assistance for rainfed farming cereals and especially for intensified farming will have to be increased substantially" (Club du Sahel 1983).

The Sahelian nations, like other arid areas in the world, have increased crop production in the last few decades by expanding the area under cultivation, rather than by increasing yields. Most of the newly cultivated marginal lands, however, are not suited to cropping, so that fertility and yields usually fall off sharply after a few years of cultivation. As a result, grain production has not kept pace with the growth in population, food self-sufficiency has fallen, and imports have risen. If present trends continue, rainfed crop production in the Sahel will increase by only 30% by the year 2000, by which time the population will have increased by 75% and each farmer will have to feed an average of 3.6 people instead of 2.8 people today. The poor performance of rainfed cropping is not necessarily the fault of the technique or those who practise it. Much of the blame lies with government agricultural policies which have favoured cash crops and irrigated cropping. It is essential to modify these policies to encourage rainfed cropping and the investment needed to increase its productivity; otherwise cereal imports will quadruple and more land will be degraded.

There are five priority areas where action could be taken to improve the productivity and sustainability of rainfed cropping:

1. developing improved drought-resistant crop varieties;
2. more extensive use of fertilizers;
3. improving cropping systems;
4. better use of water;
5. improving land-use planning.

Developing Improved Drought-Resistant Varieties

Breeding improved varieties of staple crops like sorghum and millet will allow farmers to gain higher yields than they can with traditional varieties under average rainfall conditions; although it is important to retain the drought resistance of the latter varieties so as to prevent crop failure in dry years. Development of staple crops in the Sahel has got off to a slow start, and while a number of promising improved sorghum varieties are currently under trial in Burkina Faso, millet improvement programmes have so far not been successful. Many Indian millet varieties were tested but found to be unsuited to the harsher Sahelian environment, and none performed better than the leading local variety.

One reason for the lack of progress is that local crop improvement research only started relatively recently. The International Crop Research Institute for the Semi-Arid Tropics (ICRISAT), the scientific body with responsibility for developing staple crops for drylands, was only founded in 1972 and was based in India, even though most of its member countries were African. The first ICRISAT research institute in West

Africa, the Sahelian Centre near Niamey, Niger, did not open until 1981, and scientists there found it difficult to apply the techniques that had been learnt in Asia. The first new varieties developed by ICRISAT were next to useless for local farmers since they were bred at research stations which had good soils, plenty of water and mechanical ploughs – quite unlike the conditions on most of the farms in the region. Crops were also well fertilized and weeded in ways beyond the means of the ordinary farmer. When these new varieties were tried out on real farms the yields obtained were 40–60% below those on the research stations and even lower than those of traditional varieties. Fortunately the lesson has been learnt and there is now greater awareness of the need to breed crops which will grow well under actual farm conditions, giving higher yields than traditional varieties on poor soils with erratic rainfall, and much higher yields if grown on irrigated lands treated with fertilizers.

Even if researchers are successful in developing improved varieties, poor soils and erratic climate make it rather doubtful that African farmers will be able to repeat the "Green Revolution" seen in Asia, where the soils are good and the climate more reliable. The use of selected seeds, fertilizers and pesticides has increased yields of cotton in the Sahel fivefold in the last twenty years, but this is the only rainfed crop in the region which has benefited from more intensive cropping. The picture is very different for subsistence food crops. The sowing of selected seeds of sorghum and millet is exceptional and the yields of these crops have either remained unchanged or declined. Farmers are reported to be suspicious of new varieties, particularly as regards their yields and resistance to diseases and pests, and are also discouraged by the high cost of seeds. In Niger selected seeds are used for only 3% of the area under millet and 0.3% of the area under sorghum, and in Burkina Faso selected seeds are used on less than 1% of all cereal cropland. Some progress has been made in persuading farmers to use selected seed of groundnuts and maize, but average groundnut yields in the Sahel are low and hardly rising even though most are grown with selected seeds (Club du Sahel 1983).

Sorghum is the most important cereal crop in Sudan, accounting for 75% of its cereal production and one-third of its cropland. Most sorghum is grown under rainfed conditions and only 10% of the 3.2 million ha under sorghum is irrigated. Sudan's first commercial hybrid sorghum variety was released by the Sudan Agricultural Research Corporation and ICRISAT in January 1983 after six years of research. "Hybrid Sorghum No. 1" is high yielding, early maturing and suitable for both irrigated and rainfed cropping. Four seasons of tests on Sudan's Central Clay Plains gave yields 58% higher than local varieties under irrigated cropping and 52% higher under rainfed cropping. Development of the hybrid began in 1977 with a collaborative programme involving ICRISAT, Texas A & M University and Purdue University. An Ethiopian, Dr Gebisa Ejeta, was the principal scientist (Ejeta and Woods Thomas 1984). The first large-scale trials took place on about 2,000 ha of land in 1984 during a severe drought, but yields were still impressive. Compared with the national average yield per hectare of less than 1.25 tonnes/ha in normal years, the hybrid gave 5 tonnes/ha under irrigation and 2.9 tonnes/ha under rainfed cropping.

The use of such hybrids means a change in routine for farmers since new seed has to be purchased each year instead of being saved from the previous crop. Thus, if Hybrid Sorghum No. 1 is to become widely grown in Sudan it will require a substantial local seed production facility. The US Agency for International Development (USAID) is supporting a pilot project, which began in 1983, to multiply and distribute sufficient seed of the new hybrid for initial needs and lay the foundations for a local commercial seed industry. If the new industry does become established and enough seeds are produced to convert only 25% of all sorghum cultivation to the new hybrid by 1990, then Sudan's sorghum production could be doubled.

There has been encouraging progress with another staple crop, cowpea. This leguminous crawling plant, which can grow in areas with as little as 200 mm annual rainfall, is a prolific nitrogen fixer and its stalks and leaves are also used for fodder. Improvement work began at the International Institute for Tropical Agriculture (IITA) in Ibadan, Nigeria, in 1970. The work was made difficult by the low yields of traditional varieties and their susceptibility to a wide range of pests and diseases. Crossing various strains to produce high-yielding varieties resistant to pests and diseases was therefore quite a complex task. IITA has developed varieties resistant to twelve different pests and diseases and yielding about 2 tonnes of beans per hectare, compared with the traditional 0.3–0.6 tonnes/ha. One variety, TVx 3236, has been taken up with great enthusiasm by farmers in Kano state in northern Nigeria, and is giving yields that are three times those of the traditional variety. Other improved cowpea varieties have been adapted to local conditions and introduced to farmers in Botswana, Burkina Faso, Cameroon, Central African Republic, Liberia, Somalia, Tanzania, Togo and Zambia (Harrison 1986).

Crop development work in Tunisia has focused on wheat. Five bread wheat varieties and five durum wheat varieties were developed by the Tunisian Wheat Development Programme, with funding and technical assistance from the US government, the Rockefeller and Ford Foundations and the International Maize and Wheat Improvement Centre (Anon 1983). The improved varieties were accepted by farmers and even though the programme officially ended in 1977 other varieties continue to be developed. Annual per capita production of cereals rose from 104 kg to 160 kg between 1970 and 1980. For a cost of only $3.5 million the programme saved $125.9 million in foreign exchange because the increase in production meant that less wheat needed to be imported.

Another way to improve the productivity of rainfed cropping is by introducing "new" drought-resistant crops. These are usually crops that have been cultivated in fairly limited areas of the tropics for many years but until now have not been the object of scientific improvement efforts. The US National Academy of Sciences has recommended that a number of previously underexploited crops deserve to be cultivated on a wider scale. These include grain crops like grain amarinth (*Amarynthus hypochondriacus*) and Pima-Papago 60-day maize (*Zea mays*), and vegetables like Hopi mottled Lima bean (*Phaseolus lunatus*) and Papago Cushaw squash (*Cucurbita mixta*) (NAS 1975). Persuading farmers to adopt new crops is likely to be a slow process but one which can pay dividends in terms of higher income and better nutrition.

Increasing crop yields by the use of more productive varieties will be of little benefit without good local grain storage facilities to protect the grain from pests and spoilage and to build up a reserve for drought years when crops fail. Mauritania, for example, has constructed a national network of grain silos in a $10.5 million programme co-financed by the African Development Bank and the Netherlands government through the UN Sudano-Sahelian Office (UNSO).

Increasing the Use of Fertilizers

The application of fertilizers can double the yields of even conventional varieties of millet and maize and prevent the fertility of poor soils from being depleted by intensive cropping. Phosphate fertilizers are especially valuable in the Sahel, where soils are generally deficient in phosphorus. Adding phosphorus stimulates root growth and therefore increases the effective drought resistance of a crop, since roots can reach deeper moisture levels in dry periods. The use of 15 kg/ha of triple superphosphate doubled millet yields in one area of Niger from 0.55 to 1.1 tonnes/ha, and only 10 kg/ha tripled cowpea yields (Harrison 1986).

Despite these potential gains, fertilizer use for rainfed crops tends to lag behind that for the more profitable cash crops because of the high costs involved. Thus, in the early 1970s, fertilizers in the Sahel were essentially restricted to cash crops such as cotton, groundnuts and irrigated rice, although there are now encouraging signs of a growing use of fertilizers for rainfed cereals too. The fertilization of cereals more than doubled in Senegal between 1972 and 1979; in Burkina Faso, where fertilizer use began much more recently, it increased by a factor of 24 over the same period, although still only 3% of all cereal cropland in the country is treated with fertilizer. In the Sahel, pesticides and fungicides continue to be exclusively used for cotton.

There are a number of obstacles to the wider use of fertilizers in the Sahel. Despite the abundance of natural phosphates in the region, there is only one fertilizer factory at Dakar in Senegal and so the majority of fertilizer has to be imported. While there is a good system for distributing subsidized fertilizers for cotton cultivation, low cereal prices make it uneconomic for farmers to buy fertilizers for food production and for governments to subsidize them (Club du Sahel 1983). The precarious nature of dryland farming means that yields and income are uncertain even under the best conditions, because the use of fertilizers can lead to only marginal increases in crop yields where annual rainfall is low and erratic, especially when it is under 400 mm. It costs a lot of money to buy fertilizer and seeds of improved varieties (which often require the application of fertilizers), so many farmers get badly into debt. A survey of one village in Senegal in 1980 found that in the previous year 25% of villagers were unable to repay their loans, and so could not afford to buy sufficient seed for the following season.

Views differ on the effectiveness of the "Green Revolution". There are stories in many countries of how financial overextension and crop failure forced independent farmers to lose their lands to money lenders and become either tenant farmers or casual farm labourers. Thus farmers may be unwilling to commit themselves to investing in intensified cropping without firm assurances from the government on attractive crop prices and reasonable credit terms. The situation may be different in Asian dryland areas. Surveys in India have shown that marginal farmers form the largest group of fertilizer consumers and that they use fertilizer more intensively than larger farmers (Sen 1982).

Soil fertility and crop productivity could be increased much more simply by either organic manuring or deep tillage. Restoring the traditional fallow system would mean that livestock would once again be able to feed on crop residues in return for depositing their dung on the fields. An alternative would be to plant fast-growing trees like *Gliricidia sepium* that produce plentiful foliage which can be fed to animals as fodder or turned into the soil as "green manure" (Falvey 1982). This would allow the remaining stubble to be turned under too, in the practice known as deep tillage. Farmers are reluctant to do this when fodder is scarce, even though they know it will benefit crop yields.

Improving Cropping Systems

The type of cropping system employed will also make a vital contribution to increasing yields over the longer term. Traditional dryland cropping systems, such as the four-year rotation sequence of fallow, millet, cowpeas and groundnuts traditional in the Sahel, were sustainable because they were designed to be as resilient as possible to the tremendous variability in rainfall common in these areas. However, such systems are breaking down, as fallows are reduced to grow more food for rising populations and maintain overall production as yields decline because of poor rainfall and depleted fertility. Paradoxically, the wider use of high-yielding varieties could bring about a return to traditional systems, because it would help farmers to break out of the vicious circle of continually having to reduce fallow periods and expand the area under cultivation in order to maintain production.

Intercropping, the growing of two or more crops together in the same field, has long been practised in the Sahel, and yields are superior to those of monocultures. Growing millet and cowpea together, for example, gives a total yield 50% higher than when the crops are grown separately. Intercropping can be more efficient than monoculture if the crops are carefully chosen to have different heights and root depths so that the most efficient use is made of available sunlight and water. If one of the crops is a nitrogen-fixing plant such as cowpea, then it acts as a fertilizer to the other crop, leading to higher yields. The denser vegetative cover of the multiple crops reduces soil temperature and erosion, and increases water infiltration. Weeds have less opportunity to get established, so less labour is needed for weeding.

Intercropping also provides farmers with a form of insurance against crop failure. The growing period of one crop is usually different from that of the other, so if the rains are late for one crop and reduce its growth they may arrive in good time for the other crop. According to a study by ICRISAT, the probability of a disastrous crop failure is one year in five when pigeonpea is grown alone, one year in eight when sorghum is grown alone, one year in thirteen when both are grown together in separate plots, but only one year in thirty-six when they are intercropped (Harrison 1986). African farmers are highly knowledgeable about intercropping, but further scientific studies could improve the efficiency of this technique.

Many farmers in the Sahel still cultivate the soil by hand. The use of animal traction could make a substantial improvement in farm productivity, but its introduction has been slow and confined to relatively few areas. The Club du Sahel estimates that it is only used on about 5% of all cereal lands in Burkina Faso and Niger (Club du Sahel 1983). Animal traction has become widespread in cotton-growing areas like southern Mali but even there its use for cereal cultivation is sporadic.

Making Better Use of Water

Many dryland farmers do not make full use of even the small amount of rain which falls on their lands since most of it runs off before infiltrating the soil. The productivity of rainfed cropping could be considerably increased and soil erosion decreased by the introduction on to farms of water-harvesting systems which retain rainfall for a longer time on the soil surface so that more of it percolates through to crop roots. A good example of such a system is the "stone lines" project in the Yatenga area of Burkina Faso, where years of poor rainfall had lowered the water table by up to 5 metres and the sandy soil had become so crusted by overuse that grass would not grow even during the rainy season. Local farmers traditionally used lines of stones placed along the contours to hold back rainwater before it ran down the slope. Now, with the help of the UK-based voluntary agency OXFAM and project director Peter Wright, they have made the technique even more effective. In the new technique, the lines of stones divide fields into a number of catchment bands, and water can be held back for a distance of up to 15 metres behind each stone line. The stones are placed in a shallow trench, 5–10 cm deep, and piled to form a barrier 15–25 cm high and 20–30 cm wide. Earth from the trench is placed in front of the stone line to give additional protection against water running on to it. Small gaps are left in the stone lines so that some water can pass through to the next band.

The greatest challenge was finding a cheap way to mark out the contours, for the very slight slopes (less than 2%) are difficult to assess by eye. Peter Wright devised a simple water level by using a 10–20 metre length of transparent hosepipe filled with water. The hose is moved until the water levels at each end coincide with calibration marks. Farmers are shown how to master the technique in a two-day course in which they can see scale models of the new water-harvesting system, learn from the experience of other farmers who are already using it, and actually see it operating in the field. In 1984–5 OXFAM trained no fewer than 600 farmers from thirty villages in this way.

The results of this simple technique have been amazing. Crop yields have increased by 50% and soil erosion has been substantially reduced. In the severe drought year of 1984, farmers in the village of Gourga who had stone lines on their lands were able to grow enough food to last them for three months, while farmers with no stone lines harvested nothing. With the aid of the technique the people of Gourga have been able to turn some of the most degraded land in the village into a common field. The technique is now spreading informally from farmer to farmer and from village to village (Harrison 1986).

Integrated Farm Improvement Schemes

The World Bank was for a long time hesitant about becoming involved in dryland agricultural development schemes, but in an attempt to identify the most promising techniques for improving rainfed cropping it decided in 1979 to fund a six-year experimental project at Baringo, in a highly eroded area on the slopes of the Kenyan Rift Valley, where virtually no vegetation remains at the end of the dry season and severe dust storms blow for up to a month before the rains. The result was a project which supported the development of small farms and livestock owners using simple, low-cost techniques which integrated a number of the approaches advocated above, including the cultivation of improved varieties and better use of available water.

A prominent component of the project was the construction of soil conservation works, mainly different types of earth walls (bunds) to prevent soil erosion in an area that had already lost much of its fragile soil through overgrazing and deforestation, and check dams to prevent the further growth and development of gullies, which were widespread in the area. The bunds were spaced every 1.5–3 metres, and enabled farmers to make the maximum use of the available rain (on average only 600 mm per year) for growing crops. These were planted only on the bottom part of the "terrace" behind each bund, where most of the water collected and plant growth was better than further up the slope. The bunds, together with improved structures for tapping water from rivers, led to a major improvement in the availability of water for crops. The cost of $20 per hectare was paid back within one season by sorghum yields that were two to three times those obtained before, even though only about 60% of the total area of land was planted. Once the bunds had been constructed the system required less labour than in the traditional hand cultivation of soil. Although the techniques used here are applicable to other areas they really require soils similar to the alluvial soils at Baringo. These are not hard at the start of the wet season and so rainwater can infiltrate easily instead of running off along the surface.

Farm productivity also benefited from the introduction of improved varieties of sorghum, cowpeas and tepary beans (*Phaseolus acutifolia*). Care was taken to ensure that these were locally adapted so that farmers would actually want to grow them. Two nurseries were also established to grow about 250,000 tree seedlings a year. Although both indigenous and exotic species were sold, *Leucaena*, *Prosopis* and *Cassia* were more popular with farmers, who apparently found it easier to come to terms with planting such exotics than the trees which grew naturally in the area.

The Baringo project was unusual in that it had no well-defined goals except to "see what you can do", and there was no expected financial rate of return. Perhaps the main lesson for future development projects was the importance of working with existing local institutions rather than creating new ones. The project involved the co-operation of eight different ministries of the Kenyan government (no mean feat in any country), but most of the decision making was decentralized to district level through regular district planning programmes, and project staff worked with chiefs and local leaders all the time. Management of the project was in the hands of Kenyans right from the start, with a minimal number of expatriate "experts" (all of whom spoke Swahili); this is undoubtedly one of the reasons why the project was still running two years after the last expatriate left (Republic of Kenya 1984).

Improving Land-Use Planning

Desertification is a sign that land uses have got out of balance with the land's ability to support them. Clearly something is wrong when the only way in which crop production can be increased is by expanding the area under cultivation on land that is too poor to sustain cropping for more than a few years without becoming degraded. The Club du Sahel has made the surprising but categorical assessment that the 13–14 million ha of land currently under cultivation represents perhaps only a quarter of the area which is potentially cultivable (Club du Sahel 1983). The problem is that the distribution of such unutilized lands does not match the distribution of people who need them for subsistence cropping, hence the unwise expansion of cultivation on to marginal land in some regions. So while the Club du Sahel suggests that the pressure on overcultivated land could be alleviated somewhat by encouraging people to migrate to develop new land for cropping in underpopulated areas in Mali, Chad, Senegal and Burkina Faso, it still makes the intensification of cropping systems on existing farmland its number one priority. If this aim is to be achieved, government agencies will first need to strengthen their capacity to plan and regulate land use, so that rainfed cropping is intensified on land which is suited to it, rather than on lands only fit for extensive and occasional grazing.

The Policy Imperative

Increasing the productivity and sustainability of rainfed cropping on the more fertile lands should remove the need to cultivate marginal lands, which often leads to desertification. Increases of 75–100% in millet and sorghum production are quite possible, according to the Club du Sahel, by the use of some of the techniques described above (Club du Sahel 1983). Most of the projects included in this section were undertaken simply to improve farm productivity, rather than to control desertification. The general lack of action in this field is indicative of a widespread neglect of rainfed cropping by governments, even though it is still practised by the majority of farmers in the drylands. Desertification is therefore in part a reflection of a major flaw in agricultural policy and one that must be rectified quickly, as much for the needs of local farmers and increasing food production as for controlling desertification.

Improving rainfed cropping will greatly depend on the

determination of governments to modify food and agricultural policies. Low, government-controlled crop prices may well maintain the political support of urban populations upon which governments depend for their survival, but they also discourage farmers from investing in more intensive production. Ways must be found to increase farmers' prices and their access to credit, fertilizers and selected varieties. New mechanisms for marketing crops are also needed, so governments should review the operations of their often inefficient and bureaucratic national cereal agencies, and consider alternative strategies, such as encouraging the formation of local producer associations and co-operatives. Without positive action of this kind the present stagnation in rainfed cropping is bound to continue.

Fortunately, there are signs that governments in the Sahel are at last giving their attention to food crop production. Mali has introduced programmes to increase millet and rice production; Senegal is trying to increase millet and sorghum yields in the groundnut-producing area of Sine-Saloum and to eliminate the country's rice deficit; and Niger, Burkina Faso and Chad have national programmes to expand cereals research and agricultural advice. This momentum must be sustained and governments encouraged to be far more open to new and more appropriate agricultural policies.

IMPROVING IRRIGATED CROPPING

Cause or Cure?

Irrigated cropping has been given priority over rainfed cropping in the agricultural development plans of most dryland countries in recent decades, and at first glance it does have a clear advantage. Irrigation uses water from rivers, reservoirs or underground aquifers to compensate for the lack of water in dry seasons or droughts. The greater availability of water, especially in critical parts of the growing season, can lead to yields that are two or three times those from rainfed cropping. Irrigation has helped people in the Ajmer district of the Indian state of Maharashtra, for example, where, in the Masuda irrigation scheme, groundwater was exploited by digging new wells, deepening existing ones and installing diesel and electric pumps. The scheme enabled more economic crops to be grown and yields to be increased. Small farmers in particular benefited, and because of their enthusiasm an even larger area has been brought under irrigation (Patel 1983).

Some experts would argue that more irrigation is the only way in which food production in regions like the Sahel can keep pace with population growth amid recurring droughts. Because of its higher productivity, irrigated cropping already produces 40% of all crops in the developing world on only 20% of the cropland. By concentrating food production on the most fertile lands, an expansion of irrigated cropping could, in theory, also halt the spread of rainfed cropping on to marginal lands and the desertification which results.

However, as was shown in Chapter 2, irrigation is as much a cause of desertification as a cure. Inadequate design, construction and management lead to waterlogging, salinization, alkalinization and sedimentation so that yields decline. Eventually lands turn into saline deserts and have to be abandoned. In principle, soil degradation is not an inevitable consequence of irrigation, but it appears so in practice because of the way irrigation schemes are implemented. Furthermore, the argument that more irrigation could help to concentrate food production on highly productive lands and reduce the need for rainfed cropland is only partially true. Irrigated cropping requires considerable investment and incurs high running costs, so if it is to be economically viable then irrigated lands usually have to grow profitable cash crops for export rather than food crops for domestic consumption. Even if managed perfectly, more irrigation might not therefore actually increase food production very much. It could even make matters worse if it further displaced rainfed cropping on to marginal lands, lowering yields and causing desertification due to overcultivation. There are three main ways to improve irrigated cropping and control desertification:

1. manage existing and future irrigation schemes better, so as to: (a) prevent further waterlogging and salinization; (b) increase food production (as far as possible) and therefore help to concentrate cropping on the most productive lands;
2. rehabilitate existing irrigation schemes where production

has been severely reduced by waterlogging and salinization or by mechanical problems;
3. switch from large-scale irrigation schemes, with all their management problems, to small-scale schemes which are more easily managed and have a better chance of increasing production of staple food crops as opposed to cash crops and luxury food crops.

Irrigated Cropping in Asia and Africa

Asia is the world's most irrigated region, with China, India and Pakistan accounting for almost half of all irrigated cropland. But the effectiveness of irrigation has been undermined by extensive waterlogging and salinity problems. Pakistan and India alone now account for up to 40% of all desertified irrigated cropland. Because of desertification, and poor management generally, the efficiency of irrigation in India is very poor, being less than a third of that in Israel and Jordan (FAO 1985a; Bansil 1983).

Africa, on the other hand, is the least irrigated part of the developing world. Only 5% of its cropland is irrigated, compared with 8% for Latin America and 29% for Asia. One reason for this is the general lack of underground and surface water supplies. Most of the underground aquifers in West Africa are either small or deep and expensive to tap. In Mali, for example, FAO has put the cost of groundwater irrigation at $15,400 per ha using a diesel pump, 80% higher than for a large public surface water irrigation scheme in Niger ($8,400 per ha) and more than five times the cost of a small-scale irrigation scheme in Mauritania ($3,000 per ha) (FAO 1985a).

Irrigated cropping which depends on surface water, as is the case for many African irrigation schemes, is still vulnerable to drought. Only one-third of all irrigated cropland in the Sahel has fully guaranteed water supplies in the dry season because river flows fall sharply at that time (Club du Sahel 1979). River flows in the Asian and South American drylands are more continuous because they originate from watersheds covered by forests on which rain falls throughout the year, and the "sponge effect" of the forests ensures fairly even flows. Most rivers in Africa flow instead from sub-humid uplands, like the Ethiopian highlands, which experience a seasonal drought and are highly populated, poorly forested, overcultivated and overgrazed. River flows from these watersheds would be variable even in ideal circumstances in which the watersheds had a good vegetative cover, but devegetation and chronic population pressure mean that most of the rain floods down to the lowlands in the rainy season, and little is stored in the watersheds to be released in the dry season. The soil eroded from hillsides by these floods ends up as sediment in reservoirs and irrigation canals, reducing their capacity and the overall efficiency of irrigation schemes (Freeman 1986). The giant 880,000 ha Gezira irrigation scheme in Sudan depends on the level of the Blue Nile, and therefore on rainfall in the Ethiopian highlands. In 1984 the rains were the lowest since 191. This prevented 126,000 ha of wheat being planted and led to a production loss of 150,000 tonnes.

Of the 9 million ha of irrigated cropland in Africa, half is in the five countries of North Africa and two-fifths in Madagascar, Sudan and Nigeria. The Sahel has only a small area of irrigated cropland – just 392,000 ha in the six countries in the region (FAO 1985a). In the first half of the 1970s, Africa's irrigated area expanded by 4% per annum but the rate of increase fell to only 1.7% per annum in the second half of the decade (Harrison 1986). Fewer than 5,000 ha of extra irrigated cropland are established every year in the Sahel, and of this area 4,000 ha are for rice cultivation (Harrison 1986). According to the Club du Sahel the annual increase in irrigated cropland in the region only just exceeds the area of land going out of production because of salinization and waterlogging (Club du Sahel 1979).

Improving the Management of Irrigation Schemes

Huge sums of money have been devoted to expanding irrigation in developing countries in recent decades. Investment in irrigation was running at $10–15 billion per annum in the late 1970s, and the World Bank devotes about 28% of all its agricultural lending to this sector (Carruthers 1985). In the Sahel irrigation accounts for 10% of all foreign aid (Club du Sahel 1981), in Pakistan for 10% of total public investment, and in Mexico for 80% of all public investment in agriculture since 1940 (Repetto 1986).

Much of this money goes to large-scale public irrigation schemes in which the government builds a network of canals to link a water source such as a reservoir with a large number of farms. The management of these schemes is often placed in the hands of large and extremely inefficient irrigation authorities. High-level responsibility is usually shared between a number of government ministries: lack of co-ordination between them is another source of inefficiency. Poor management means that water and other inputs are not delivered to farms on time, and many farmers with plots at the end of the irrigation canals often do not receive their full allocation of water because farmers at the top of the system have used more than their share. In some Indian projects as much as half of the total area is without proper water supplies (Walach 1984).

In the $400 million Rahad irrigation project, covering an area of 4,000 sq km in eastern Sudan, cotton yields dropped by more than 50% over the first two years of operation. Responsibility for managing the highly mechanized project, which began in 1972, was vested in a bureaucratic corporation which, according to Paul Harrison:

> [The corporation] took all decisions on crops to be grown (cotton and groundnuts without fallow), the type and quantity of inputs and the nature and timing of operations. It controlled directly no less than 90% of all the tenants' costs. Tenants who did not comply with the corporation's directives faced fines or eviction. The rationale was that experts must know better than illiterate farmers. The experience proved that the opposite is true. Corporation services such as tractor ploughing, seeding and weeding were often late, due to breakdowns, shortages of spare parts and fuel, lack of trained drivers and mechanics. (Harrison 1986)

Large-scale schemes are also often too inflexible to respond to market forces as quickly as individual farmers would. In Sudan the corporation persisted in growing cotton, even though falling world prices made it increasingly uneconomic and farmers wanted to grow the more profitable sorghum instead.

Even judged on its own merits, overlooking the harmful environmental impacts, irrigated cropping does not look very attractive from an economic perspective and hardly justifies the investment allocated to it. Initial crop yields never seem to reach projected levels, and then often decline over time as poor management reduces the efficiency of irrigation, and salinity and waterlogging rise. Bad management and the harvesting of only one instead of two crops in Mali, for example, has meant that rice yields have been only 1.7–2.6 tonnes/ha instead of the expected 5–6 tonnes/ha (Club du Sahel 1979).

P.K. Joshi and A.K. Agnihotri, of the Central Soil Salinity Research Institute at Karnal, compared yields immediately before and after the start of six Indian irrigation projects in the 1960s and 1970s. In only two projects did rice yields exceed 1 tonne/ha after the start of irrigation and in only one does the yield now exceed 2 tonnes/ha. Yields initially increased by 30–40% in two projects and by 10–20% in another two, but in one project the yield actually fell by 4% following irrigation. In three of the six projects yields are now lower than they were at the beginning, in two yields have remained about the same, and only in one project have yields actually increased (Joshi and Agnihotri 1984).

The construction of large-scale irrigation schemes frequently runs over time and budget. Nine recently completed major irrigation projects financed by the Asian Development Bank took 72% longer to complete and cost 66% more than originally planned. This is not untypical of projects generally (Bhuiyan 1985). To add to the problem governments find it extremely difficult to collect all the water charges owed by farmers participating in these schemes. In Pakistan, states Robert Repetto of the World Resources Institute, "gross receipts represented only about 13% of the fiscal cost of public irrigation services" (Repetto 1986). The less money is collected the less is available for looking after recurring costs, such as those of maintenance.

Improving the Maintenance of Equipment and Canals

There are many reasons why existing irrigation schemes fail to meet expectations, but the lack of maintenance of machinery has been stated by the Club du Sahel to be "the first problem encountered in all Sahelian states without exception" (Club du Sahel 1979). This is caused in part by a shortage of trained personnel and leads to the frequent breakdown of pumping and control machinery. Most of this is imported, so when breakdowns do occur, it is often difficult to make repairs even if personnel are available, because a lack of foreign currency delays the purchase of essential spare parts, sometimes for years. Fuel shortages occur for the same reason. Thus pumping stations often have very much shorter lives than originally planned.

There is no simple solution to the maintenance problem, bearing in mind the economic difficulties which many dryland countries now face and are likely to continue to face in the future. One solution would be to promote the domestic manufacture of pumps to save on foreign-exchange needs, but in the first few years as the industries establish themselves their products might not be as reliable as desired. Alternatively, pumps could be imported from another developing country. A government-controlled irrigated wheat-growing scheme in a remote area of northern Mali, supported by funds from USAID, used diesel pumps made in India to save money, but that particular model of pump was subsequently discontinued, spare parts were not available in Mali, and it was difficult to find skilled personnel to service the pumps (Moris 1984). Certainly the training of more technicians is absolutely essential. An irrigation project in Nigeria (described on p. 189) used Japanese pumps, but trained motorcycle mechanics to service and repair them.

Irrigation canals also suffer from poor maintenance, and a large proportion of the water never reaches any field because of seepage through unlined canals, holes in canal banks and sedimentation. Seepage was causing the loss of more than 20% of irrigation water in the Indian state of Punjab before canals were properly lined (Gupta 1982). Canals become infested with weeds like water hyacinth, cat-tails, water lettuce and water fern. The weeds obstruct the flow of water, provide favourable environments for the hosts of waterborne diseases such as dysentery, cholera, malaria, filariasis, schistosomiasis and bilharzia, and result in a substantial loss of water by evapotranspiration before it ever reaches the fields. In one irrigation scheme, the Gandak command area, it was estimated that the amount of water transpired by weeds would be sufficient to irrigate 117,000 ha of wheat and 467,000 ha of rice (Joshi and

Agnihotri 1984). Determining who is responsible for maintaining and operating irrigation canals can cause endless disputes both within the irrigation authority and between the authority and farmers. When farmers are not closely involved in the design and management of large irrigation schemes it is probably inevitable that they should not feel the same sense of responsibility for looking after canals as they would in a smaller project.

Better Project Design

The design of irrigation projects needs considerable improvement. The omission of adequate drainage from project specifications leads inevitably after a few years of operation to waterlogging and salinization and to a decline in yields. Numerous examples of this were given in Chapter 2. Some projects are even constructed without adequate preparatory studies to assess whether sufficient water will be available to service the needs of farmers.

Greater attention also needs to be paid to social aspects of project design. Preliminary studies of key sociological factors which will have a bearing upon the success of a project are not always carried out. Large-scale irrigation schemes are usually "imposed" on farmers with no consultation or regard to their knowledge of local conditions. Ensuring the widest possible distribution of benefits from irrigation schemes is also important, since such schemes have been criticized for being of little benefit to poorer people (Stryker and Gotsch 1981).

Another vital part of social design is to ensure that both project staff and farmers receive adequate training. Poor technical expertise on the part of both farmers and management staff has been blamed for the disappointing performance of three government-funded schemes in the Morogoro, Tanga and Arusha areas of Tanzania, dating back to the late 1950s and early 1960s and covering almost 2,500 ha (Mrema 1984).

Greater Farmer Involvement

The greatest possible involvement of local farmers is crucial to the success of irrigation projects, and can pay great dividends since their knowledge may prevent costly mistakes in both technical and social design; such involvement is, however, seldom obtained (Barnett 1984). In large-scale irrigation schemes, farmers are usually only allowed to be tenants and so become justifiably alienated when their land is compulsorily purchased with little or no compensation (Famoriyo 1984). This perhaps explains why farmers are often disinclined to maintain properly parts of the system for which they are responsible. In the Rahad scheme, described on p. 000, village production councils were supposed to give the farmers the opportunity to take part in the decision-making process but, according to Mary Tiffen of the Overseas Development Institute in London, "they are agents for the Corporation's orders and not channels for feedback" (Tiffen 1984). In Tanzania, most of the small-farm irrigation schemes initiated and constructed by governments have been unsuccessful, largely due to the lack of active participation by farmers (Mbwala 1980).

Paying attention to farmers' needs is essential. Peasant farmers must be assured that benefits will accrue to them, for without a specific political commitment by the government it is unlikely that they will have the same access as larger farmers to the benefits of irrigation projects (Metral 1984; Edwards and Vincent 1983). If farmers' needs are ignored, then projects may not be fully implemented. In some areas only one crop is grown every year, instead of the two needed to make a scheme economic, because grain prices set by the government are too low to give farmers sufficient incentive to buy expensive seed and fertilizer. A World Bank survey in Mexico found that farmers were still growing traditional low-yielding maize varieties on irrigated lands and that yields were therefore only half of those expected (World Bank n.d.). On the other hand, the Ghab Regional Pilot Project in Syria, implemented by government agencies between 1976 and 1979, was mainly profitable for local peasants; the acquisition of land and water was guaranteed, financial and technical support was provided, and government intervention did not stifle competition between farmers.

Rehabilitation of Existing Irrigation Schemes

It is clearly in the interest of governments to try to maximize the returns on their investments by making existing irrigation projects work as efficiently as possible. That is why the Club du Sahel, USAID and other agencies are stressing that the priority in future must be on the rehabilitation of existing irrigation schemes that have fallen into disrepair and are not working at full capacity, rather than on the construction of new ones. There are an estimated 25,800 ha of irrigated lands needing rehabilitation in the six Sahelian countries alone (Club du Sahel 1979). In India and Pakistan, whose agricultural sectors are heavily dependent on irrigation, there is simply no alternative to large-scale rehabilitation if yields are to be improved and the worrying rate of desertification is to be halted.

In some areas, especially in Africa, rehabilitation may only require the repair of pumps, canals and other system infrastructure, and better overall management of the system, but elsewhere extensive reclamation of saline and waterlogged land will be necessary. The techniques for reclaiming salinized land are well known, and include the installation of pumps and drainage works to drain the soil and reduce waterlogging; leaching salts from the soil with fresh water; the use of gypsum to help in such leaching; improving soil cover with vegetation or mulch to suppress capillary rise and surface evaporation; cultivation practices such as deep ploughing or loosening to improve the depth of the rooting zone; and the preliminary planting of highly salt-tolerant trees and shrubs to restore nutrient cycling and drain the soil (NAS 1974). Alkaline soils are more difficult to reclaim than saline soils (see p. 32).

Draining the soil is the first priority because inadequate drainage is usually the original cause of waterlogging and salinization. Open drains are the simplest to construct but typically occupy 15% of the total area of the cropland which they serve, and so present a major cost to farmers in terms of income forgone from cropping. If poorly maintained they serve as breeding grounds for mosquitoes and the snails which are vectors for bilharzia. Tile drains are less costly than open drains, especially if plastic piping is used. Vertical tubewells are in theory much cheaper than horizontal drains, where the groundwater level is appropriate, but in practice they have proved difficult to operate (Carruthers 1985).

Pakistan began a series of Salinity Control and Reclamation Projects (SCARPs) in 1961, improving drainage channels and sinking tubewells. So far, forty-two of these projects have been completed at a cost of over $2 billion, but they have not been very successful. Drainage initially reduces waterlogging so that crop yields increase, but when more water is applied to increase yields further, the operation and maintenance of the schemes decline and waterlogging returns (Repetto 1986). According to the World Bank, waterlogging in 1981 was at critical levels in 36% of all SCARP areas covering a total area of 4.3 million ha (World Bank 1984). Salty groundwater has severely reduced the average life of tubewells from the initial estimate of forty or fifty years to only twelve years. Average crop yields are only about 80% of those projected at the start of the project, and because the SCARP programme has difficulty in collecting the full water charges from farmers the government is effectively subsidizing the programme by about $22 million a year. The economics of the programme were severely affected by big increases in energy costs (Johnson 1982).

The World Bank and the government of Canada are attempting to reverse this cycle of failure by funding yet another SCARP project at Mardan in north Pakistan. The area is intensively cultivated but yields are low because 60% of the irrigated croplands are waterlogged or salinized. The project aims to rehabilitate 29,000 ha of affected lands by building new drains and widening and deepening 320 km of canals. The emphasis is on using gravity, rather than electric power, for drainage, since Pakistan's 20,000 existing tubewells are estimated to account for one-quarter of the country's total electricity consumption. This project is a new departure because it uses horizontal drains rather than vertical tubewells for drainage (Johnson 1982). The problem with vertical drains is that any salts washed out of the soil stay in the system. The only way to ensure that salts do not return to the soil is to flush them out to rivers and oceans by horizontal drains and canals. Initial reports are positive, and water tables are falling. However, it remains to be seen whether this SCARP project avoids the mistakes of previous ones (CIDA 1986).

India has considerable experience in the reclamation of saline lands. Salt-tolerant trees are being planted in a number of states to reduce salinity and waterlogging, and this technique is especially cost-effective when large areas of barren land need to be treated. Before tree planting, gypsum is normally added to the soil, and the increased transpiration of water which accompanies tree growth helps to lower the water table where it is too high. The shade of the trees reduces surface evaporation, thereby retarding the upward movement of groundwater and the deposition of salts near the soil surface. Fallen leaves increase organic matter and improve soil structure, and the penetration of roots opens up the soil, improving permeability and assisting the leaching of salts. A system developed in Haryana state uses *Prosopis juliflora*, *Acacia nilotica* and *Eucalyptus* spp. to reclaim highly saline barren land. These multi-purpose trees are attractive because they can also be cropped for fuelwood, fodder and other products.

Governments will have to bear a considerable proportion of the cost of drainage and reclamation of irrigated lands, but determining how much of the cost has to be borne by farmers is bound to cause political difficulties. Even without government assistance, some private farmers in Egypt are reported to have been digging deep open drains on their farms, an activity made possible by low labour costs at certain times of the year and subsidized energy prices (Carruthers 1985). Cost-benefit studies in both India and Sri Lanka have shown that the value of potential increases in crop yields make reclamation of saline and alkaline lands not only socially desirable but also economically feasible (Joshi 1983; Herath 1985).

The Small-Scale Alternative

The sheer complexity of large public irrigation projects makes them a challenge even for the most skilled manager (Ali 1980), so perhaps the only solution to the problem of poor management is to switch the emphasis from mammoth schemes to small-scale projects, especially those which build on local people's knowledge and expertise. In sub-Saharan Africa, for example, an estimated 2.5 million small farms irrigated using traditional labour-intensive techniques such as those based on shallow wells in river basins. Improving the productivity of these farms could have a major impact on food production. Small-scale schemes can be administered by local representative committees, without the need for large bureaucracies; the produce can be marketed locally; training farmers in irrigation skills is easier because fewer people are involved in each scheme; and starting small allows farmers to learn by experience

(Makadho 1984). This kind of decentralized approach to irrigation can lead to more efficient farming, and is less wasteful of water because the management of tubewells and pumps is in the hands of only a few people. Organizational problems are likely to be fewer, and water can be applied by the farmer as and when required, rather than by the irrigation authority. In the Indian state of Uttar Pradesh, a survey found that farmers using private tubewell irrigation had significantly higher agricultural production, cropping intensity and income compared with users of public tubewell and canal irrigation (Thakur and Kumar 1984).

There are 2.2 million ha of land with potential for irrigation in the major river basins of the Sahel; more than half of this area is in Chad (FAO 1985a). The Club du Sahel thinks it vital to increase the area under irrigation to 1.2 million ha by the year 2000 so that all the region's rice, wheat and sugar requirements, as well as a substantial amount of corn, millet and sorghum, can be produced locally. Intensifying rainfed cropping is also vital but will not be sufficient to keep up with demand. Rehabilitation of existing irrigation projects, while extremely necessary, will not have nearly as much impact on food production as it will in South Asia because of the much smaller irrigated area in the Sahel. On the other hand, since the past performance of irrigation in the Sahel has been so poor, it would clearly be wrong to continue with the same types of projects in the hope that these would lead to an increase in food production.

Small-scale irrigation may therefore be the only way in which the Sahel can realistically meet these targets. The Club du Sahel has found that: "In most irrigated projects where good yields have been obtained and where maintenance is good, such projects are of modest size and producers are closely associated with their management." To meet the target, therefore, the Club du Sahel and USAID are advocating a new approach which involves farmers far more closely in the design and management of projects. These two agencies intend to focus in the future on small-scale projects which benefit a wider cross-section of people and are less likely to fail for social reasons or on account of poor equipment maintenance.

A good example of a successful small-scale project is the FAO/UNDP village irrigation programme on the Senegal River near Matam on the border between Senegal and Mauritania.

Farmers have traditionally planted crops on land close to the river which is naturally irrigated by flood waters. The aim of this programme was to help the Halpulaar people to use water pumped from the river to make productive land which otherwise would not be cropped because the annual flood did not reach it. The emphasis is on growing subsistence crops of rice and maize, so the plots allocated to each family are relatively small. The programme began in 1974 on an experimental basis with just two schemes of 7 and 8 ha. Soon other farmers came to visit these schemes, decided to follow their example, and the area cropped grew rapidly. Membership in the programme is voluntary, and if a village wishes to join it is allowed to choose the site in consultation with technical advisers from SAED (Senegal's national irrigation authority). Villagers clear the site themselves and are responsible for operating and maintaining the scheme and for collecting money from participants to pay fuel bills and the dues to SAED.

Local people are enthusiastic about the programme: the productivity of their rainfed and flood recession cropping has been badly affected by the drought, and they prefer this small-scale approach to participation in the larger irrigated rice projects directly managed by SAED. By 1983 the programme had attracted 400 schemes covering about 8,000 ha on either side of the river. Two crops are grown per year on each plot. In the lower part of the valley both crops are rice but in the upper valley rice alternates with maize. Rice yields average between 4 and 5 tonnes/ha and maize yields are about 2.5 tonnes/ha (Diemer and van der Laan 1983).

Ellen van der Laan of the Centre for African Studies in Leiden, Holland, has identified five main reasons for the success of the programme. First, there was a convergence of interests between farmers and the government, since farmers wanted to be less dependent upon erratic rainfall while the government was concerned about the increasing shortages and imports of food. Second, plenty of uncultivated land was available near the river so that there was no competition between rice and the traditional crops of millet and sorghum. Third, women constituted an untapped reservoir of labour: traditionally women have been responsible for sowing millet and sorghum, guarding crops against birds and helping with the harvest, while men take care of the weeding. Because the small-scale programme was new it was not governed by tradition and so women (and children) were free to undertake the weeding. Unfortunately, there is an overlap between the peak labour times for traditional crops and the new schemes, so that rice rather than millet is given priority and millet yields are lower than they otherwise would be. Fourth, many households had money available to pay for the costs of the scheme, such as diesel oil, repairs and the wages of the pump driver. The money comes from the wages of members of the household who leave the valley to take jobs in Dakar and elsewhere. Households without migrant members would find it difficult to pay the costs of the scheme since only 5–10% of the total rice harvest is sold. Fifth, village institutions already existed which could be used as a basis for managing the scheme on behalf of the community. There were well-defined lines of authority and mechanisms for decision taking, collective work on canals and dikes could be carried out as it would be for any other village task, and several joint saving funds were already operating, so that the collection and management of money for the irrigation scheme did not present an insuperable problem (van der Laan 1984). The only reservation concerns the future economic viability of irrigated cropping once the programme ends, since the pumps, which cost £10,000 each, are given as a free grant to each village and the dues paid by the village to SAED are at present insufficient to cover replacement costs (Diemer and van der Laan 1983).

There is considerable potential for expanding small-scale irrigation projects in Africa but they need the positive support of government through technical assistance and funding. Too much government involvement, on the other hand, can be detrimental because it tends to inhibit farmers' initiative and motivation and so may lead to disappointing yields, low rates of new development, high capital costs and low-grade maintenance (Abernethy 1984). Schemes should therefore be designed to maximize the amount of freedom which farmers have to make decisions so that they can "learn by doing" and improve their skills in using irrigation.

This small-scale approach has also been adopted in projects funded by non-governmental organizations. Farmers on low-lying valley bottoms in Niger traditionally planted vegetables in seasonal ponds after the water has evaporated or drained away. Then they learned that by sinking a shallow well in the centre of the "pond" they could hand-draw more water for drinking and gardening in the dry season. However, such wells only last a year or two before collapsing, often more than one well is needed for one garden, logs to line the wells are not easy to find in many areas, and the gardens are vulnerable to invasion by livestock. The staff of Lutheran World Relief showed the farmers how to construct permanent wells with concrete linings made in sections 1 metre high and 140 cm across by pouring the concrete around steel rods in a simple mould. Between six and ten linings are normally required for each well, although some wells have been as deep as 18 metres. The sandy soil is dug out from inside the rings so they sinks slowly into the ground. Lutheran World Relief trains local people to make the linings and provides the cement and steel rods. Farmers repay the cost of about $300 over a three-year period. The gardens are protected from livestock and the *Harmattan* by "living fences" of *Prosopis juliflora*, a fast-growing thorn tree which can also be pruned for fuelwood. Even people who were doubtful about participating in village woodlot schemes planted living fences to protect their gardens. On finding that the fences could be cropped for fuelwood they gained a whole new awareness of the importance of trees.

There are twenty-eight such projects in Niger and more than 3,200 wells have been dug. Each project is managed on a co-operative basis with the involvement of government officials. Lack of enthusiasm by these officials and corruption among the leaders of some co-operatives have led to problems, however, with some members afraid to pay money to the co-operative for fear of it being "lost". The gardens have been used mainly to grow vegetables like lettuce, onions and tomatoes rather than staple cereals or root crops. Nevertheless, the gardeners earn a useful income since most have a surplus of food to sell, and sales vary between $400–2,000 per hectare. The market for such vegetables is large, and merchants use donkeys or trucks to transport produce to markets 40–80 km away. The gardens have increased the income of local people and the availability of food in the dry season, and together with the fruit trees and hedges have changed the dry-season environment out of all recognition. The cost of each well is only a third of the cost of wells made by the government (Harrison 1986; Cottingham

1987). A similar kind of project in Mauritania is promoting the creation of small market gardens near thirty-six deep wells which were drilled as part of another project jointly funded by the UN Sudano-Sahelian Office (UNSO) and the African Development Bank.

The World Bank has used a "small-scale" approach to irrigation as part of its Kano Agricultural Development Project in northern Nigeria. Farmers on 60,000 ha of valley-bottom land previously grew dry-season crops in small basins 2 metres square, obtaining water from shallow wells or rivers using a shadoof (a large calabash on a pivoted pole with a stone as counterbalance). The project has increased the area of the basins to 4 metres square and given the farmers access to higher-yielding improved seeds, pesticides and fertilizers. Cheap shallow tubewells costing $900 were built and the slow shadoof replaced by a cheap Honda petrol pump costing $500.

The start-up costs of up to $1,250 per hectare for this approach were much higher than for the Lutheran World Relief project. Although much cheaper than the $19,000 per hectare of the previous large-scale river basin irrigation schemes in Kano state, the project could still be criticized for favouring the richer farmers. On the other hand, a farmer can earn between $1,800 and $3,500 per hectare per year, thus easily covering his initial investment in the first year of operation. This is an important consideration. The start-up cost can be reduced to $500 on sandy river-beds where the water is close to the surface and tubewells are not necessary. A simple $150 tubular screen (called a washbore) through which water, but not sand, can pass will suffice. One of the potential weaknesses of the project is that it depends upon imported pumps which, as already mentioned, can fall into disrepair because of poor maintenance and lack of foreign exchange to buy spare parts.

Both the Lutheran World Relief and World Bank projects have, for relatively low cost outlays, built upon existing practices instead of replacing them. Small-scale schemes need not be communal, and in other countries there has been a rapid expansion of private irrigation schemes. However, small-scale irrigation is not without its problems. Farmers, lacking adequate technical support, may choose pumps which are too powerful for their needs, and the proliferation of private wells without proper government control may deplete groundwater reserves (Vincent 1982; Bottrall 1982).

THE ROLE OF CROPPING IN DESERTIFICATION CONTROL

Making rainfed and irrigated cropping more productive and sustainable on the most fertile lands will play a crucial part in bringing desertification under control. Because this will also increase food production in the drylands, a policy goal high on the list of economic priorities for many African and Asian countries, it is a strategy which is politically feasible as well as technically desirable. Nevertheless, it will demand a number of shifts in agricultural policy. First, high priority must be given to the improvement of rainfed cropping by allocating more funds for the development of high-yielding varieties and more productive cropping systems, and ensuring that crop prices are sufficiently attractive to give farmers the incentive to crop more intensively. Second, new large-scale irrigation schemes should only be introduced after the most careful preliminary social and technical evaluation has determined their feasibility. Third, existing irrigation schemes that have fallen into disrepair because of either waterlogging and salinization or mechanical problems should be rehabilitated. Fourth, the emphasis should be on small-scale irrigation schemes that are likely to be of greater benefit to farmers, more easily managed, and therefore less susceptible to waterlogging and salinization. Fifth, land-use planning techniques and skills in dryland areas should be strengthened so that the lands best suited to cropping can be identified and cropping is intensified as far as possible only on those lands where it can be sustainable. Farmers on other lands should be given incentives to practise less intensive types of farming which are not as likely to cause desertification. Controlling desertification in these ways is likely to be economically beneficial to most of the parties involved in dryland agriculture, and this demonstrates once again that sustainable development and environmental conservation do not have to be conflicting objectives.

6. Controlling Desertification by Improving Livestock Raising

Nomadic herdsmen received much of the initial blame for desertification in the Sahel in the early 1970s because it was felt that they had let their herds grow too large during good rainfall years and that this had led to overgrazing. However, it was recognized later that overgrazing was as much the result of external factors as of the dynamics of pastoralism, because the area of pasture available for grazing had been reduced by the expansion of rainfed cropping on to marginal rangelands normally used for grazing, and by the increased cultivation of cash crops like groundnuts on fallow lands traditionally grazed by nomadic herds in the dry season. Simply trying to regulate grazing in isolation from other agricultural sectors will therefore not be sufficient to prevent overgrazing and control desertification.

While curbing overgrazing is one priority, another is to restore self-sufficiency in animal products in drought-affected areas and supply the food needs of growing populations. Since the start of the drought in 1969 the Sahel's meat exports have fallen, and domestic consumption of meat is now only three-quarters of the pre-drought level. Milk imports have increased sixfold so that the region is now a net importer of milk and milk products. Imports of all livestock products to sub-Saharan Africa reached about $2,000 million in 1980, twice the value of all foreign aid to the livestock sector over the past fifteen to twenty years (Anteneh 1984). Rising demand for livestock products could cause overgrazing to become far worse than it is at present, if herd sizes increase sharply and the area of pasture available for grazing remains restricted by the continued cropping of marginal and fallow lands and by desertification. Action to improve the productivity of cropping should help to reduce or even reverse the rate at which it is encroaching on to grazing lands but clearly the productivity of livestock raising will also have to increase. There are five ways by which this could be achieved, although not all experts agree on the value and feasibility of each of them:

1. improving the quality of animals by disease control and selective breeding;
2. increasing the offtake (the number of animals sold each year for slaughter) so that stock levels fall and become more in line with sustainable carrying capacity;
3. improving rangelands by measures including reseeding, allowing time for regeneration, planting new fodder/forage crops and other measures;
4. improving infrastructure by digging wells, making other improvements along routes to market and establishing feedlots and abattoirs;
5. changing the organization of pastoralism by taking measures to restrict or regulate grazing in certain areas, encouraging the sedentarization of nomads and their herds, establishing ranches, and developing regional livestock-raising schemes by stratification or other means.

IMPROVING ANIMAL QUALITY

Raising the quality of livestock by disease control and selective breeding will increase the average yield per animal. During the 1960s and 1970s the production of meat, milk and other products in the Sahel grew less rapidly than in previous decades, and most of the increase was due to higher livestock numbers rather than to increased productivity per animal (Anteneh 1984). Disease control is the oldest component of livestock programmes in many countries and is readily accepted by herdsmen. Yet vaccination and other steps to reduce disease often lead to livestock overpopulation, overgrazing and – when drought strikes – disaster. In the 1960s, for example, the World Bank supported a livestock disease control programme among the Karamojong people of Uganda. The programme was also intended to improve the marketing of animals and thereby counteract the increase in the overall number of livestock which was inevitable if their health improved. However, the marketing scheme failed, the herds increased in size and overgrazed the land, and the Karamojong ended up worse off than before the programme started.

Other animal health programmes have attempted to increase cow fertility, milk production and calf survival rates, but since the methods used have depended largely upon vaccinations, mineral and vitamin supplements and concentrates, they are expensive, and often beyond the means of owners once external payments stop. A much more basic approach is now being taken in southern Ethiopia, where herders from the Borana tribe are being trained as "barefoot vets" to give basic injections and cure simple health problems. Experts from the International Livestock Centre for Africa, based in Addis Ababa, are also introducing fodder legumes to help improve calf nutrition, lowering the mortality rate and raising the weaning weight. The legumes, grown as part of the crop rotation on farms which the Borana have established to cope with the drought, are fed to the cattle mixed with maize stalks. The lower mortality of male calves means that more animals can be sold for meat while leaving the level of milk production unaffected. This is important because the diet of the pastoralists depends far more on milk than on meat (Harrison 1986).

Selective breeding has received rather less attention than animal health. Considerable research has been directed towards the breeding of a tsetse-tolerant cow that would open up to livestock raising large areas of the more humid regions of West Africa that are currently unsuitable because of tsetse infestation. The hardy zebu cattle common in the Sahel, for example, are highly susceptible to the disease trypanosomiasis carried by the tsetse fly when they are moved south to the higher rainfall zones.

Some exotic sheep breeds which have been introduced into the Sahel have failed because the animals could not adapt to the environment, while other apparently more suitable breeds have not been introduced. These include the long-legged Kabashi desert sheep from Sudan, which is an excellent walker, produces good-quality meat and needs watering only every three or four days; and the small Moroccan ewes bred at the Royal Farm in the Souss region which have a high twin pregnancy rate but have not been introduced very widely.

Goats have received little attention from improvement schemes so far, despite their prominent role in livestock husbandry. There were more than 143 million goats in Africa in 1978, mostly in Chad, Ethiopia, Kenya, Mali, Mauritania, Niger, Nigeria, Senegal, Somalia and Sudan (Wilson 1982). Goats are usually found in areas with 350–550 mm annual rainfall at stocking levels of 20–25 head per square kilometre. They account for 31% of all meat production but only 16% of total liveweight biomass. One of the main reasons for their neglect is that goat meat is mainly consumed domestically and is not as significant as beef in international trade.

Another neglected animal is the camel. Daniel Stiles of UNEP has suggested that helping nomads to herd more camels and fewer cattle may be in the interests of both the pastoralists and the environment (Stiles 1983). As demand for camels grows, so will the need for better-quality herds. Camels are far better adapted than cattle to dry degraded lands. They eat grass but prefer browsing shrubs and trees for leaves, twigs and berries, and so do not remove the soil's protective grass covering as would a herd of cattle. Camels spread out when they browse and their soft hooves do not cause as much damage to the soil as do the hard hooves of cattle. Camels can go for more than a week without water and even in dry periods continue to produce 5–10 litres of milk a day compared to less than half a litre for a cow. Milk is the main food for many nomadic peoples and so camels provide them with an insurance against drought. Because camels are much more efficient at converting vegetation into milk a family of six will only need a herd of twenty-eight camels to provide its subsistence, instead of sixty-four cattle (Stiles 1983).

Camels cost more than cattle, so a project funded by the West German government is helping the Samburu people of northern Kenya to buy camels. The Samburu exchange their cattle for camels which the project buys in Somalia, and the money from the sale of the cattle goes to buy more camels. Camels suffer from a number of diseases and, like cattle, they are susceptible to trypanosomiasis carried by the tsetse fly. Some new initiatives are in hand to tackle such problems. The Food and Agriculture Research Mission, a non-governmental organization, has begun a "flying vet" service for camels in Kenya, while in 1986 UNEP held a camel workshop so that the Samburu could learn from peoples like the Rendille who have herded camels for hundreds of years and developed techniques for preventing and curing diseases (Hilsum 1987).

REDUCING STOCK LEVELS

It could be argued that if pastoralists decreased the size of their herds there would be more food and water for the remaining animals, productivity would increase, and overgrazing and desertification would decrease. The average proportion of all animals sold each year was estimated to be around 9% for Burkina Faso, Niger and Mali in the late 1960s, a few percentage points higher than in the early 1950s but very low by western standards (Jerve 1981). Raising fewer animals would smooth out the dramatic swings in livestock numbers that result from large herd sizes during good rainfall years which are then reduced by high mortality rates in times of drought.

Estimating the sustainable carrying capacity of a region is difficult, however. One suggestion is that livestock numbers should never significantly exceed those which could be maintained during the worst years and seasons and that it is best to stock at 80% of average capacity in order to keep at least a 20% forage reserve for drought years (USAID 1980). On this scale the 1982 Sahel herd size of 23 million was 5% above the safe sustainable level. Such exercises, however, remain largely academic, for reasons discussed in Chapter 2, and attempts to reduce stock levels by increasing the number of animals sold each year (the offtake) always seem to lead to problems. "Stock reduction schemes do not work because they operate on fiat, creating resentment and antagonism among a traditionally independent people", according to Walter Goldschmidt of the University of California at Los Angeles (Goldschmidt 1981). Nomads will fiercely, and if necessary violently, defend their right to retain their mobility and to determine herd sizes, the two cornerstones of nomadic pastoralism.

Any plans to increase offtake will therefore remain theoretical unless they fit in with the herdsman's point of view, which is that herds are a means of storing wealth, rather than so much potential beef to be sold, and are more a form of capital than a source of income (Jerve 1981). The need to conserve such capital, especially as an insurance against bad years, is paramount. Thus, suppose that we dispense with the notion of a fixed sustainable carrying capacity, in favour of a more rational one which is related to the cyclical change in pasture production as annual rainfall falls and then (in normal circumstances) rises. We might suggest to herdsmen that, to prevent overgrazing, they should reduce the size of their herds as soon as a drought begins. However, the herdsmen would probably still object to such a strategy because it would reduce the chances of a viable herd remaining after the drought has ended.

Herdsmen barter their livestock, animals and milk in exchange for grain from farmers. Cash need not, and often does not, change hands. In recent years the exchange rate for livestock and grain has changed to the disadvantage of the herdsmen, and more animals are now required to obtain the same amount of grain. Conditions have changed for the farmer too, as some of his grain is now sold on national and international markets, leaving less available for barter.

As far as herdsmen are concerned, therefore, the herd represents the basis of an alternative form of economics. It is also central to all social negotiations – recording kinships and creating patterns of alliance, dependency and support. "A herd is the monument and obituary of its owners, and a close reading of the animals in the kraal is a record of the major social interactions of a herder's life", writes Walter Goldschmidt, describing the Sebei people of Uganda. The same is true for most pastoral societies. Cattle are commonly given as bridal payments, and the Sebei calculate lineages of cattle to express social rather than biological continuity. "A Sebei wants to retain at least one representative of important sources of such cattle lineages, e.g. one from each aunt's and sister's and daughter's bride payment" (Goldschmidt 1981).

Such attitudes are quite logical to the herdsmen but difficult to represent in mathematical equations of livestock supply and demand; hence the problems experienced in introducing a more rational offtake system based on estimated carrying capacity. More mundane reasons can also lie behind limitations on offtake. Herdsmen who look after livestock belonging to farmers receive their payments largely in the form of milk from those animals. This leads to overmilking, high calf mortality and low rates of reproduction, and so fewer animals can be sold for slaughter.

IMPROVING RANGELANDS

If there were fewer livestock, then the valuable perennial grasses and browse shrubs would have a chance to regenerate. Reducing livestock numbers in the Badiya El-Sham area in Syria, which receives only 150 mm rainfall a year, resulted in the re-emergence of desirable perennial grasses and shrubs. In Gabes, in southern Tunisia, which receives 195 mm annual rainfall, controlling stock numbers and rotating the areas of rangeland used for grazing led to improved vegetation cover. This reduced the area which a ewe needed to graze every year from 7 to 2 ha. Similar results have been achieved in Jordan by protecting rangelands and allowing only occasional light grazing (FAO 1985b).

In the absence of natural regeneration, the improvement of rangelands will require seeding with desirable fodder plants. The introduction of two new forage grasses was one of the major accomplishments of the huge $103 million seven-year Drought Prone Areas Project to stabilize and increase agricultural production in six drought-prone areas in west and central India. The project, which ended in 1981, was funded by the World Bank. The two grasses, *Stylosanthes hamata* and *Stylosanthes scabra*, come from Australia and have the potential to increase fodder production substantially, providing a low-cost way to improve rangelands rapidly. *Stylosanthes* can be established on the better wastelands just by broadcasting seed, while even on badly eroded sites only simple furrows are required. The plant sets seed quickly even under low rainfall and, because its early growth is unpalatable, premature grazing is discouraged (World Bank, personal communication). Tunisia is creating 138,000 ha of pastoral reserves in the governorates of Gabes and Medenine by planting forage species and protecting them from overexploitation (UNEP 1985). Reseeding rangelands has been successful in areas of the USSR, Australia and Pakistan receiving less than 300–400 mm rainfall but other countries have experienced difficulties (FAO 1985b).

Trees are another valuable source of fodder and one which until recently was sadly neglected in livestock development projects. One of the lessons learned in the course of the Indian project was that the reintroduction of traditional fodder trees and shrubs like *Prosopis* and *Zizyphus* into rangelands was just as important as establishing perennial forage grasses. The potential of many of these so-called "multi-purpose" trees, which can be cropped for fodder, fuelwood and a variety of other products, has only really been recognized in the last ten years, but now their use is being encouraged by prominent organizations such as FAO, the International Livestock Centre for Africa and the US National Academy of Sciences. Research to develop this potential is one of the most rapidly growing fields in tropical forestry.

The forty-four species of the genus *Prosopis* are typical multi-purpose trees. They are highly drought-resistant and saline-tolerant, and as members of the Leguminosae family they bear pods containing up to 14% protein and 44–55% carbohydrate. Most species are nitrogen fixers. *Prosopis* can grow in areas receiving less than 100 mm of rainfall per annum where even the highly drought-resistant *Acacia tortilis* could not survive. Quick-growing roots reach as deep as 80 metres beneath the surface to tap underground water, but the tree can also grow in hyper-arid areas where hardpans prevent roots from reaching groundwater; there is some evidence that *Prosopis* can take in water through its leaves. Because of its saline tolerance *Prosopis* can be used to make salinized lands productive again.

Both the foliage and pods of *Prosopis* trees can be used as fodder. In the arid and semi-arid lands of India *Prosopis juliflora* and *Prosopis cineraria* trees are lopped for fodder every winter without harming future growth. The leaves and pods are also browsed by sheep and cattle, and while *Prosopis* should preferably be used as part of a mixed diet, in Chile the pods have formed the sole diet of sheep without ill-effects and supported more than ten sheep per hectare (Pedersen 1980). Plantations of *Prosopis nigra* and *Prosopis alba* have been established in Chile and Brazil, giving an annual fodder production of 1–2 tonnes from 50–105 trees per hectare. Animals are excluded from the plantations for the first five years, and afterwards their browsing has to be carefully managed. *Prosopis* species are also being planted in India and Pakistan for this purpose (FAO 1985b).

Acacia albida is another valuable fodder tree. It is highly drought-resistant and in some dryland areas may be the only greenery remaining in the dry season. It is also a soil-improving, nitrogen-fixing tree, with the additional advantage of a relatively open canopy during the crop-growing season that allows it to be grown together with field crops like millet. Far from having any detrimental effects, crop yields under and around *Acacia albida* have been found to be equivalent to those in fertilized fields. Another outstanding fodder tree is the carob, *Ceratonia siliqua*, which has a long history of use in countries around the Mediterranean and in other semi-arid and sub-humid areas where it will grow on the thin soils of degraded lands and rocky hillsides. An evergreen tree growing up to 10 metres in height, the carob's pods are poor in protein but rich in sugars; besides being harvested for local livestock they are also used in commercial feeds (Winer 1980). Two equally versatile multi-purpose trees widely used as sources of fodder in sub-humid areas are *Leucaena* (which requires more than 600 mm rainfall per annum) (Blom 1981) and *Gliricidia* (more than 800 mm required) (Falvey 1982).

Fodder trees and shrubs need to be protected from uncontrolled browsing if their benefits are to be available over the long term. The concept of "fodder reserves" adjacent to boreholes has been advocated, but without proper protection the trees would probably be overbrowsed or even cut down for fuelwood. However, the feasibility of such reserves is increasing now that agencies like the World Bank and USAID are funding the establishment of pastoral associations which can take responsibility for managing specific boreholes and the land around them (see pp. 211–13).

Another possibility would be to restore, if only in part, some of the village fallow land lost to cash-crop cultivation. More and more farmers are switching from cash crops to staple crops because of falling returns, and fallowing could be stimulated by appropriate incentives. The land could be replenished with fodder grasses, trees and shrubs, and the traditional symbiosis between pastoralists and farmers restored. Another possible way to improve animal feed value and the productivity of the herds could be by using the oilcakes produced by farmers as supplementary feed (Khogali 1983).

Long-term management of fodder trees and shrubs is perhaps easier in India where the keeping of animals in villages is widespread and there are communal grazing lands, but protecting these lands from unauthorized browsing is still a major problem. In the state of Gujarat a large-scale social forestry scheme has planted multi-purpose fodder trees and shrubs on marginal village lands (see Chapter 7). Fodder is often a more attractive product to farmers than fuelwood, and in both India and Nepal it has been found that people are more easily motivated to plant trees if they know they can harvest fodder from them. Villagers still need to agree on the use of fodder trees and to protect them from outside herds, but the problems, though substantial, are not as great as in areas like the Sahel where pastoralists are nomadic and do not stay in one place long enough to manage fodder reserves. Pastoral associations could overcome this problem since, while the herds continue to move around a region, the responsibilities of herders are well defined.

Many pastoralists, like the Borana in Ethiopia, have been forced by dire circumstances to grow crops, while others have been pressured by the strong hand of government to become totally settled (sedentarized). In both cases the possibility exists to intercrop valuable fodder plants with regular food crops. Sedentarization has led to severe overgrazing around some villages because of the high concentration of livestock that were previously scattered over large areas of rangelands. In such cases high-yielding fodder trees and other sources of forage may be the only way to support intensive livestock production without causing further environmental degradation. It should be noted, however, that in Africa many villagers are still sceptical about keeping animals. Efforts to widen the use of animal-based ploughs instead of hand-wielded hoes have made little headway, despite the economic advantages.

From a broader perspective, as mentioned in the previous chapter, the pressure on rangelands could be alleviated by intensifying field cropping on existing arable lands (Khogali 1983). A more radical measure, that of controlling the encroachment of cropping on to traditional rangelands, is being tried in Syria, where the government is restricting rainfed cropping in semi-desert grazing areas (albeit with opposition by farmers) and encouraging a co-operative approach to the control of grazing. The programme also supplies supplementary feed and is establishing a national emergency fodder reserve. In Niger cultivation is prohibited by law north of latitude 15°N, although in practice this simply means that only farmers south of this line can sue herders for compensation if their crops are damaged by livestock.

MORE WELLS?

Considerable effort has been devoted to trying to improve the facilities available for herders and their livestock, principally by sinking new wells. The benefits of such projects are debatable, however, because they have often contributed substantially to desertification by encouraging the growth of livestock beyond the number which can be supported by available feed supplies. Animals can starve when local supplies of forage are exhausted, even if plenty of water is available from the wells. According to Walter Goldschmidt: "I have no indication of any instance in which the use of wells has had any positive effects" (Goldschmidt 1981). (He refers to wells in the arid and semi-arid zones, not village wells for domestic use.)

The phenomenal growth in the number of watering points, especially along routes to markets, has concentrated herds in limited areas along routes and around wells, causing degradation and desertification through trampling, browsing and grazing. Although herdsmen clamour for the construction of new wells, they are often aware of such dangers and prefer privately-owned wells to which access is limited. The Bororo herdsmen from the Bernmou region of Niger have urged the government not to sink new wells in their traditional grazing areas since this would attract outsiders and overload rangelands. They have suggested instead that wells be sunk in Tuareg country to the north, so that the Bororo could use them in the dry season but would not have to reciprocate by sharing their own wells. On the other hand, the Tuareg also appreciate the threats which wells pose to their rangelands. In the early 1970s, the Illabakan Tuareg of Niger petitioned to have the pump at one well turned off because it had attracted so many outside herds that pressure on rangelands was severe and relations between the Tuareg and nearby tribes were becoming very strained.

Despite the great profusion of such examples and the clear warnings at UNCOD of the dangers of too many watering points, Sahel governments, assisted by aid agencies, continue to sink new wells. Since the mid-1970s, for example, almost $7.5 million has been channelled through UNSO alone to projects of this kind in Gambia, Mauritania and Burkina Faso. Walter Goldschmidt despairs: "Planners do not learn from their own mistakes. To see governments plan to make elaborate installations of waterholes or to launch stock-reduction programmes after these have been repeatedly branded as failures makes one wonder why writing was ever invented" (Goldschmidt 1981). "The indiscriminate sinking of boreholes with no parallel grazing programme should cease", says Randall Baker of the University of Indiana.

A moratorium by aid agencies on projects to sink new wells may therefore be advisable. Assistance in the immediate future should be given instead for the rehabilitation of existing projects, trying out a number of techniques. Groundwater studies could be undertaken, where they were not made before the well was sunk. The output from wells might be reduced by replacing pumps with more traditional hand-drawn and animal-draught methods for raising water. The watering point and adjacent areas could be redesigned to regulate and decentralize access; regenerate overgrazed rangelands; and incorporate shelterbelts of trees. Reserves of native pasture grasses and fodder shrubs and trees could be planted along the major market routes to reduce the pressure of livestock on areas adjacent to water-holes. Finally, the crowding of animals could be alleviated by ensuring that watering points are not sited near villages.

Once a watering point has been established, regulating its use can prove very difficult and in some cases can lead to violence. According to Peter Hopcraft of the World Bank:

In areas where bore holes have been installed, attempts have sometimes been made to turn off the pumps before grazing destruction sets in. The results of these attempts have generally been such anger that pumping equipment is deliberately destroyed. This has led to proposals that pumps and equipment be removed to the next block (grazing area), again forcing the livestock to move to where there is water. In general, as anyone who has been thirsty and without water in a hot and arid environment knows, tampering with what is regarded as a possible water supply is an explosive activity.

There are some encouraging signs of a change in approach to wells. UNSO is financing a project to rehabilitate degraded rangelands surrounding boreholes in Gambia. Villages are given responsibility for taking care of boreholes and rangelands in their vicinity, and in return they receive the produce from orchards of cashew and other trees which have been planted close to the boreholes. Cattle from co-operating villages are circulated around a number of boreholes to lighten the pressure on rangelands (UNSO, personal communication).

REGULATING NOMADS

Nomadic grazing is a sound technique for using scarce pasture resources in dryland areas, where rainfall fluctuates widely from place to place, and mobile herds have freedom to search for areas where rain has fallen and grass has grown. For example, Maine-Sora in southeast Niger has an average rainfall of 432 mm per annum. In 1949, it received 230 mm of rain, but only 67 mm fell at Diffa, 50 km further east. Herds would have suffered if government grazing regulations had prevented them from moving out of Diffa that year.

Mobility also means better nutrition: if herds can arrive in a place when the grass is still green, digestibility is of the order of 70% compared with 43% for dry grass. Without mobility, herds would gain ever poorer feed value from the pasture as its digestibility decreased throughout the season. Cattle which annually trek more than 1,000 km between the Malian Delta and the Mauritanian Sahel return in better physical shape than the milking cows and calves which remain in the village throughout the year.

Despite the advantages of nomadic grazing there have been continuing attempts by governments and aid agencies to regulate it, mainly by reducing herd mobility. Some attempts arise from a desire for a more "scientific" approach to livestock raising which, it is hoped, will prevent overgrazing. Others result from a determination by governments to bring these wandering peoples more under their control, whether for altruistic reasons such as providing education and health facilities or merely to tax them more easily. There have been four main types of interventions: 1) grazing controls; 2) sedentarization; 3) establishment of ranches; 4) stratification.

Grazing Controls

Despite strenuous efforts over the last thirty years, most attempts to control grazing have failed. This is largely due to the ignorance of "experts" or the neglect of traditional grazing controls exerted by pastoralists. It may be imagined that because grazing lands are communal they are also unregulated, but this is not usually the case. Michael Horowitz, a social anthropologist at the State University of New York at Binghampton, writes: "There is an emerging awareness among scientists ... that strict regulation of access to scarce resources (water and grass) and limits on herd size may well be the rule rather than the exception among herding societies in semi-arid lands" (Horowitz 1979).

Why do governments wish to control what has previously been self-regulated by nomadic peoples? "With the exception of Mauritania in the Sahel and Somalia in East Africa", writes Horowitz, "the ruling elites of these states are drawn from groups which are not only not pastoral, but which have historically viewed pastoral peoples with ambivalence at best, and often outright hostility." The dominance of the farming peoples in such countries was confirmed by the colonial powers, forcing the herdsmen to retreat in the face of constantly expanding cultivation. "The final insult was the implantation of deep wells open to all comers, leading to chaotic competition for grazing land" (Horowitz 1979).

The imposition of grazing controls by governments has therefore aroused resentment and resistance. One type of control is the grazing fee, which, while not limiting the herd's freedom of movement, has the motive of forcing the herdsmen to dispose of non-productive animals. In the view of Peter Hopcraft:

> Enthusiasm for the notion of a grazing fee should not be
> anticipated, especially from the owners of relatively large
> herds of cattle. Paying for a resource that has such a long
> and culturally embedded history of being freely available

(or at least free, even if not always available) is an innovation that is likely to have its legitimacy seriously questioned. Grazing fees were an integral part of a number of grazing schemes in the colonial period, and they undoubtedly generated considerable antipathy.

Another type of control is the division of a region into blocks of rangelands. These are then used successively as part of a rotational scheme in which each block is left for a fallow period during which pasture regenerates. One way to manage such a scheme would be to turn off or physically remove the pumps from water-holes in a particular block where forage is approaching exhaustion. This, as has been seen above, could provoke violent reactions on the part of the herdsmen. When tried out in Kenya among the Poktot and Samburu, the block system failed dismally. Like enforced sedentarization, it limits freedom of movement, one of the key components of the nomadic life.

Sedentarization

The sedentarization (settlement) of nomadic herdsmen is one option for improving livestock raising in semi-arid areas, but in practice it is difficult to consider it in isolation from the strong political pressures against nomads (described above and in Chapter 2) which in some countries have led to forcible sedentarization. This is not to say that sedentarization is always forced or even planned. Many nomads, such as some of the Fulani people in Mali, settled spontaneously in the aftermath of the 1968–73 phase of the Sahel drought. They saw it as a temporary measure until their herds built up again, but many have not returned to their former way of life.

Sedentarization has a number of disadvantages in practice. The increased concentration of livestock around a village often degrades nearby pastures. This degradation may be exacerbated by an increase in herd size as the settled nomads tend to become more dependent upon grain, require less meat and milk, and therefore kill fewer animals. Resettlement schemes in Sierra Leone, Burkina Faso and Nigeria have not been very successful because animals have suffered health problems in the new locations and conflicts have arisen between the nomads and local peasant farmers owing to planners' failure to recognize the previous range of uses of project sites. Nomads used to travelling longer distances have experienced particular difficulties in adjusting to being settled in one place, and some pastoralists resettled by the Senegal government after the 1968-73 drought began to neglect their cattle, even though tubewells were provided, with the result that the herds increased in size and became vulnerable to disease (Oxby 1984; Santoir 1983).

One of the few really successful examples of sedentarization is that involving the Fulani in central Nigeria, who combine livestock raising with cropping. Each household crops an average of about 0.9 ha, with sorghum and maize (either alone or together) accounting for 70% of the total area. Forage crops are grown and the fields are manured by cattle, although most Fulani also use some chemical fertilizers. Crop residues are another valuable source of fodder. However, because the area being settled was previously underpopulated, and the climate is less arid than in other African countries with large nomad populations, the example might not be generally applicable (Powell and Taylor-Powell 1984).

Nomads may suffer a drop in income when they are sedentarized, according to Mustafa Mohamed Khogali, of the Department of Geography at the University of Khartoum in Sudan, who has estimated that the average annual returns of (unsettled) nomads may exceed those of rainfed cultivators by at least 30-50%. Khogali thinks that nomads should be encouraged to settle if they want to, but that planners who urge them to become cultivators without animals have failed to understand the nomad mentality since settlement may require just too much of a break from past traditions (Khogali 1983).

Ranching

The introduction of intensive livestock ranching to developing countries has rarely been a success. After World War II, "development" usually meant transferring western technology without adaptation to developing countries, and many aid agencies essentially tried to take Texas to the Sahara. The President of Niger and other Sahelian heads of state talked of exporting beef all over the world. Reality has proved less rosy.

According to the Club du Sahel: "Results on the whole have not been satisfactory: the investments called for were too heavy; the farms, placed in regions poor in resources, resulted in mediocre productivity; and finally, marketing of production was hampered because the ranchers were too far from the large commercial centres" (Club du Sahel 1980).

Under colonial rule, Europeans appropriated land and established commercial ranches, showing little regard for the welfare of the indigenous people. The ranches tended to be effective when established in areas with favourable soil and climatic conditions. After countries gained independence, their governments sometimes replicated the actions of the colonizers. In the 1960s, the World Bank promoted commercial cattle ranching in southwest Uganda with the construction of over 100 ranches, each of several thousand hectares. The political élite of Uganda took control of the lands and succeeded in establishing themselves as absentee landowners of large tracts of Uganda's grasslands.

The "group ranch", so-called because it is run as a co-operative by several families, was supposed to be more democratic and give a wider spread of benefits than a strictly commercial ranch. The government of Kenya established fourteen group ranches in the late 1960s, each averaging 19,000 ha and 100 families, settling the people and registering land ownership. The Maasai agreed to participate in these ranches, less out of enthusiasm than in fear of what the government might do if they refused. As one elder said: "If there is rain in Kenyawa and people have ranches there, I cannot move my cattle into that place." Some of the Maasai arranged to have family members registered in different ranches so that they could still move to more favourable lands when drought threatened their own grazing lands. The experiment was not very successful. The co-operative has improved marketing and the availability of cattle dips, sprays and the like, but the ranches are still run basically on traditional subsistence lines and the lands have often been invaded by outsiders, leading to armed clashes.

In Tanzania the government has tried for many years to integrate pastoralists into its Ujaama (socialist community) villages. After various incentives and even coercion had failed, in 1973 the government tried (with the help of USAID and the World Bank) to establish co-operative (group) ranches, but with communal herds. Members could bring their own cattle if they wanted, but people have been loath to do this and unwilling to accept government-subsidized communal cattle because this ran contrary to their basic belief in self-reliance.

The government of Botswana tried to curb overgrazing near villages and increase beef production by opening up new rangelands in the western Kalahari to grazing. It established twenty-five ranches near Ncojane in Ghanzi district, drilled boreholes and erected fences, and leased them to the owners of large cattle herds from adjacent villages. The project ran into problems because most of the new owners were absentee ranchers who only visited their herds occasionally and did not have much interest in taking care of their rangelands (UNEP 1985).

Game ranching, in which wild game rather than traditional domestic livestock are raised, has been widely advocated as a both profitable and more ecological form of livestock husbandry. However, one evaluation of the relative merits of game or cattle ranching on Kenyan savanna lands receiving less than 600 mm of rainfall annually found that, at current costs and prices, neither activity would be economically viable if projects had to start from scratch and all land, facilities and equipment had to be purchased (McDowell et al. 1983). Higher prices would be needed to make game ranching profitable, and it was unlikely that it would improve the protein intake of local low-income families because they would never be able to afford to buy game meat.

Stratification

Stratification is a far more elaborate form of reorganization of livestock management than the types already mentioned. Proposed for West Africa by the World Bank and FAO since the end of the 1960s (Mayer et al. 1983), the basic idea was that calves would be reared, as at present, on arid rangelands in the north but then moved south to the more humid pastures for fattening, either by peasants looking after small numbers of animals or on large intensive feedlots where animals would be kept in stalls and given concentrated feeds. Finally, they would be sold in the cities and on the coast. This was the basis of the SOLAR project (Stratification of Livestock in Arid Regions), which was adopted as one of six transnational projects that were part of the UNCOD Plan of Action.

With a few very local exceptions, stratification has been a failure. The lifestyle and culture of nomadic pastoralists severely limit the rate of offtake from the herd, and herdsmen have not been willing to sell large numbers of young stock for fattening. The planned reduction in herd sizes has not been achieved and there has been no improvement in living conditions for nomadic people. On the contrary, the majority of nomads involved in stratification schemes have probably suffered a reduction in their quality of life because the schemes did not take account of crucial aspects of nomadic cultures. Stratification requires nomadic pastoralists to make a severe cultural transition, and considerable efforts will be necessary in future schemes of this kind to prepare them for a change whose psychological and social implications may not always be fully appreciated by planners from the developed world (Baumer 1981).

Nor has stratification proved to be an economic proposition so far. Dozens of abattoirs in the more humid "fattening zone" of sub-Saharan Africa have either been abandoned because they were unprofitable, or operate at only a fraction of their maximum capacity. It has been estimated that the cost of feed and supplements needed to fatten the cattle in this grain-deficient region is two or three times the value of the gain in livestock weight. Intensive fattening would increase the price for which livestock could be sold, but since beef is imported into West Africa, linking the Sahel with world markets, there is an upper limit on local livestock prices.

Some success has been obtained with attempts to stratify the Kenyan beef industry by breeding herds and producing store cattle on dryland ranges and then finishing animals for slaughter on intensively managed pastures on better lands or in special feedlots (Creek 1984). The feedlots increased income and employment, but because they were owned by large farmers or state enterprises the benefits were not widespread, especially since European cross-breeds were used for fattening. When it was shown the Boran cattle raised in northern Kenya could be economically fattened in feedlots, pastoralists became convinced of the potential benefits. High-quality beef was produced and the project helped to stimulate more government support for foot-and-mouth disease control programmes. However, the rate of growth of feedlots has been disappointing, mainly due to government controls on meat prices.

THE PROSPECTS FOR LIVESTOCK DEVELOPMENT

Can livestock raising in the Sahel ever be improved? Walter Goldschmidt concludes: "The picture that emerges is one of almost unrelieved failure" (Goldschmidt 1981). Michael Horowitz talks of "the almost unblemished record of project non-success in the Sahelian livestock sector" (Horowitz 1979). Ibrahima Toure, a UNESCO range management expert based in Dakar states: "There have been many livestock projects since the drought, but few very positive results. Things go wrong in implementation. Planners don't have an overall view of the problem."

The lack of success is not surprising. Nature does not run like clockwork in the Sahel or other arid areas, and it would be difficult if not impossible to devise a block grazing scheme which was flexible enough to give levels of productivity and insurance against poor rainfall similar to those of the traditional system. Michael Horovitz recommends that: "Range management interventions can and should be based, where possible, on the system of controls already practised by the people" (Horowitz 1979). Instead, schemes are devised on the basis of inappropriate and inadequate policies, and very often it is these, rather than the actions of pastoralists or poor project administration, which result in failure (Haldeman 1985).

Such experience has left its mark. Planners and overseas aid agencies now feel they cannot do anything to improve livestock raising. Only 5% of all development aid to the Sahel goes to this sector, even less than the 8% for rainfed cropping (Club du Sahel 1981). Decades of failure have also made pastoralists suspicious of all government attempts to intervene in their practices. For their part, governments tend to label nomads as stubborn and inflexible. There are elements of truth in both views.

Even if perfect management could be assured there is no guarantee that the results obtained would be any better. A study by a team from the Institut d'Économie Rurale in Bamako,

Mali, and the University of Wageningen in the Netherlands concluded that failure to appreciate the effects of low soil fertility on pasture growth in the Sahel has meant that the potential for improving livestock production has been overestimated and traditional pastoral systems have been greatly undervalued (Breman and de Wit 1983).

In principle, it should be possible to improve livestock production in ways which do not result in overgrazed rangeland and which are compatible with the traditions and wishes of the pastoralists. However, major problems still await solution in terms of achieving the right balance between conservation of the environment and the constraints of land tenure; balancing the individual and communal interests of pastoralists; overcoming the cultural and economic conflict between pastoralism and commercial beef production; and deciding what is the best path for the "development" of the livestock sector (Haldeman 1985).

PASTORAL ASSOCIATIONS: THE LAST HOPE?

A promising new initiative now being tested in the Sahel gives some cause for hope. A number of projects in Mali, Senegal and Niger, funded by the World Bank and USAID, are promoting the formation of pastoral associations. Instead of trying to regulate pastoralists from outside, these projects try to adapt the traditional self-regulation of nomadic herders to modern conditions. Groups of closely connected families are encouraged to form a pastoral association which is given responsibility for managing a borehole or watering point and the surrounding rangelands, and is eligible to receive benefits such as low-cost drugs, vaccinations and other animal health measures.

The idea of pastoral (or herders') associations in West Africa developed in the early 1970s from the concept of group ranches tried previously in East Africa. USAID sponsored an extensive study of pastoralism in central Niger between 1978 and 1983 before formulating the practical second phase of its project, but as a result of the good relations which project personnel formed with local people ten pastoral associations were established before the first phase ended. The follow-up (second phase) project, covering 4 million hectares in an area bounded by the towns of Agadez, Tahoua and Tanout, began in January 1984 and was intended to increase the number of associations to at least 110. Instead, it failed dismally, though more as a result of poor project implementation than of any fault in the basic concept. In the transition from the first to the second phase, project personnel with whom the pastoralists were acquainted left the area. This, together with a variety of difficulties in relating the project to local government bureaucracy, made the pastoralists disillusioned with the scheme and prevented it from being extended as planned (Aronson 1982; Swift 1984; Abdou *et al.* 1985).

World Bank experts are cautiously optimistic about the furture of their projects. In Mali up to ten pastoral associations, each consisting of thirty to fifty families, have been formed around existing watering points. In Niger eighteen associations have been formed in an area to the southeast of the USAID project in the departments of Maradi, Zinder and Diffa. The most promising example of pastoral associations is in Senegal, where Fulani herders who are already well settled have formed fifty-two associations. The concept is now also being tested in Burkina Faso and Mauritania (Sihm, personal communication).

Experience in Lesotho, on the other hand, casts some doubt on the effectiveness of pastoral associations, suggesting that self-managed associations of herders may not be able to take unpopular decisions and that some external influence may be required after all. A grazing association established in the Sehlabathebe area to improve range management and animal health covers eleven villages and accounts for 96% of all livestock in that area. Farmers already perceive benefits in the form of better rangelands, animal quality, and facilities for wool shearing and dip tanks. The association enforces the grazing and livestock regulations, and is managed by a committee of village chiefs together with two elected members from each village. However, a detailed evaluation of the project has stated that:

> because of the sensitivity of the issues involved, grazing association members often cannot act decisively or agree among themselves .... Field technicians often decide for the grazing association, thus relieving its members of the responsibility of making unpopular decisions on controversial issues .... Although the local chieftainship structure is incorporated into the grazing association, the

project represents externally imposed decision criteria for exercising systematic control over resources in ways not previously done. The project becomes a kind of adversary, forcing unpopular but necessary decisions that grazing association leaders, by virtue of their identity with village society, find it impossible to impose. At present there is no institution that can play this adversarial role after the technical assistance team is withdrawn.

The report concludes that for this reason the grazing association may well not last beyond the end of the project unless there is some external leadership which can take the blame for unpopular decisions (Warren *et al.* 1985).

THE LIVESTOCK DILEMMA

Technically, it would seem quite simple to control desertification caused by overgrazing, either by reducing the number of animals or by improving the quality of rangelands. All attempts to do this, however, have so far met with little success, and these experiences illustrate the importance of social controls over resource use and what can happen when attempts are made to replace or circumvent them. Whether the new pastoral associations will put Africa's livestock sector on a sounder footing remains to be seen, but in general projects which work with the people should have more chance of succeeding than those – all too frequent in the past – that work against them. Reducing the extent of overgrazing will also require a greater recognition by governments of the importance of pastoralism in dry areas, and corresponding action to prevent the incursion of rainfed cropping on to marginal rangelands which reduces the area available to herds and is a major indirect cause of overgrazing.

7. Controlling Desertification by Planting Trees

Tree planting plays a vital role in establishing more productive and sustainable land uses in dry areas. Trees protect land from soil erosion, reclaim land that has already been degraded, and help to prevent further deforestation by providing sustainable supplies of fuelwood, fodder, food and other products. Present rates of afforestation are small in comparison with the clearance of 4 million ha of woodland each year in the world's drylands, two-thirds of it in Africa (Table 2.1), (Lanly 1982). Afforestation data are not available for dry areas specifically, but the Sahel and the Sudano-Sahelian region (the latter including humid tropical areas) had average afforestation rates of only 4,800 and 42,000 ha per annum respectively between 1976 and 1980 (Grainger 1986). Nevertheless, tree planting and soil conservation account for the largest proportion of projects currently being implemented to control desertification.

Trees are being planted in a wide variety of ways. In early projects, forest departments established large-scale fuelwood plantations, usually on land in government forest reserves, to counter the growing scarcity of fuelwood, the major source of energy in the drylands. Because planting rates could not keep pace with rising demand for fuelwood, the focus then switched to social forestry schemes in which trees were planted on village lands (either by forest department personnel or by villagers themselves) and on farmlands. This chapter looks at progress in establishing fuelwood plantations; social forestry; the growing importance of non-governmental organizations, which seem to have had more success than forest departments in inspiring local communities to plant trees; and the spread of improved wood and charcoal stoves to make the most efficient use of wood from remaining forests and the new plantations. The next chapter focuses on farm forestry and the potential for managing natural woodlands instead of establishing plantations.

THE FUELWOOD CRISIS

In the mid-1970s, while people in the developed world were experiencing the OPEC oil crisis, they became aware that their fellow human beings in developing countries were suffering from shortages of fuelwood in what Erik Eckholm called "the other energy crisis" (Eckholm 1976). Based on FAO data, it is estimated that the deficit of fuelwood in the drylands in 1980 was equivalent to the sustainable production of almost 26 million ha of intensive fuelwood plantations (an area the size of West Germany). If present planting rates are not increased the deficit could grow to the equivalent of nearly 62 million ha of plantations (an area the size of Spain and Greece combined) by the end of the century (FAO 1981; Grainger 1988). Just to make up the shortfall by the year 2000 in all African drylands south of the Sahara would need the establishment of plantations which, if collected together, would form a forest belt 6,000 km wide and 34 km deep across the Sahel from Senegal to Ethiopia. The required planting rate of about 1 million ha per annum dwarfs the present rate (see p. 214) and is equivalent to the current rate of afforestation for all tropical countries combined.

In response to these growing shortages of fuelwood, projects got under way to improve supplies by establishing plantations of fast-growing trees. Forest departments used two main approaches: first, planting large blocks of trees, often on forest reserve lands not currently forested; second, encouraging and assisting farmers and villagers to plant trees on private and public lands. The second approach, called social forestry, was necessary because of difficulties in requisitioning land being used for growing crops or raising livestock, and because in many countries there were not enough foresters even to look after existing forests properly.

LARGE-SCALE FUELWOOD PLANTATIONS

The initial response of forestry departments in the Sahel to rising fuelwood shortages was to establish large-scale plantations of fast-growing tree species using conventional forestry techniques on government-owned land. These projects have not been very successful, mainly because of the slow growth rates of trees in areas with poor soils and low rainfall. This has led to a general pessimism about the feasibility of large-scale fuelwood plantations in the Sahel.

In a project funded by the World Bank in Niger, 779 ha of fuelwood plantations were established between 1979 and 1981 within the boundaries of forest reserves. This was 10% higher than the target area and was achieved in two years rather than the three years originally intended. Neem (*Azadirachta indica*) accounted for 71% of the area and *Eucalyptus camaldulensis* for the remainder. However, growth rates were lower than anticipated because the soils were so poor – no preliminary soil survey seems to have been carried out (World Bank, personal communication).

The Mali Forestry Department established fuelwood plantations in the La Faya and Monts Mandingues forest reserves between 1980 and 1985, in an attempt to keep up with demand for fuelwood in the nearby capital city of Bamako, which was expected to rise from 0.3 million cubic metres in 1982 to 1.5 million cubic metres by the end of the century. *Gmelina arborea* plantations accounted for over 70% of the total area, and the remainder consisted of neem (*Azadirachta indica*) and *Eucalyptus* spp. in both pure and mixed stands. However, the project (also funded by the World Bank) failed to meet its target. The 2,273 ha of plantations established were only two-thirds of the intended area.

A project in Senegal funded by USAID was supposed to establish 3,000 ha of fast-growing *Eucalyptus* plantations in the Bandia Forest area between 1979 and 1983 to provide much-needed fuelwood to the cities of Thies and Dakar. Again, less than two-thirds of the target area was planted. Mortality was high because of low rainfall, and annual growth rates were only 1.5 cubic metres per hectare instead of the 10 cubic metres per hectare expected, mainly because the species used were not appropriate to the site (USAID 1983a).

One way of improving growth rates is by irrigating the plantations. Another part of the Niger project described above established 400 ha of experimental irrigated fuelwood plantations at Namarde Gongou, 35 km from Niamey on the lower terraces on the left bank of the Niger River. These plantations were expected to be ten times as productive as rainfed plantations, to produce fuelwood on a three-year rotation, and be a model for future fast-growing plantations; but these hopes were not fulfilled. Many difficulties were experienced in establishing the irrigation system, in which each tree was to receive water separately by drip irrigation. Equipment was late in arriving, and the irrigation system had to be changed in the middle of the project to a type more suitable for the soil and terrain. Consequently, only 240 ha were planted, the initial plantations were not properly irrigated, and many trees suffered from drought and had to be replaced. Despite the faster growth rates, the high cost of pumping water makes irrigated plantations difficult to justify economically unless cheap fuel is available or there are low-cost sources of water nearby. Suitable locations include land on valley bottoms close to rivers, or adjacent to existing irrigated cropland where the trees could benefit from unused irrigation water and provide windbreak protection to the crops (World Bank, personal communication).

SOCIAL FORESTRY

Social forestry, in which local people plant trees outside regular forest areas, seems to be the only viable long-term strategy to meet afforestation targets, given rapidly rising demand for fuelwood and the fact that national forestry departments are already stretched by their present modest afforestation commitments. Even if government forestry departments were able to plant the required numbers of trees, without wide popular support the young trees might be lopped or felled for fodder and fuel by farmers and nomadic herdsmen. Until recently, most foresters have treated plantations as places from which local people must be excluded. They now have to learn how peasants view trees and forests, and to develop new practices from this understanding. For forest officers, who were previously little more than policemen, to change overnight into patient, inspirational "tree gurus" is quite a challenge. Some of the following examples show how foresters are meeting that challenge.

China and South Korea

In the developing world, the most spectacular examples of afforestation carried out largely by ordinary people have been in China and South Korea. Between 1950 and 1983, China's forest area increased by 60% to 122 million ha. This required a phenomenal afforestation rate of 1.5 million ha per year, and included massive projects like the "Green Wall" against the Gobi Desert, a huge shelterbelt 1,500 km long and 12 metres wide which 700,000 farmers established in just two seasons. Although China was able to mobilize large numbers of people to plant trees, the programme has not been a total success. When vast areas of plantations are established over such a short period they may not receive sufficient care in the crucial early years; consequently the plantations established in the 1950s and 1960s had high failure rates. In 1987 large areas of forest, including part of the "Green Wall", were destroyed by fire. Preliminary estimates spoke of almost 700,000 ha being lost in Heilongjiang province alone but no data were available for other affected areas such as those in Inner Mongolia (Thomson 1987). Nevertheless, China aims to increase its forest cover from the present 12.7% to 20% by the end of the century and ultimately to reach 30%.

When South Korea was faced with a shortage of fuelwood, more than 20,000 village forest co-operatives were mobilized to plant over 1 million ha of fruit, fuel and timber trees in just five years, from 1973 to 1977 (Arnold 1983; Gregersen 1982). South Korea is a highly capitalist country but like China has a tradition of organized communal activity. There already existed a national movement to improve rural living (the Saemaul Undong), and elected Village Forestry Associations were established in each participating village. The government enacted legislation to make land available for the new woodlots, and this, together with the existence of a grassroots organization which the forest service could advise and provide with seedlings and other materials, had a great deal to do with the success of this programme.

In other countries the social conditions for afforestation are not as favourable as in China and South Korea, so it is unlikely that they will be able to achieve the same rates of afforestation. Nevertheless, social forestry schemes may well prove to be the only practical option if forest cover is to be restored. There are two main components in most social forestry schemes: tree planting on village lands by forest department personnel and hired labour, and true community forestry in which it is the villagers themselves who plant and tend trees. Often these schemes also promote farm forestry in which trees are planted on private farms.

Social Forestry in India

Gujarat: the pioneering state
Social forestry was pioneered in India in the western state of

Gujarat. In 1969 the state forest department set up a separate Extension Forestry Wing to plant trees outside regular forest boundaries, and its first activity was to establish multi-row windbreaks alongside roads and canals, with shade trees nearest the road and faster-growing trees for fuelwood and timber behind them. The forest department owned the land and planted the trees, but nearby villages had grass-cutting rights, protected the saplings, and in return took a share of the profits. These "mini-plantations" were not planned or planted with much public participation but, according to B. K. Jhala of the forest department: "They marked a critical psychological turning point. The new plantations also gave many people their first positive impressions of the forest department. They saw that our activities could benefit them directly." By 1982, trees had been planted along 14,750 km of roads and canals in Gujarat, equivalent to 29,500 ha of forest. The trees were to be harvested on a 7–10 year rotation and the benefits shared equally between the forest department and local people (Karamchandani 1982; Foley and Barnard 1984).

The next phase began in 1974 with the "Supervised Wood-lots" scheme. A number of villages were persuaded to establish small plantations of about 4 ha on communal grazing lands. Various timber, fuelwood, fodder and fruit species were supplied free of charge by the forest department. The villagers protected the plantations from grazing and fuelwood poaching, and had an equal share with the forest department in the eventual returns from harvesting the trees. The response to this scheme was enthusiastic. By 1982, 28,000 ha of plantations had been established in a fifth of the 20,000 villages in the state, although some plantations have suffered heavily from livestock grazing. Now the forest department is contacted by village councils who have seen how neighbouring villages have benefited and who want to do the same themselves.

In 1980 the forest department launched a new "Villager Self-Help Scheme". This was intended to be a truly community forestry scheme in which the forest department would provide free seedlings and technical support but villagers would take responsibility for all the tree planting and in return would share all of the benefits from the sale of the harvested trees. The general response to this scheme has, however, been poor.

The programme has a number of other components. Prominent among these is the highly successful farm forestry scheme, described in the next chapter, in which farmers received free seedlings for planting on their lands. By 1983 10% of all farmers in Gujarat had participated in the scheme and 150,000 ha of plantations had been established (Spears 1983). Landless people were not neglected, and a special scheme was evolved whereby landless families were each assigned 37.5 ha of degraded forest to plant at a rate of 2.5 ha per year over fifteen years. In return for forty person-days of work per month each family received 250 rupees, free material to build a hut, rights to all minor forest products (i.e. those other than wood) on the land, and 20% of the net profit when the forest is harvested after fifteen years.

Gaining the co-operation of local people is vital if trees are to be protected from grazing and illegal felling, but physical protection is still necessary. This might entail barbed-wire fencing; cattle-proof trenches, up to 0.9 metres wide and 0.75 metres deep, with the soil piled up on the inner boundary as a bund on which thorn trees like *Acacia nilotica* and *Prosopis juliflora* are planted to form a living fence; small circular trenches, 3.6 metres in diameter, to protect individual trees, often with an additional living fence or thorn basket; and, in rocky areas, stone walls, about 1 metre high for individual tree protection (Gupta 1985). Barbed-wire fences are often used to protect high-risk roadside plantations but are twice as expensive as cattle-proof trenches and bunds which are more popular and also generate employment for unskilled people.

Much of the success of social forestry in Gujarat stems from the decision to set up a new section within the forest department consisting of staff specially trained for the job. In social forestry, the ability to deal with people is at least as important a qualification as technical knowledge of forestry, because personal contacts are fundamental to gaining the co-operation and involvement of farmers and villagers. Another key ingredient was the identification of some of the early farm forestry plantations and village woodlots as demonstration centres which farmers and villagers could visit to see what could be accomplished. A great deal of attention was paid to public relations: a forestry officer with the grade of Deputy Conservator of Forests was appointed to work full time in liaising with newspapers, radio and television; pamphlets and posters were widely distributed; two mobile exhibition vans toured villages, towns, schools and colleges with slide shows and documentary films; and slides on forestry with suitable slogans were screened during normal programmes in almost all cinemas in the state. Finally, children were also closely involved. Primary schools were encouraged to establish their own tree nurseries, the seedlings raised were purchased by the forest department, and the proceeds went to benefit school activities. Both primary- and secondary-school students took part in nature education camps in forest and wildlife reserves (Karamchandani 1982).

Social forestry in Uttar Pradesh

While social forestry was spreading in Gujarat, a similar project began in the state of Uttar Pradesh in 1976, receiving World Bank support from 1979. The initial aim was to plant 48,600 ha of forest on public and village lands, mainly in forty-two districts which had a per capita forest area of only 0.01 ha but also about 500,000 ha of underused or unproductive lands. All but 14% of the target area was to be planted by the forest department using hired labour. This included 4,741 ha of village lands that would be turned over to the forest department for planting but returned to the villages after five years when the first trees were ready for harvesting. Another 3,000 ha were to be planted on village common lands by community self-help schemes. The project would provide technical assistance, seedlings and other materials to those villages which asked to participate in the scheme. It would also distribute 8 million seedlings for farmers to plant on 4,000 ha of private lands (Murthy 1985).

Project achievement exceeded targets in all major categories where the forest department undertook the planting, although only 3% of all available village land was afforested. By 1984, 59,700 ha of plantations had been established, of which 9,700 ha were on village land, 18,558 ha on degraded forest land, and 31,397 ha on the sides of roads, canals and railways, exceeding targets by 100%, 40% and 40% respectively (Seetheraman and Singh 1985). The area planted along canals was far less than intended because of irrigation department restrictions.

The amount of self-help planting, that is, true community forestry, was very small. No reliable data on achievements in this vital category exist because no systematic monitoring took place, but the area has been estimated variously as only 133 ha (Foley and Barnard 1984) and 200 ha (Murthy 1985). Nor are there any data on the extent of farm forestry plantings but one conservative estimate is just 1,300 ha of plantations, with a further 1,300 ha planted on the bunds bordering fields (Singh 1985). According to another source the area might have been far higher because the actual number of seedlings distributed to farmers was up to twenty-four times the target number. K.R.S. Murthy blames the forest department for the poor record in community forestry: "The social forestry division has yet to capture the essence of the social part of social forestry in its operations. It has to choose whether to remain in the mainstream forestry tradition or . . . to fit itself in organizational terms with the local people" (Murthy 1985).

Social Forestry in Africa

Slow progress in sub-Saharan Africa
Social forestry projects in the Sahel date from about 1980, when it was becoming clear that large plantations would not give sufficient yields. Some progress has been made with projects in countries such as Burkina Faso and Niger in which people are paid to plant trees, but community-based projects in which villagers work together to plant trees on village lands for joint benefits have been less successful. Church groups and other non-governmental organizations in the Sahel are often more active than governments in promoting afforestation and have financed a number of small-scale projects.

The most popular social forestry measure in the Sahel is the "greenbelt", a shelterbelt of hardy, drought-resistant trees which protects a town or village from harsh winds, swirling sand storms and encroaching sand dunes, and which can be harvested for fuelwood, poles and fodder. Examples include the 500 ha greenbelt around Ouagadougou (capital of Burkina Faso) and the expanding 300 ha belt around Niamey (capital of Niger). Smaller green belts reach into the centre of the towns, providing shade in every street. The town of Bouza in Niger is now famous for its cool, tree-lined streets.

Tanzania began a community forestry programme in 1967 but ten years later little progress had been made. The programme took time to get started, and between 1973 and

1978 only about 2,000 ha were planted each year on average (Skutsch 1985; Mnzava 1980; Kilahama 1980). One African village woodlot programme that has been successful is in Lesotho where, since 1983, 260 woodlots covering 3,650 ha have been planted in a nationwide effort to grow wood for fuel and small poles. The programme is funded by the UK government's Overseas Development Administration, the Anglo-American Corporation and the government of Lesotho. The majority of the trees in the woodlots are species of *Eucalyptus* and *Acacia*. Trees are planted at altitudes of up to 2,000 metres, and coniferous trees have grown faster than *Eucalyptus* although they do not coppice and are susceptible to fire (UNEP 1985).

Reforestation by refugees in Sudan
The Eastern Refugee Reforestation Project in Sudan, which contains both paid and voluntary planting components, provides a good example of the difficulties faced in African social forestry. The project area receives between 250 and 500 mm of annual rainfall and is located on the clay plains of Kassala province in eastern Sudan that were almost totally deforested by the rapid expansion of mechanized sorghum cultivation after 1960, which led to extensive soil erosion and gully formation. In addition, an estimated 400,000 refugees have streamed into the province from Ethiopia, and made it their home. The major aims of this multi-faceted project, funded by USAID and the US-based voluntary agency CARE, are to restore tree cover to improve the environment and quality of life in the area, provide renewable sources of fuelwood to local people, increase employment opportunities, and encourage the adoption of sustainable agroforestry practices. The project, intended to benefit both refugees and rural Sudanese, includes large-scale fuelwood plantations, smaller village woodlots, windbreaks and tree planting in house compounds.

By mid-1985, two years into the project, progress was mixed. More than 1,260 ha of fuelwood plantations had been successfully established, despite the ravages of drought in 1984 which left many seedlings dead. This meant that a lot of replanting had to take place in the following year when heavy rains caused additional hazards. Of the 93 km of windbreaks planted, only 28% were successfully established, owing to incorrect seeding, competition from weeds, and inundation by heavy rains in

1985. The plantations were to be cropped for fuelwood on an eight-year rotation. The native species *Acacia seyal* and *Acacia senegal* were the main trees used for the plantations and windbreaks. Another 250,000 trees, mainly *Acacia mellifera*, *Azadirachta indica* (neem), *Leucaena leucocephala* and *Eucalytpus camaldulensis*, together with various fruit trees like lemons and limes, were distributed for planting in house compounds to provide shade and living fences; an estimated 50–70% survived. Over 500 improved charcoal stoves were produced and sold commercially in just nine months in an attempt to increase the efficiency of wood use.

The voluntary planting component was less successful. The extension programme certainly led to many trees being planted in house compounds, but this diverted the efforts of project staff away from their main purpose of introducing farmers to the benefits of agroforestry and windbreaks. According to an evaluation of the project by a USAID team, the focus shifted from agroforestry because of the inexperience of extension staff, and the fact that some project officers were unwilling to discuss openly the difficulties they faced in implementing agroforestry systems in the area. The large number of trees planted in house compounds could turn out to be counterproductive because many people felt that they were under pressure to plant trees, accepting them as yet another relief measure, the most important of which was food. Only a few people were accustomed to tree planting, and even those who were enthusiastic were often unhappy about having to buy water (not freely available in this dry area) and the brushwood necessary for fences to protect the seedlings. Such costs were often beyond the means of the poorer people, and so widespread tree mortality was inevitable. On the other hand, the richer members of the community were more able to afford to establish compound trees and, in some cases, private woodlots.

The village woodlot scheme was a failure. Only one woodlot was planted, and the trees suffered from a shortage of water and from grazing by local goats. Wire fences were insufficient to keep out the goats and the villagers were unwilling to supply volunteer guards for the woodlot. There was a general lack of interest on the part of villagers in managing the woodlot for their common benefit. These people had long been used to gathering fuelwood wherever they wanted; suddenly expecting them to begin to plant and then care for a plantation of fuelwood just as they would an agricultural crop was asking a lot. There were endless disputes over who was to guard the woodlot and benefit from its harvest. Villagers may not necessarily be familiar with the concept of co-operation and equal sharing, since in each village "rights", even on supposedly communal land, are determined by wealth and family ties. Most people felt that trees in the plantations and woodlots belonged to the state. One way to obtain more participation by local people in future afforestation and plantation management activities might be to allow them to crop the fuelwood plantation using an agroforestry system. Thus farmers could grow sorghum on a replanted area for a few years until the tree canopy closed and then move on to another area to repeat the process. The system would be even more attractive if farmers were to benefit from the sale of wood on the land before planting crops and seedlings.

Some of the social aspects of the project were successful. Refugees accounted for 90% of the casual labour force and benefited enormously from being employed to plant trees in the windbreaks and plantations. However, they were confused by the tree-planting programme, since there did not seem to be a clear distinction between the "relief" and "development" aspects. Many refugees accepted trees for planting in their compounds just as they accepted relief food, rather than out of any strong commitment to afforestation, and were subsequently surprised at how much it cost them to buy water to irrigate the seedlings.

The project has also laid excellent foundations for future activities through its school programme. Extension workers visited schools in the area to inform students of the value of trees to the environment and to encourage them to plant trees in the school grounds. The importance of school extension was vividly explained by a teacher: "Most of the children here will be farmers or will live in the rural areas, and so learning to plant and care for trees as part of their environment is just as important as other school lessons." One school established its own nursery and a small plantation with over fifty trees which reached a height of more than 7 feet in just two years. In another school, which did not have the benefit of its own water supply, the students brought little jugs of water to school each morning to water the trees they planted (USAID 1983b; Resch *et al.* 1985).

One of the problems with social forestry in African countries is that, in contrast to India, it has largely focused on growing trees for fuelwood in the belief that this corresponds to a real local need. Unfortunately, as Tom Catterson, Fran Gulick and Tim Resch have pointed out, "the fuelwood model accounts imperfectly for the disappearance of trees in the desertification equation" and community forestry will be unlikely to succeed unless it is supportive of local people's main priority: growing food. This means planting trees that will provide food, fodder and shelter as well as fuelwood. Village woodlots can therefore exacerbate the distinction between agriculture and forestry whereas what is needed is a much greater integration between the two practices (Catterson *et al.* 1985). This topic will be discussed in more detail in the next chapter.

DOES COMMUNITY FORESTRY HAVE A FUTURE?

For the last ten years or so there has been a general assumption that community involvement is essential if countries in the Sahel and other dry regions are to plant enough trees to control desertification and satisfy the need for fuelwood. The ultimate goal of most social forestry programmes has therefore been to achieve true community forestry by: (a) mobilizing a larger number of people to plant trees than would be available from the forestry department alone; (b) ensuring that trees once planted are protected from damage and premature harvesting; and (c) improving the livelihoods of local people by increasing the productivity of land and widely distributing the benefits obtained thereby. So far, few community forestry projects have been totally successful, especially in Africa. In areas receiving less than 800 mm rainfall per annum community projects are subject to the same limitations on tree growth as the large plantation schemes mentioned on pp. 215–17 but the reasons for lack of success seem to be more social than technical. Does this mean that the experiment has been a failure or does it merely reflect the difficulties involved in properly designing and implementing such projects?

Generating Awareness

There is no doubt that community forestry projects face numerous obstacles. It may take years for a local community's awareness of the need for tree planting to reach the point at which they ask to join a scheme. This will require, among other things, exposure to general publicity on posters and in the media, patient work by social forestry officers in discussing matters with villagers, and visits to other villages that have already joined the scheme to see tree planting in practice. This preliminary phase, although protracted, can well determine whether or not the project is likely to be successful even before it has begun. For it is during this period that social forestry officers learn about the particular characteristics of a village and the wide range of local requirements which need to be taken into account in the design of the project.

Foresters may limit the success of projects if they only consult with the men in a village, for it is women who usually grow food crops and have to walk for hours to collect fuelwood which could easily be grown locally. They also know far more than their husbands about gathering essential minor (non-wood) forest products (see Chapter 8). Many households are effectively headed by women because their husbands are away working in cities, sometimes even overseas. While these women might wish to join a social forestry programme they may be prevented from doing so because they lack rights to the land which they farm and find it difficult to get credit. According to Gill Shepherd of the Overseas Development Institute in London: "Women are the very category who have most to gain from tree growing for subsistence purposes and arguably for cash too" (Shepherd 1985). Shepherd makes the interesting point that since women already tend to co-operate with each other for subsistence cropping purposes, involving them more closely in social forestry projects could greatly increase their success rate.

Ensuring a Wide Distribution of Benefits

Ensuring that as many sections of the community participate in and benefit from community forestry projects is not a desirable extra but absolutely essential. Suppose, for example, that a project contains a component which protects young trees – by fences, guards or other means – from browsing by animals or felling by humans. Technically such a component is understandable and necessary, but socially it is not sufficient, because access to resources is closely associated with equity, that is, the way in which project benefits are shared between all sections of the community. Without careful consultation and planning, some people are likely to feel aggrieved or dispossessed by the enclosure of land for a forestry project, and they may try to continue to use that land for grazing or other purposes. All sections of the community will generally have established rights and privileges on village lands, but often the people most affected are those in the poorer sections of the community who depend heavily upon village common lands for grazing their animals or gathering firewood and fodder, both for their own use and for selling at local markets.

Equity may also have a major role in determining whether or not social forestry projects are accepted by the wider community at home and overseas. Social forestry projects in India, for example, have been criticized by development economists and non-governmental organizations there who claim that they benefit the rich and disadvantage the poor. This is a common objection to any government policy or programme in developed as well as developing countries. Perhaps in this context it arises from a combination of idealism and false expectations. The terms "social forestry" and "community forestry" may give the impression that they will help to redress the economic imbalance between rich and poor inside developing countries, but it is doubtful whether such an interpretation was ever intended by foresters in the early development of social forestry. They thought that the terms simply described a new type of forestry which involved ordinary people rather than professionals, and probably the most they expected was that local people would derive some benefit in terms of improved earnings, supplies of wood and other materials, and a better environment. The idea that social forestry would also be a form of social engineering was probably far from their minds, and there is no reason why the forestry sector should be any better at redistributing wealth than the agricultural sector.

Social Structures: Ideas vs. Reality

Some of the critics of social forestry expect more altruistic behaviour from people in developing countries than is evident in their own. The reality is that the success of projects can be threatened by political or economic schisms inside the village which are totally unconnected with tree planting. According to Erik Eckholm, Gerald Foley, Geoffrey Barnard and Lloyd Timberlake:

> In few if any villages do all the residents have identical interests and priorities. Communal programmes must be based on compromises on matters of deep concern to all. Rigid social stratification, such as India's caste system, makes any communal action extremely difficult. Communal programmes may presuppose the existence of competent and representative local organizations that can plan and implement projects and distribute benefits fairly. But often villages have no organization with such capabilities. The main lesson from the past is not that communal programmes should be abandoned; rather their design and implementation should be improved. (Eckholm *et al*. 1984).

Another complication arises in India, where village forests are often *panchayat* forests. A *panchayat* is the administrative body in charge of usually more than one village. There are endless possibilities for one village to feel aggrieved about a project if it feels disadvantaged relative to others. In one *panchayat* in Madhya Pradesh, for example, the grass from a young plantation area was distributed among all people in the *panchayat* at a nominal price but the money raised was used to electrify just one of the five villages in the *panchayat*. People from another village felt aggrieved and released their cattle into the plantation.

The Economic Incentive

It is unrealistic in most cases to expect villagers to co-operate merely for the sake of a communal ideal. As Tom Catterson remarked: "The attractive solution of shared work for shared benefits in pursuit of a solution to a common problem, the *raison d'être* of village forestry, remains an elusive, utopian goal" (Catterson 1984). This helps to explain why the types of social forestry which have been most successful so far have either been those with a low degree of community involvement or projects on private farms where the prime motivation is profit rather than some hazy ideal of improving the life of the "community". "The commercial incentive is often the strongest stimulant to tree growing" (Foley and Barnard 1985). Indian farm forestry has been criticized on the grounds that the richer farmers benefit most and that such schemes lead to a fall in employment, but while richer farmers may well be the first to establish woodlots, they are often the only ones able to take risks with a locally unproven concept. The smaller farmers will follow their example later when they see that the larger farmers are making money. Certainly the latest studies of farm forestry in Gujarat show a fairly wide spectrum of participating farmers.

Political Support is Vital

Political support by government leaders can mean the difference between success and failure in social forestry projects. Tree planting has reached the level of a mass movement in Kenya, and this is due in no small measure to the advocacy of government officials from the president downwards and to frequent coverage of the issue by newspapers, radio and television. The numbers of nurseries and seedlings raised are growing all the time but are still unable to keep up with the demand for tree planting on farms and house compounds (see Chapter 9). Every district commissioner, chief, secondary school and branch of KANU (Kenya's only political party) is now required by presidential order to have a tree nursery. In 1983, 83 million seedlings were produced by 1,300 government tree nurseries. This is in addition to seedlings produced by nurseries belonging to voluntary groups (Harrison 1986).

Conflicts Over Land

Lack of available land is often a constraint on community forestry projects. While a village may have a sizeable area of degraded grazing lands, individual members of the community will have different views on how to use it. This is the classic problem of the "commons" which recurs throughout tropical land-use questions (Hardin 1972). Some experts have come to regard it as intractable. "Proposals to use communal grazing land for trees will be perceived quite differently by those who presently use that land for grazing their livestock than by those who do not", says Michael Arnold of the Oxford Forestry Institute. "Such conflicts may be very difficult to resolve in communities which do not have homogeneity of ethnic, economic or social interest, or which lack social cohesion, or where there is lack of confidence in the community leadership" (Arnold 1984).

A Matter of Trust

Community forestry projects sometimes fail to live up to expectations because of the limitations of those responsible for implementing them. Often, the foresters deputed to organize such schemes have been trained originally in the technical aspects of forestry but not in dealing with people. However, even a well-trained social forester could fall victim to a long history of antipathy between foresters and villagers. Gerald Foley and Geoffrey Barnard argue that the most crucial factor in determining the success of a community forestry programme is the degree of trust between villagers and forest department representatives. In one village in Gujarat, for example, it took five years before full trust was established. Foley and Barnard ask whether forestry officers are best suited to community forestry schemes, bearing in mind that the popular perception of the forest department may not be a good one. Thus before a community forestry project began in Pakistan there were more than 50,000 cases of forestry offences pending in the local courts, with one family in six involved in a legal conflict with the forest department (Foley and Barnard 1984). Once such suspicions have finally been overcome it is vital not to rekindle them. When villagers in a number of West African countries did not receive the benefits to which they were entitled after woodlots had been harvested, it naturally became unlikely that they would participate in any future projects (Weber 1982).

Villages which already have successful communal farming

projects are more likely to establish communal woodlots, according to a study by Margaret Skutsch of Twente University, Enschede, of the participation (or lack of it) of eighteen villages in Tanzania's unsuccessful community forestry programme. She found that some of the usual explanations for poor performance, for example villagers' dislike of trees and lack of knowledge about tree planting, were not supported by the evidence, and the only significant contributing factors were the absolute quantity of fuelwood used per villager and their degree of trust in their own Village Councils (Skutsch 1985). Other studies have drawn attention to the fact that villages may be unwilling to participate in social forestry schemes if the tree-planting period coincides with the time when farmers need to plant food crops or leave the farm to work elsewhere for cash wages (Skutsch 1985; Foley and Barnard 1985; Hoskins 1982).

Is Fuelwood a Prime Need?

Village woodlot projects can fail for a much more fundamental reason: local people require other products far more urgently than fuelwood. "Fuel is seldom the only product [people] get from trees," says Michael Arnold of the Oxford Forestry Institute, "and often not the most important .... Community level [af]forestation therefore needs to provide for the multiple functions of trees and forests in the life of rural people" (Arnold 1984). "The fact that outside observers are able to calculate a shortfall in fuelwood supplies in a particular region", say Gerald Foley and Geoffrey Barnard, "does not mean that local people will be prepared to plant trees for fuel" (Foley and Barnard 1985). According to Henry Kernan and Tim Resch, previous projects have been primarily concerned with wood supply and the environment, and "although they are very important to society in general, they are not the primary concerns of the rural people upon whose interest and cooperation the success of the projects depends. These primary concerns are increased yields from crops and livestock. To have any significant effect on the wood supply or environment the forestry project must take the farm family as the basic unit and its welfare the basic objective". (Kernan and Resch 1984) Trees should therefore be able to enhance farming and yield products other than wood. The success of the Majjia Valley windbreak scheme in Niger (described in the next chapter), which relied only on farmers' voluntary labour, shows that community forestry can work when its support for agriculture is directly evident.

An even more radical explanation for the lack of interest in planting woodlots has been proposed by Gerald Foley of the Panos Institute in London. By means of theoretical models applied to a typical village in Mali and the capital city Bamako, Foley argues that talk of a fuelwood crisis at rural level may be mistaken and that the rise in urban fuelwood deficits may not be as rapid as some have suggested. Foley's point is that in the absence of adequate forest resource data, previous projections of fuelwood supply and demand have had to rely upon models that are in fact too simple to explain the real situation. In particular, they have ignored wood supplies from fallow lands. While Foley admits that it would be wrong to use these conclusions as a justification for complacency in tackling the problems of the Sahel, he states that: "There is no point in using scarce resources to solve energy problems which do not exist or are being dealt with quite adequately by local people." If Foley is right about the situation in rural areas, then he substantiates the view of Gorse and Floor quoted in Chapter 2 (p. 100). All models are necessarily approximations to reality, and his is no exception, but what is clearly needed is far more research into fuelwood consumption patterns and the resources available to satisfy them. If, as Foley states, "at present energy policy making in the woodfuel area is taking place without an adequate factual or analytic basis" (Foley 1987), then it is obviously misguided.

Time Delays Before Villagers Benefit

Even if a social forestry project can yield a desirable mix of products, the delay before harvesting them might be too long to satisfy the needs of local people. Projects should therefore be designed in such a way as to ensure that the short-term costs of enclosing village land while trees become established are kept as low as possible, and that short-term benefits, for example wages for planting and protecting trees, are maximized. Grass may be collected from inside the enclosure for fodder instead of animals grazing it as before. After a few years livestock can often be allowed to return to the plantation at certain times of the year when the trees are not at risk. The provision of paid, rather than voluntary, employment, with either wages or food for work, may be a significant contributing factor to the success of "directed" village forestry schemes. It could also explain the relative lack of success of truly community-based projects in which a major part of the costs is borne by the village in return for long-term benefits.

A Challenging Future

Community forestry projects are likely to become gradually more successful in the future, but only after social foresters become more skilled in their jobs, projects are designed to take more account of the needs of local people, and the communal activities of villages grow as part of the general social and economic development process. In the meantime, the continuation of other types of social forestry, including farm forestry, will increase general awareness of the value of tree planting. Eckholm, Foley, Barnard and Timberlake probably summed up the present state of community forestry best when they said: "The main lesson from the past is not that communal programmes should be abandoned; rather their design and implementation should be improved" (Eckholm *et al.* 1984). A major part of the problem lies in the social design of projects, and this can only be overcome by improved training of social foresters and detailed (but cost-sensitive) research and consultation prior to the start of projects. Existing communal institutions may not be suitable for initiating and managing a forestry project and it is possible that new institutions may need to be developed with the assistance of non-governmental organizations active in the area.

THE ROLE OF NON-GOVERNMENTAL ORGANIZATIONS

Non-governmental organizations (NGOs) already play a key role in rural tree planting and afforestation, and have the potential to become even more prominent in this field in the future. Also known as voluntary, non-profit or charitable organizations, NGOs are dedicated to a wide range of community service activities at home and overseas, taking advantage of voluntary help and donations from their supporters and the general public. They range in size from small church groups active only within a single town in a developing country to huge organizations like OXFAM that support development and humanitarian work all over the world. What NGOs lack in funds and technical knowledge they make up for in enthusiasm and grassroots contacts that sometimes result in quite extensive tree-planting campaigns (Grainger 1984b). A number of projects involving NGOs, such as CARE, have already been referred to. Although some NGOs have been established specifically to promote or undertake tree planting, often the most prominent organizations in this field are church groups, women's organizations and youth clubs for which tree planting is just one of a number of activities. Gujarat's forest department, for example, co-operated with a variety of NGOs in its social forestry programme (Karamchandani 1982). According to Fred Weber, the success of NGO afforestation schemes in the Sahel, compared with the poor performance of village woodlot schemes organized by government forest departments, suggests that foreign donors could make more effective use of their money by working through NGOs (Weber 1982).

The National Council of Women of Kenya (NCWK) has thirty-five affiliated women's organizations, and is a powerful example of what an NGO can achieve. In June 1977, it founded a national Green Belt Movement to help communities establish greenbelts of at least 1,000 trees on open spaces, school grounds and roadsides. The first greenbelt was planted at the time of UNCOD, in September 1977, but the number of greenbelts has now risen to more than 1,000. This does not include the many smaller greenbelts which farmers have been inspired to plant on their own properties. Communities who ask to join the Green Belt Movement form a committee of local leaders which chooses the site and is responsible for organizing the planting and management. In the early days all seedlings were supplied free of charge by the Ministry of Natural Resources, but demand outstripped supply and member groups had to establish sixty-five new nurseries to increase seedling production. The practical job of tending the greenbelts and supervising the nurseries is undertaken by Green Belt Rangers. The NCWK raises money for its work from member groups, local companies and individuals all over the world (Harrison 1986; Mathai 1985; Anon. 1982).

Until recently the actions of NGOs in this field, while not unnoticed, were mainly unsupported and undocumented. Information on the extent of afforestation work by NGOs based in developing nations is therefore very sparse, even though in some countries it is known to exceed that performed by government forestry departments. In Senegal, for example, 5,100 ha of trees were planted by community efforts in 1983 compared with 4,700 ha planted by the government forest service (Sene 1984). The involvement of NGOs in tree planting was given a fresh focus in 1985 when, as part of International Youth Year, the Tree Project was established at the United Nations under the auspices of the UN Liaison Service for Non-Governmental Organizations (NGLS). The International Tree Project Clearinghouse was set up to collect and disseminate information on NGO tree-planting activities, and later published the first detailed directory of African NGOs involved in this field (NGLS 1986).

The Tree Project's Directory for Africa listed 211 NGOs (182 if the numerous overseas branches of NGOs based in developed nations are omitted) (Melamed-Gonzalez and Giasson 1987). Twenty-one NGOs indigenous to dryland countries and for which specific data are available are shown to have planted more than 5.5 million trees. Some organizations have planted as few as 500 trees, and others as many as 1 million. Annual planting rates also vary widely: some plant only 200 trees a year, others 20,000 or more. In many respects, numbers are much less important than the impact on local lands and communities, which can be quite considerable. India is another country with a wealth of NGOs working in rural development, although of at least ninety-three such organizations only a small proportion, like the Millions of Trees Clubs and the Ranchi Consortium for Community Forestry, are active in planting trees (Basu 1984).

Indigenous NGOs often receive funding for their work from NGOs based in developed nations. The contribution of these NGOs to development assistance as a whole is quite substantial – estimated at one-tenth of the development funds given by the governments of donor nations on a bilateral or multilateral basis in 1980 (OECD 1981; World Bank 1981). The number of NGOs based in developed nations which have a significant forestry component in their programmes is still small, mainly due to a lack of awareness of the need for action in this field. In a survey of 1,700 NGOs by the Organization for Economic Co-operation and Development (OECD), only seven indicated that they were involved in forestry (OECD 1981), although the actual number is probably at least twenty-five (Grainger 1984b). Several examples of projects funded or implemented by NGOs based in developed nations are described in this book.

NGO involvement can do much to help to motivate local people, but the success of projects may still be affected by organizational, technical and financial constraints. There is often poor co-ordination between the NGOs on the one hand, and governments and intergovernmental organizations like FAO on the other. This can mean that NGOs are not consulted about or involved in the planning of national and international programmes, and receive only limited access to necessary technical information. The collection of international statistics on their activities is also given a low priority (Grainger 1984b).

NGOs also encounter considerable technical constraints on their activities in afforestation. Thus, an NGO might be enthusiastic about a project, and succeed in motivating local people to participate in it, but its successful completion could be threatened if NGO staff at headquarters and at project level have insufficient technical knowledge and receive inadequate advice from consultants. Many projects are not designed by trained foresters, and the consultants who are brought in at the planning stage may not be available to deal with problems which arise during implementation. Much that is currently being undertaken in agroforestry and social forestry is very experi-mental, because applied research in these fields is poorly funded, and so the activities of NGOs suffer accordingly. Poor-quality seed is another handicap. NGOs are often dependent for seed upon national forest departments which are mainly experienced in obtaining seeds of industrial wood species. Seed of multi-purpose species therefore has to be imported. It tends to be expensive and of unreliable quality, and if obtained through informal channels may not be subject to the kind of quality control that is desirable.

The role of NGOs in afforestation will become increasingly significant in the coming years and it is therefore vital to take steps to ensure that their activities are as effective and successful as possible. Those NGOs currently active in this field should do much more to publicize their work, since this will help to convince other NGOs of the need to include forestry in their programmes, and will also increase the flow of funds from the general public for tree planting. Better co-ordination between NGOs, national forest departments and intergovernmental agencies like the FAO is essential if NGOs are to be more closely involved in planning and policy formulation and are to receive more technical information. More co-operation between NGOs themselves is also needed: this would allow the sharing of experiences, the production of more comprehensive information on their forestry activities, and an improvement in technical support and seed supplies. NGOs would benefit from better links with forestry research and educational institutions, which could, for example, hold special training courses for NGO personnel, second staff and students to NGO projects, and receive funding from NGOs for appropriate research projects (Grainger 1984b).

There are already some encouraging examples of NGO coordination. In 1981, NGOs in Kenya concerned with renewable energy and community development issues, formed a network called KENGO (Kenyan Energy Non-Governmental Organizations) which assists member NGOs with technical matters, training and information, and provides a mechanism for members to exchange information and experiences among themselves. KENGO has over 130 local member bodies, issues a newsletter, and has sponsored the publication of a field manual listing sources of tree seeds in the country. Senegal has a similar organization, called CONGAD (Conseil des Organisations Non-Gouvernementales Appui au Développement) to coordinate NGOs concerned with national development issues in that country (Williams 1985b). Finally, the African NGO Environment Network (ANEN) was formed in 1985, with the assistance of UNEP, to link NGOs throughout Africa concerned with environmental issues. It currently has 230 affiliated NGOs.

IMPROVED STOVES

Tackling fuelwood deficits involves more than planting trees to increase wood supplies. Many families cook their food on inefficient open fires or simple stoves; consequently, since the mid-1970s a variety of projects, sponsored by both large aid agencies and smaller NGOs, have tried to persuade villagers to use more fuel-efficient stoves that in theory could burn up to 50% less wood. The wider use of such stoves could reduce deforestation and make the most effective use of wood produced by fuelwood plantations. However, the majority of these projects have not been very successful because they did not take into account the needs of the people who were to use the stoves, which were usually abandoned after a subsidized project ended.

The Improved Stoves Project in Niger

Recent projects offer some hope that lessons have been learned from past experience and that a more effective approach to stove design and dissemination has been found. The Improved Stoves Project in Niger takes a novel approach to solving Africa's energy crisis, based on the belief that it has mainly been caused by the growth in demand by urban centres. This leads to what project co-ordinator Willem Floor of the World Bank has called "urban shadow" – ever-widening circles of degradation around cities after trees have been cut down for fuel.

A two-pronged approach is used in the project: first, changing demand by encouraging the adoption of improved stoves while at the same time promoting substitute fuels such as kerosene; and second, changing the supply side by raising the price of wood. Low fuelwood prices encourage growth in demand, inhibit the protection of remaining fuelwood resources and the establishment of new tree plantations, and give people no incentive to buy more efficient stoves. So the project increases prices by levying a tax on all fuelwood passing through the gates of Niamey, the capital of Niger. Part of the revenue raised goes back directly to rural people, who are given the sole right to harvest fuelwood in their area. They bring it to "bush markets" to sell to entrepreneurs who then market it in the city. This system helps to prevent the wholesale felling of trees over large areas by gangs paid by city-based dealers.

The project is funded jointly by UNDP, the World Bank and the German and Dutch governments. It is only two years old but has already distributed 40,000 stoves – twice its target number – in urban areas. The stove, called the Mai Suki, is small, portable, made out of scrap metal, and costs about $3, three times as much as the traditional, less fuel-efficient variety. It is made by a number of local manufacturers to a standard design adapted by the Association Française de Volontaires du Progrès (AFVP) from another improved stove being promoted in Burkina Faso. The Mai Suki gives the same 50% saving in fuelwood achieved by the earlier design, but is more stable and has a lower production cost because it is made from scrap metal. The promise of higher profits and the absence of royalty payments give metalsmiths an incentive to switch from making traditional metal stoves. After attending a training course they are officially licensed to make the Mai Suki and receive the standard templates. Quality is maintained by stove inspectors who test each stove and give it a seal of approval if satisfactory. Special sales points are set up initially by the project but most stoves are later sold through traditional outlets (UNDP/World Bank 1986a; Floor, personal communication).

Advertising the stove to the general public and to potential manufacturers accounted for about 80% of the $500,000 project budget during the first two years. The first priority is to publicize the advantages of the stove and the reasons why people should pay three times as much for a stove which is twice as efficient as the traditional model. Then women need to be shown how to use the stove correctly, with the right size kettle or pot, cleaning the grate to improve airflow and efficiency, and only using two sticks of wood in the stoves instead of the traditional four. Two teams tour Niamey neighbourhoods in a Volkswagen bus, giving cooking demonstrations in which they compare the improved and traditional stoves. They announce their presence with loudspeakers and drummers and the prior distribution of posters and pamphlets.

A popular song composed specially for the project is played on radio and can be bought on cassette tape. Special short promotional "commercials" are shown regularly on television, and T-shirts printed in French and the two local languages display the messages: "With my Mai Suki I save on wood", "A modern stove – a friend of trees", and "With a modern stove, no more waste of wood". Restaurants are highly visible to the general public, so the project persuaded many of them to buy the new stove. As an indication of the enthusiasm generated by the project, 5,000 people each paid a hundred francs (CFA 100) to attend an "Improved Stoves Festival" including sketches and dances in Niamey in November 1986 while another 4,000 filled an overflow arena.

Only time will tell whether the project is a real success; that depends on whether local businesses continue to sell the improved stoves after the project has ended. So far, the signs are good. A representative sample of 1% of those who bought the stoves indicated that 94% of the people are still using them – and using them properly. Preliminary evidence from one town points to an average energy saving of 28%. The project is now expanding to include other towns in Niger and could well serve as a model for projects elsewhere in the region. Two more projects are already under way in Senegal and Burkina Faso (UNDP/World Bank 1986a; Floor, personal communication).

Promoting More Efficient Stoves in Gambia

Another recent project in Gambia, funded by UNSO, is training people to produce three different types of stoves: a stove made from scrap metal; a cheaper portable pottery stove; and a large mud stove made from clay combined with other ingredients like cow dung and straw which, while not portable, can be made by villagers themselves at low cost. The metal stove sells at $4 for the medium size and $5 for the large size. Early stove designs were tested first at the UK laboratories of the Intermediate Technology Development Group, and then in family compounds in Gambia, to see how they responded to the needs of "real" people, and to identify problems arising during use which had not been anticipated in the design process. Manufacture and distribution of the stoves started in September 1985, but adoption of the new stoves has been rapid and by October 1986, 10,000 metal stoves had been produced for use in the Greater Banjul area alone, where a survey of over 800 households in May 1986 found that almost three-quarters possessed metal stoves. There are twenty workshops producing the stove in Gambia and its use is spreading.

The three stoves have short lives. The pottery stove lasts for just one or two years but is cheap to replace. The scrap metal stove has a similar life and while using imported metal can extend it to three or four years, this makes the product far more expensive. The mud stove, in contrast, lasts for a mere 1–1.5

years, and only reduces wood use by 30%, compared with up to 40% for the metal stove and up to 50% for the pottery stove. In practice its actual efficiency is often lower. Many mud stoves were made during training programmes but, because they take two weeks to build, villagers are unlikely to go on making them after the project has ended. Mud stoves are not portable, which is a disadvantage since certain meals in rural areas are normally cooked outside the house and others inside.

Gambia does not suffer from severe fuelwood shortages, but to prevent the rapid growth in fuelwood consumption that has led to destruction of forest resources in other African countries the project has a component to encourage the use of fuelwood substitutes, particularly briquettes made from groundnut shells, of which the country has a large surplus. Briquettes are suitable for both household use and for local industries like breweries. The project has developed a portable metal stove that burns both wood and briquettes efficiently, and a plant established in Kaur with Danish funding could produce enough briquettes to supply half of Banjul's needs for domestic cooking fuel (Bennett 1984; Joseph and Loose 1983; UNSO, personal communication).

The Potential for More Efficient Charcoal Stoves

Increasing attention is also being paid to improving the efficiency of charcoal stoves. The rapid growth in consumption of wood-based fuels (particularly charcoal) in urban areas presents the biggest threat to dryland forest resources, so improving the efficiency of charcoal use could have a significant impact on future rates of deforestation and desertification. Most of the 30 million charcoal stoves used worldwide are poorly designed and wasteful of wood according to a study by UNDP and the World Bank. The problem is not the efficiency with which water is boiled, but that after boiling it is difficult to reduce the heat from the stove sufficiently to simmer the water for cooking (UNDP/World Bank 1986b).

Tree Planting or Improved Stoves?

An interesting policy question for governments and development agencies is whether planting trees or promoting improved stoves is the best way to sustain forest resources under pressure from fuelwood cutting. US forestry consultant Fred Weber has calculated that, if 50% of all deforestation in dryland areas is caused by overcutting for fuelwood and 85% of fuelwood is burnt for cooking, then the introduction of a new stove which reduces fuelwood consumption by 30% and is used by half of all households in the country to cook 70% of all their meals will only reduce deforestation rates by 4.4% (Weber 1982). If this is correct, then improved stoves are not a panacea, and tree planting and the use of substitute fuels such as kerosene are still necessary to reduce the pressures leading to deforestation.

FUTURE TREE-PLANTING STRATEGIES

Experience so far suggests that large intensive plantations established and managed by forest departments are not the answer to shortages of fuelwood and other types of wood in the drylands. The productivity of rainfed plantations may be reduced by poor soils, and returns do not seem to justify investment costs, except in areas receiving at least 800 mm rainfall per annum. Indeed, one estimate, by forester Kjell Christopherson (Weber 1982), is that growth rates of at least 12 cubic metres per hectare per year are needed just for projects to break even in the Sahel. This is far higher than the yields actually achieved hitherto. Irrigated plantations are more productive, but still may not produce sufficient returns to justify the much higher investment required. Both rainfed and irrigated plantations will suffer from low market prices for fuelwood as long as merchants can cut trees without charge in natural woodlands. At present, social forestry seems to work when trees are planted on community lands by the forestry department ("directed" social forestry), but not when villagers are expected to do all the work themselves. Farmers in India and elsewhere, on the other hand, seem to have taken very readily to establishing woodlots on their lands.

Future tree-planting strategies should therefore be modified to take this experience into account. Large plantations, especially near to cities, will probably still be needed to some extent, but a World Bank assessment of forestry in the Sahelian zone has concluded that, "it is logistically impossible for government plantations to be established on a scale large enough to ensure future fuelwood requirements". "Directed" social forestry planting will continue, and the benefits from such projects will give local people greater confidence in applying for self-help schemes. Forestry departments may find that they could achieve much higher planting rates by working through non-governmental organizations than they would by undertaking all projects themselves. However, if true community forestry is to become popular, it will require much more detailed study of the constraints facing it and how they may be overcome, and of the kinds of products which would make community forestry desirable to villagers (World Bank 1986b). Although village plantations might be more acceptable if they consisted of a variety of fruit, fodder, fuel and timber trees, preferably with as many multi-purpose trees as possible, an alternative strategy to guarantee an attractive diversity of produce might be to manage local natural woodlands instead of establishing new plantations. The possibilities for doing this will be discussed, together with the potential for planting more trees on private farmlands, in the next chapter.

8. Farm Forestry and Natural Woodland Management

Increasing the tree cover in dryland areas will help to control desertification by protecting soil from erosion and reducing the pressure on existing forest resources for fuelwood and fodder, but in most countries there are simply not enough personnel in government forestry departments to undertake all the afforestation which is desirable; hence the need for social forestry schemes that involve varying degrees of participation by local people. The last chapter showed that communal tree planting (true community forestry) has so far met with only limited success. The most successful kind of social forestry has actually been farm forestry, the establishment of plantations on private farmlands for profit. Farm forestry is but one of a number of agroforestry techniques which combine agriculture and forestry, and make drylands more productive and less vulnerable to degradation. The first part of this chapter describes progress made in farm forestry, and puts it in the wider context of other agroforestry techniques, such as silvop-asture (mixed grazing and forestry). Instead of tree planting, another strategy may be to prevent deforestation from ever taking place. The second part of the chapter, therefore, looks at recent experiments in managing natural woodlands in dry areas, previously considered to be much less productive than plantations. Natural woodlands might actually supply just as much wood as plantations, and many other useful products besides, so it is possible that they will receive much greater support from local people than plantations which only produce fuelwood.

FARM FORESTRY

Farmers in India and other dryland countries are becoming increasingly interested in growing trees on farm plantations. In the Indian state of Gujarat both farm forestry and community forestry were promoted as part of the wider social forestry programme beginning in 1970 (see Chapter 7), but farm forestry has been by far the more successful of the two. Khalidas Patel used to grow cotton and tobacco on his 79 ha farm near Ahmedabad, but agreed to a forest department suggestion that he use some of their *Eucalyptus* seedlings to establish a trial woodlot. The first trees were ready for felling within five years of planting, and their rate of growth so impressed Mr Patel that he converted over 80% of his property into a tree farm. The power of his example has been astonishing. By 1977, eight-six farmers in the state had asked the forest department for seedlings and technical advice on how to establish their own plantations. In 1978 the number rose to 300. Gujarat then asked for a $37 million World Bank loan to expand its farm forestry programme. The number of trees planted more than doubled from 48 million to 100 million between 1979 and 1981, and nearly doubled again to 195 million by 1983, by which time at least 10% of Gujarat's farmers had participated in the scheme and over 150,000 ha of trees had been planted (Spears 1985).

This programme is not without its critics. Some argue that farm forestry benefits only the richer farmers. The Centre for Science and Environment, New Delhi, has alleged that 10,000 farmers in one Gujarat district converted prime, irrigated agricultural land to *Eucalyptus* plantations to grow wood mainly for pulp and rayon production and for poles to be sold in cities, rather than for fuelwood to benefit the rural poor. The Centre also argues that tree plantations require little labour and that as a consequence the poor are thrown out of work (Anon. 1982). While the programme did begin by concentrating on large farmers, their success inspired many other farmers to follow their example, and of all seedlings distributed in 1982, 19% went to farmers owning more than 4 ha, 44% to farmers with 2–4 ha, and 37% to farmers owning less than 2 ha. To spread participation even further the programme has a separate scheme targeted specifically at the poorest farmers.

Another criticism concerns the value of *Eucalyptus* trees. The Centre for Science and Environment quotes one critic who describes *Eucalyptus* as an "ecological terrorist" which harms the soil without providing fuel, fodder, green mulch or shade

(Anon. 1982). Mr S. A. Shah, one of the pioneers of social forestry in Gujarat, now retired from the Indian Forest Service, has defended the use of *Eucalyptus* on community lands in an advanced stage of erosion on the grounds of its rapid establishment and growth, and resistance to browsing: "I know of no indigenous tree with comparable survival and growth mechanisms!" says Shah. He argues that so far there is little scientific evidence to link *Eucalyptus* with the contamination and drying of soil and consequent desertification. On the contrary, without the use of fast-growing *Eucalyptus* plantations the harvesting of fuelwood and industrial wood would place such a heavy pressure on India's remaining natural forests that the result would be severe environmental degradation. On the use of *Eucalyptus* in farm forestry, Shah asks: "Traditionally farmers in India have grown crops which are the most profitable to them. So far nobody has raised even an eyelid against this practice. Why then against *Eucalyptus*?" He then points to the crucial role of *Eucalyptus* in encouraging popular afforestation: "The most difficult task in India today has been to make people plant and care for trees. Now that people have taken to planting trees, nothing would be worse and antinational than to create a climate against planting *Eucalyptus*" (Shah 1985).

Farm forestry has two other advantages over large plantations. First, it is more profitable because costs are lower, typically $250 per hectare for establishment and maintenance over one rotation, compared with $800–1,300 for a large plantation. Second, higher yields can be obtained by pollarding or coppicing trees to maximize the production of total wood volume, instead of only stem volume as in normal forestry practice. (Coppicing differs from clear felling because trees are cut near their bases so that one or more shoots can spring up from the stool and later be harvested as poles. In pollarding the cut is made higher up the tree.) It has been estimated that if pollarded, thirty neem trees could supply an average family's wood and fuelwood needs in Niger on a sustainable basis, but ten times as many trees would be needed if the plantation were managed in a conventional way (Fishwick 1965; Spears 1983).

AGROFORESTRY

Farm forestry is but one of many agroforestry techniques that combine agriculture and forestry on the same plot or on adjacent plots of land. Persuading farmers to stop growing crops or raising livestock in areas subject to desertification would be almost impossible because it would threaten their livelihood, but agroforestry can restore tree cover while agriculture continues. Farmers soon begin to receive indirect benefits in the form of higher crop yields and reduced soil erosion because of the protective effects of trees. In a few more years a useful crop of fodder, fuelwood and poles can be harvested from the trees. High priority has been given to agroforestry and social forestry since major policy shifts in 1978 at FAO and the World Bank (FAO 1978b; World Bank 1978). In the same year the International Council for Research in Agroforestry (ICRAF), now based in Nairobi, was established to promote and co-ordinate research in agroforestry.

Agroforestry is a new word but not a new practice, having been part of traditional agricultural systems for many years. For example, the role of *Acacia senegal* (the source of gum arabic) as fallow in a rotation system including millet and sesame in the Kordofan region of Sudan has been mentioned in Chapter 2. In West African countries such as Senegal, Mali, Burkina Faso and Niger, *Acacia albida* is grown with millet, adding nitrogen and organic matter to the soil and providing a valuable source of forage in the dry season. A similar system using *Prosopis cineraria* instead of *Acacia albida* is found in the Indian states of Rajasthan, Gujarat and Haryana and in Pakistan's Sind province (FAO 1985b).

Multi-Purpose Trees

At the heart of any dryland agroforestry schemes are multi-purpose trees, for example species of the genera *Acacia*, *Leucaena* and *Prosopis*, which can be grown quickly on poor soils to yield food, fodder, fuelwood, building timber and other products. Interest in multi-purpose trees has been increasing since the mid-1970s, with the encouragement of bodies like FAO and the US National Academy of Sciences.

The multi-purpose tree *Acacia albida* (also called *Faidherbia albida*) is intercropped with sorghum, millet and other field crops in semi-arid regions of Africa and is unusual in that it comes into leaf at the end of the rainy season and maintains its foliage during the hot dry season. This has two advantages. First, field crops growing beneath it are not hindered by shade in the rainy season. Second, its foliage shades the soil and also acts as a valuable fodder reserve in the dry season. The tree also fertilizes the soil by increasing levels of nitrogen and phosphorus: crop yields beneath the tree are as high as in fertilized fields. In Senegal yields of millet near *Acacia albida* trees have been shown to be two and a half times the yields in open fields. Cattle benefit from the foliage and from the large crop of pods full of protein and carbohydrate. At Zalingei in Sudan each tree was found to produce an average of 135 kg of pods per annum. In the village of Lama-Lama in Chad's Bongor region, where *Acacia albida* grows naturally on farms, farmers keep their cattle in a compound during the rainy season and feed them the fruits harvested from the tree. They then collect manure from the compounds and deposit it on their fields as fertilizer.

Recognition of the value of *Acacia albida* is shown by a remarkable example of tree protection in Niger, reported by Michael McGahuey: "Protecting trees was provided in a ... draconian fashion by a Sultan of Zinder about a hundred years ago. By his edict, anyone caught cutting an *Acacia albida* tree would lose his head. A possible legacy of this measure is the large number of farm fields with field trees in the Zinder Department of the Republic of Niger" (McGahuey 1985). The US voluntary agency CARE has been assisting farmers in Chad to plant *Acacia albida* in fields of cereal crops since 1975, although the project was interrupted by the civil war between 1979 and 1982. Funding for the project comes from USAID. Survival rates as high as 70% in some areas were achieved by the dedicated care of local officials and workers. Elsewhere only 10% of trees survived because many fields were abandoned during the fighting and trees fell prey to bush fires and browsing by livestock. The growth rates of *Acacia albida* have been impressive, and in the Guelendeng region the trees grew as tall in eight years as the more conventional plantation tree *Eucalyptus camaldulensis*, and had larger boles. One of the disadvantages of *Acacia albida* is that farmers do not start to receive benefits until eight years after planting and so are sometimes unwilling to plant the tree unless they receive government incentives (McGahuey 1985).

In India, *Prosopis juliflora* plays a similar role to that of *Acacia albida* in Africa. It is one of the species of the leguminous genus *Prosopis*, most of which fix nitrogen in the soil (Pedersen 1980). *Prosopis* trees are very drought-resistant, having both deep and shallow roots that can collect water from different levels. They also tolerate soil salinity to varying degrees, thus *Prosopis juliflora* will grow in soil with a pH of 9.5–10.0 and a soluble salt content of 0.54–1.0%. The most remarkable member of the genus is *Prosopis tamarugo*, which grows in Chile in soils covered by a layer of salt 1 metre thick (Pedersen 1980). It has been suggested that the tree can take in water from misty air through its leaves in the extremely arid desert areas on the coast of Chile (Sudzuki 1969). The importance of *Prosopis tamarugo* was underlined by the holding of an international conference devoted entirely to it in Chile in 1984 (FAO 1985d). Some *Prosopis* species spread as weeds and secrete phytotoxins that prevent the establishment of other vegetation nearby but this is not a general characteristic of the genus. The use of pods and leaves as animal fodder is generally advantageous as long as they are part of a varied diet.

Multi-purpose trees also find use as "living fences". These quickly establish themselves, either from small cuttings stuck in the ground or from planted seedlings, and are more effective than fences made out of dead thorn tree branches which require many days to construct and repair each year. At Darda in Chad, 60 km south of N'Djamena, living fences of *Prosopis juliflora* were established between 1976 and 1978 to protect sorghum fields. They provide fruit and fodder from the leaves as early as the fourth year after planting and are also harvested for poles and fuelwood.

Windbreaks

Windbreak design

Windbreaks have proved to be a popular type of agroforestry because they protect soil, crops and livestock from hot desiccating winds. Windbreaks reduce wind speed by 50% or more to a distance of 5–10 times the height of the windbreak (commonly abbreviated as 5–10H) and by 20% or more to a distance of 12–20H. Soil is only lifted up from the ground and blown away when winds exceed a certain threshold speed, so windbreaks which reduce wind speed below this level can reduce soil erosion (World Bank 1986a). Experiments in India have shown that cropped fields lost 72 tonnes/ha in just twenty days during April; the establishment of multi-row windbreaks cut the wind speed by 40% and reduced soil loss to just 3 tonnes/ha (Sur 1986). Long multi-row windbreaks are often called shelterbelts but for convenience the term windbreak will be used generally in this section to refer to both types.

Crops grown in fields sheltered by windbreaks give higher yields than unsheltered crops. Windbreaks protect crops from mechanical damage by winds, for example the flattening of crops, the blasting of foliage by the sand particles carried by the wind, and the stress due to the high transpiration rates which are induced. Windbreaks also improve the general conditions for plant growth, and in particular lead to higher soil moisture and ambient temperatures. The surfaces of sheltered crop plants are less subject to cracking by winds and so lose water less easily. The lower wind speed can reduce plant water loss (evapotranspiration) by up to a third at distances of 3–5H and by 8% even at 20H (Grace 1986). Windbreaks also reduce the evaporation of water from the soil surface and therefore increase soil moisture. On the other hand, temperatures are higher in the shelter of the trees. For example, in Sudan temperatures were found to be typically 2°C higher for distances of up to 12H from the windbreak (World Bank 1986a), and this promotes the evaporation of water from the soil. However, the overall effect of windbreaks is to increase soil moisture.

The design of windbreaks is all-important if they are to give optimum benefits to crops and soils. Effective protection extends for a distance of up to 15–20 times the height of the trees, depending upon the species used. The aim of introducing windbreaks is only to reduce the speed of the wind; blocking it out entirely would lead to eddies that would be just as damaging to crops as if the windbreaks were not there at all. Thus the spacing of trees, the way they are pruned, and the types of shrubs and grasses planted between them determine the proportion of wind which passes through the windbreak; this is usually recommended to be of the order of 40–50% (Kort 1986).

Some farmers initially object to windbreaks because they take up space that could be used for growing crops, and because crop yields close to the windbreaks are lower due to shade and competition for water. Such effects can extend for distances of up to three times the height of the trees. Still, overall the use of windbreaks will give net benefits to farmers. It was estimated, for example, that using 8,000 km of three-row windbreaks to protect 560,000 ha of cotton in the large Gezira Irrigation Scheme in Sudan would save 380 million cubic metres of water, enough to irrigate both the windbreaks and an extra 33,600 ha of cotton (World Bank 1986a). Windbreaks can therefore make money for farmers. An FAO study has shown that, excluding the cost of land and any income from the sale of fuelwood and other products, a 10% increase in crop yields results in positive financial rates of return on investments in windbreaks (World Bank 1986a).

Farm windbreaks

The first large-scale demonstration of the value of farm windbreaks came in the 1930s when, following the tragic soil erosion and crop losses which turned the Great Plains of the USA into the infamous Dust Bowl, almost 30,000 km of windbreaks comprising 217 million trees were planted in the states of North Dakota, South Dakota, Nebraska, Kansas, Oklahoma and Texas between 1935 and 1942. Interest in windbreaks in the USA has waned somewhat in recent years, as many farmers began to think of tree planting as an "old technology" made redundant by the use of irrigation (Arnold 1986). Fears that farmers were removing windbreaks led to a survey of the five Great Plains states by the US General Accounting Office, which reported that there had been a trend to reduce the width of windbreaks and that 1,857 km of windbreaks were removed between 1970 and 1975 (US Soil Conservation Service 1980). Nevertheless, the length of new windbreaks exceeded this loss by 1,098 km. Indeed, since 1942 an average of 3,220 km of new windbreaks has been established each year compared with 3,743 km per annum in the initial period from 1935 to 1942.

Windbreaks are now being planted to protect farmlands throughout the developing world. In India's major arid state of Rajasthan an estimated 1,500 km of roadside windbreaks were planted between 1983 and 1988, and they consisted mainly of tree species which could be harvested for fodder and fuelwood. The normal practice was to plant three rows of trees on either side of a road, following an aerodynamic design calculated to achieve maximum benefits. Tall trees, like *Azadirachta indica* (neem), *Albizia lebbek, Acacia tortilis, Cassia siamea* and various species of *Eucalyptus* were planted nearest to the road; species of *Acacia* and *Tamarix* were in the second row, and trees like *Acacia nilotica, Prosopis juliflora, Parkinsonia aculeata* and *Zizyphus mauritiana* were used for the third and smallest row next to the fields. This double wedge-shaped design gently lifts the wind away from the field without causing harmful eddies.

Neem (*Azadirachta indica*) is an excellent windbreak tree, growing well in areas with low rainfall and poor soils. The tree can be harvested for firewood, and since its leaves are not highly palatable to animals it is less likely to be browsed while becoming established. Another protective device is that the seeds and leaves contain a natural pesticide said to be effective against more than 100 species of insects, mites and nematodes. In India stored grains are actually protected from insect infestation after harvest by being mixed with neem seeds and leaves, and a water emulsion made from the neem seeds can be sprayed on to plants as an insecticide (Benge 1987).

Neem has become very popular in Africa and was the main tree species used in Niger's Majjia Valley windbreak scheme. Funded by the US voluntary agency CARE and USAID, this scheme established 325 km of windbreaks between 1975 and 1985 in an area near Bouza (500 km northeast of the capital Niamey) where the annual rainfall is 350–400 mm. The fertile valley (the river only flows after the monsoon rains) is home for about 33,000 people in twenty-seven villages, who grow millet, sorghum and other crops, and raise livestock. However, the farms are at the mercy of the fierce *Harmattan* which causes soil erosion and crop damage. Paula Williams has described how:

> From the plateau to the north of the velley, the Majjia looks bleak and desolate, with few trees in evidence. ...
> The strong winds that blow through the Majjia threaten the villagers' livelihood. During the long dry season, from November to May, little vegetation covers the ground: the *Harmattan* blows almost incessantly, carrying away valuable topsoil. During the rainy agricultural months, the wind continues to blow, drying out young sorghum and millet plants. (Williams 1985a)

The windbreak scheme was conceived by a local forester, Daouda Adamou, and a Peace Corps forester, Don Atkinson

Adams, as a way of protecting humans, animals and crops from these fierce winds. Daouda had already won the trust of the local people, whom he had previously encouraged to plant private woodlots for pole production, so he managed to persuade them to work together to plant windbreaks that would benefit all their crops. Co-operation was vital, not only for planting the trees, but because the long straight lines of windbreaks would cut across the lands of many smallholders, to the detriment of some more than others. The windbreaks are up to 2 km in length, and consist of trees spaced 5 metres apart with 100 metres between windbreaks (Persaud *et al.* 1986). The windbreaks were constructed from both ends – like a typical alpine tunnel – starting in the north of the valley near the three villages of Garadoumé and in the south from Taboé, some 20 km away. Apart from neem, other species used in the windbreaks included *Acacia seyal*, *Acacia scorpioides*, *Prosopis chilensis*, *Prosopis juliflora* and *Eucalyptus camaldulensis*.

Local people are already reaping benefits from the windbreaks. They appreciate the lower wind speeds and the higher millet yields. Some estimates have suggested a mean rise of almost a quarter in the average millet yield per hectare (Thomson and Hoskins 1984; Delehanty *et al.* 1985; Long 1985; Dennison 1986). (Over the whole project area the difference between mean yields with and without windbreaks was not statistically significant, because of the wide variation in yields on different farms, the small number of farms sampled, and the large fluctuations in rainfall in recent years. Nevertheless, the increased yield was sufficiently large to indicate a net positive benefit from planting the windbreaks.) Only a third of the farmers questioned in a survey thought that the scheme had any disadvantages; among those mentioned were the reduction in the area of cropland and the fact that the trees attracted birds (Delehanty *et al.* 1985). Perhaps the people most disadvantaged by the windbreaks are the pastoralists whose animals were banned from the planted areas for three years. They are allowed to collect crop residues and grasses from the protected area, but can be given heavy fines if their animals are caught there (Thomson and Hoskins 1984).

Majjia Valley is undoubtedly one of the most successful afforestation projects in West Africa, but can its success be sustained? The long-term future of the project depends upon the continued support of local people. Although the recent survey indicated extremely positive attitudes towards the project, this was apparently not the case at the beginning, and some observers have criticized the project staff for not consulting villagers sufficiently before the project began (Thomson and Hoskins 1984). Nevertheless, it did provide a solution to a deeply felt local problem, the villagers trusted Daouda, and he had the backing of local officials. The project started small and did not expand until perceived benefits had been obtained. Indeed, it seems unlikely that the project could have been implemented unless local people were strongly supportive of it. The project could still be vulnerable to whatever bad feelings remain, especially on the part of pastoralists, and so it is vital to increase the participation of local people in future extensions of the scheme and to widen the range of benefits, for example by planting more fruit trees.

The first crucial test for the project came in 1985, when it was necessary to decide how to distribute wood from the first controlled harvesting of the windbreak trees. With careful management many different products can be harvested from windbreaks without diminishing their effectiveness, so they benefit the local economy directly as well as indirectly (in the form of shade and protection). US forestry consultant Fred Weber estimated early in the project that 250 km of windbreaks could produce 2,500 cubic metres of fuelwood every year with a local value of about $80,000. However, most people still did not believe assurances by project staff that they owned the wood harvested from the trees. So it was important that, after the first experimental harvesting of wood at Garadoumé under the direction of CARE foresters, the villagers were given the wood free of charge and the responsibility for determining how it would be distributed. Each farmer eventually received about $15 per hectare from the sale of wood harvested from the windbreaks, and this radically changed local perceptions on windbreak ownership (Delehanty et al. 1985).

In the course of the project, studies were carried out to determine the best way of harvesting fuelwood from the windbreaks so as to maximize fuelwood production without impairing the protective effect of the trees (Persaud et al. 1986). The preferred method was partial pollarding, only those branches overhanging crops being removed from both rows of trees in the windbreaks. This maintained at least a 30% reduction in wind speed for most wind directions. Less effective was coppicing, or cutting back the entire tree to a stem 2.5 metres high. Complete coppicing of only one of the two rows or one in every four trees along two rows gave reductions in wind speed of up to 20%, while complete coppicing of both rows of trees severely reduced wind protection for one year.

The next hurdle which project staff have to overcome is to find a way for the project to become self-supporting after CARE financing has been removed. To achieve this, some of the revenue from the sale of harvested products will have to be used to pay for the management of nurseries so that new seedlings can be raised to replace old windbreak trees, and for the wages of the guards needed to prevent damage to the windbreaks by animals and humans. One possibility being investigated is the formation of a farmers' co-operative.

Farming in the Majjia Valley is also threatened by water erosion, following the deforestation of watersheds for cropping, livestock raising and fuelwood cutting. Streams now flood down from the hills into the valley, forming gullies, deepening the bed of the seasonal river so that its waters do not spread out as widely as before, and reducing soil moisture (and thereby crop yields) in some fields. Also possible is a fall in the valley water table. This would threaten not only farming but also the successful establishment of future windbreaks. CARE has therefore now expanded the project, planting trees and building terraces and bunds on the valley sides to control this additional source of soil erosion.

Town greenbelts

The other type of windbreak project that has proved popular in Africa is the planting of a greenbelt of trees to protect towns and cities from wind and dust. Mauritania is a country the size of the US states of Texas and California combined, and 70% of its territory consists of Sahara Desert. By the mid-1970s Nouakchott, the capital, was afflicted by sand storms and threatened by sand dunes encroaching on its outskirts. A ring several kilometres wide around the town had been stripped of trees and shrubs by nomads fleeing from the drought-stricken interior of the country and desperate for material for housing, fencing and cooking. As a result, the area became almost completely desertified.

Lutheran World Relief saw the need for a greenbelt to stop the formation and movement of the dunes and to reduce the intensity of sand storms around Nouakchott. The aim was to plant 700 ha of drought-resistant *Prosopis chilensis* trees on the northeast side of the city within four years. Work began in 1975 with the establishment of a nursery and by 1976 the first seedlings were ready for planting. The rains in 1975 were kind, 190 mm compared with an annual average of between 100 and 150 mm, but only 76.5 mm fell in 1976 and just 2.7 mm in 1977. The annual planting rate had to be reduced from the original target to allow time for proper watering of the trees; by 1980 only 360 ha had been planted. Despite the low rainfall 80% of the trees survived.

Tree seeds are planted in pots in the nursery each September and transplanted into the field the following year at the end of the rainy season. They are then watered twice a week during their first dry season, the area being so dry that it is necessary to water even the hardy *Prosopis* tree. This increased project costs, and so in 1980 cuttings of a local shrub *Euphorbia balsimifera* were planted in the *Prosopis* plantations to produce a denser and more diverse vegetation. The shrub can achieve a 70–80% survival rate without any watering at all and has previously been widely used by Mauritanians as a living fence.

The Nouakchott Greenbelt Project has been funded and implemented by Lutheran World Relief and the Mauritanian Red Crescent Society, and managed by two Mauritanian foresters from the country's Forest Service. Although the People's Volunteer Movement plants *Eurphorbia* cuttings in the plantations on Sundays, this is not a true community forestry project because some 160 workers have been hired to help in the nursery and to plant, protect and water the trees. When there is insufficient rain to achieve planting targets in the greenbelt, surplus seedlings are donated to individuals for planting around homes, schools and factories throughout the country. There is now a greater awareness of the need for trees and tree planting, by both the people of Nouakchott and the Mauritanian government. This has so far been the only forestry project in the country, but greenbelt areas are planned for other parts of the city and former nomads resettled around Nouakchott are required to plant four trees in the courtyards of their new homes (Lutheran World Relief, personal communication).

Silvopasture

Another type of agroforestry, known as silvopasture, integrates trees and pastures to make livestock raising more productive. Animals are often let loose in woodlands throughout the drylands to feed on grasses. This can cause considerable damage to trees, many of which are attractive food sources for the animals. Silvopasture turns unregulated grazing into a managed and sustainable system to improve livestock raising without harming trees and the environment generally. If tree crop plantations are undersown with pasture grasses, for example, cattle and sheep can graze the pastures, feed on the highly nutritious tree pods and foliage collected by the herdsmen, and also benefit from the shade. Experience in Niger has shown that 18–20 cattle may be grazed per square kilometre where *Acacia albida* grows, compared with 10 cattle/sq km of open pasture. The protein-rich pods of *Acacia albida* may be dried and stored for later use.

Natural stands of *Prosopis juliflora* are important food sources for both animals and humans in the arid lands on Peru's Pacific coast (Peck 1984). Pods from the tree are collected for fattening pigs and cattle in feedlots, and are also eaten by free-ranging sheep and goats. To counter the loss of large areas of *Prosopis juliflora* woodlands by deforestation, plantations of the species were established on barren wastelands, using water from irrigation canals flowing through the area to serve other crops such as cotton, rice and sugar cane. Pod production begins after three years and the 6–7 tonnes of pods produced in each hectare can be sold to feedlots for about $30 per tonne.

Prosopis is also a valuable source of fodder in India's dryland areas, with *P. juliflora* and *P. cineraria* the most common species. Tirath Gupta, of the Indian Institute of Management in Ahmedabad, Gujarat, has shown that a silvopastoral system combining pasture grasses and trees for fodder, fuelwood and small timber, gives far higher yields than conventional crop growing and livestock raising on an equivalent area (Gupta 1983). A survey of seventy farm households in five villages in Rajasthan in 1977–8 found that net annual returns per hectare from silvopasture were more than three times those from annual cropping and at least 50% greater than those from livestock raising.

Another tree with potential for more widespread use as a source of fodder in dry areas is the carob tree (*Ceratonia siliqua*), which grows well on the rocky, degraded lands common in Spain, Greece and other countries around the Mediterranean. Carob pods have a relatively low protein content but are rich in sugars and are therefore a valuable source of livestock feed. Cyprus, for example, has long exported most of its carob pod production to the UK for use in commercial compound animal feeds. Carob pods are also widely used in confectionery, and an industrial gum extracted from the seeds is used in convenience foods, textiles and paper manufacture (Winer 1980).

Tree Crop Plantations

Acacia senegal, the source of gum arabic, is a popular tree crop in the Sudano-Sahelian region. It is extensively grown in Sudan, which is the main world supplier of gum arabic, accounting for over 80% of all production. Gum gardens have historically been the basis of an ecologically stable agricultural system in which the *A. senegal* trees occupy agricultural land in the fallow periods between crops (see Chapter 2). The system helps to conserve soil, and gum arabic is a major cash crop for local peasants. Gum arabic currently accounts for 25% of Sudan's agricultural income and is its third largest export commodity, providing 10% of all foreign-exchange earnings (Beshai 1984). However, during the first phase of the drought in the early 1970s, the productivity of gum trees fell sharply. Fallow periods were shortened to increase food production as yields declined, there was considerable livestock encroachment into the gum gardens because of poor pasture growth in the Sahelian zone, and large numbers of trees, weakened or killed by drought, were blown down in the wind. Sudan's gum production plummeted and overseas importers turned to substitutes. Production slowly increased with the return of the rains, but by 1980 it was still only 22,000 tonnes per year, half of the average production in the 1960s, and it fell again when the drought reached a new level of intensity in 1984.

In 1981 the Sudan government, aware of the economic

importance of gum arabic and the way in which the "Gum Belt" forms a natural buffer zone between the desert and the main agricultural zone to the south, launched a $5 million scheme to restock 144,000 ha of gum gardens in the provinces of North Kordofan, Northern Darfur and White Nile. The scheme relies on the involvement and co-operation of large numbers of individual farmers. By the end of 1986, 7,472 farmers in 206 villages were taking part in the project, although data on the exact area of gum gardens restocked are not available. The farmers receive new tree seedlings for transplanting from a number of well-run government nurseries or they sow trees directly from seed. Survival rates vary widely, and are often less than 60%. Many seedlings died in the severe drought year of 1984, but the loss of many livestock during the drought temporarily reduced the grazing pressure on gum gardens. A few of the trees established at the start of the scheme began to yield gum in 1986 (UNSO 1986).

The future success of the scheme will depend on whether many more farmers can be persuaded to restock their gum gardens and ensure that these remain as productive as possible. Farmers now have two powerful incentives to join the scheme. First, a food-for-work scheme, in which both farmers and labourers receive food rations equal to half their needs, encourages farmers to remain on their farms and to continue some food cropping. Second, the price paid to farmers for gum was more than tripled in 1985. Previously the Gum Arabic Company, the monopoly in which government holds 30% of the shares, retained too large a proportion of the international market price and some farmers found it more profitable to fell their trees for charcoal production than to keep them for gum tapping.

Economic incentives to farmers will have to be continued if gum production is to be maintained. One possible new initiative is the formation of co-operatives to eliminate the need for a middleman between farmers and the Gum Arabic Company. Farmers' skills also need to be increased through training courses and frequent farm visits by extension workers. Up to 1986 the number of training courses organized was far smaller than needed, and while farmers received visits from extension workers at planting time and over the following three months, the visits tended to tail off afterwards. So far, however, progress has been very encouraging, and a UNSO report has stated that: "There is little doubt that the project is contributing substantially to the control of desertification and to the rehabilitation of degraded lands" (UNSO 1986).

Palms are another major type of tree crop in the drylands. In Algeria and Morocco the date palm (*Phoenix dactylifera*) is under threat from a fungus called bayoud (*Fusarium oxysporium*) which reduces palm growth rates and fruit production. The National Research Centre for Arid Zones (CNRZA) in Algeria is selecting varieties resistant to bayoud, testing various chemical and biological control methods, and studying the biology of date palms to find ways of increasing productivity. A similar selection programme is taking place in Morocco at the central station of desert agronomy of the National Institute for Scientific Research (INRA). Resistant exotic varieties introduced from the USA are among those being tested (UNEP 1985).

Integrated Land-Use Schemes

The ultimate form of agroforestry is one in which trees, cropping and livestock raising are managed over large areas in an integrated and planned way which is sustainable over the long term. This approach was the basis of the two huge transnational greenbelt schemes, consisting of mosaics of forests, shelterbelts and well-managed croplands, that were proposed at UNCOD to stretch along the northern and southern fringes of the Sahara, from the Atlantic to the Red Sea. The project on the southern fringe of the Sahara has been tacitly abandoned since UNCOD. Algeria, Libya and Tunisia actually signed a co-operative agreement for the North African Green Belt in 1979 but so far this has only resulted in joint training programmes, studies and seminars. Ironically, it is in this region that large-scale plantation programmes directed by forest departments have resulted in impressive achievements. Algeria established 267,500 ha of new forests between 1978 and 1981, although tree mortality after planting was high; Morocco afforested 19,600 ha in the Rif region between 1977 and 1980; and good progress has also been made in Libya. In Tunisia, on the other hand, protective afforestation has not been very successful, owing to a combination of poor climate, population pressure and inappropriate afforestation techniques.

The Algerian effort is part of an ambitious programme in which the army will plant a 20-km-wide forest wall, called the "Green Dam", to stop desertification on degraded soils in the Algerian steppe. It is planned to stretch eventually for about 1,200 km from east to west along the Sahara Atlas and to cover a total of about 3 million ha. The early plantations consisted almost entirely of *Pinus halapensis* and the needs of local people were ignored; but now twenty-three other species, such as *Atriplex* spp., *Pistacia atlantica* and *Lygeum spartum*, are being planted to reduce the susceptibility to fire and disease which threatened the former monoculture. New grazing areas are also being created and fodder trees and shrubs planted to replace those lost to local people when the plantations were established (UNEP 1985).

Developing Multi-Purpose Trees

Considerable research and development are needed if multi-purpose trees are to attain their full potential in making drylands more productive. Until recently, research on multi-purpose trees has been quite limited in scale, and undertaken in an unco-ordinated way by scattered research institutions for very specific objectives, such as the analysis of fodder potential. The tree which has received the most research attention, *Leucaena leucocephala*, has been subsequently introduced in many countries, but has been more successful in moist than in arid climates (NAS 1977). Awareness of the need for wider development of multi-purpose trees spread after UNCOD in 1977, and the Eighth World Forestry Congress in Jakarta the following year, when the new emphasis of FAO and the World Bank on "forestry for people" was unveiled.

The first stage in expanding research was an admirable effort by the US National Academy of Sciences (NAS) to publicize the potential of multi-purpose trees and encourage research activities and trials of various species in different countries. The NAS followed up its very popular publication *Underexploited Tropical Crops* (1975), which included some multi-purpose trees with potential for development, with two other reports, *Tropical Legumes* (1979) and *Firewood Crops* (1980, 1983), whose coverage was more comprehensive. These reports achieved extensive worldwide distribution and the NAS can claim a lot of the credit for bringing multi-purpose trees to the attention of the general public. In 1980 a specialist journal, the *International Tree Crops Journal*, was established to communicate the results of the latest research in this field.

Research on tree crops is badly needed because they are still at a very primitive level of development compared with agricultural crops. Often little information is available about the actual origins of seeds and grafting wood used for introducing exotic tree crops to new locations. Future development work must therefore be far more rigorous. The first priority is to gain more detailed knowledge about the natural geographic varia-tion of a particular species or genus, together with sources of already improved varieties, for only when such work has been done will it be possible to design systematic expeditions to collect seeds from distinct areas and test their growth and yields by trials at research stations in a variety of locations.

Development must proceed hand in hand with conservation. The rapid degradation of vegetation in dryland areas represents a serious threat to plant genetic resources in these areas, and makes identification of significant natural populations of multi-purpose species an urgent priority so that effective conserva-tion programmes can be designed for trees which are of both current and potential economic importance. In Somalia, for example, *Cordeauxia edulis*, a leguminous multi-purpose shrub growing in arid areas of the country, is threatened with extinction. The tree is useful for firewood and browse and has a nut-like seed eaten by both humans and animals, but the tree has been overexploited during long periods of drought and war, and although it constituted 50% of the woody vegetation of the region in 1929 it is now very scarce (Resch 1981).

The International Board for Plant Genetic Resources (IBPGR), based at FAO in Rome, identified in 1977 four multi-purpose tree genera for arid or semi-arid lands whose develop-ment was an urgent priority: *Acacia*, *Azadirachta*, *Eucalyptus* and *Prosopis* (IBPGR 1977; FAO 1977). The current status of research on *Prosopis* (see p. 250) gives an example of activities in the field generally. The genetic centre of *Prosopis* is in South America, where such species as *P. alba*, *P. chilensis*, *P. juliflora*, *P. nigra*, *P. pallida* and *P. tamarugo* are found. Several species have been introduced and naturalized in Asia, Africa and Australia, especially *P. juliflora*. Three species are native to Asia and Africa: *P. cineraria* (India and Pakistan), *P. farcta* (North Africa and Southwest Asia) and *P. africana* (Sahel region) (Pedersen 1980; Pedersen and Grainger 1981).

In 1980, IBPGR initiated a co-operative pilot programme (IBPGR 1980) involving nine countries – Senegal, Sudan, India, Israel, People's Democratic Republic of Yemen, Argentina, Chile, Mexico and Peru. Institutions in these countries were to establish their own priorities within the guidelines laid down by IBPGR. Concerning *Prosopis*, Chile promised some collection of *P. tamarugo*; Yemen a limited collection of *P. cineraria*; (but not from a natural stand) India was not interested in either *P. cineraria* or *P. juliflora*; Mexico was interested in collections of *Prosopis* spp. but had done little in this field before; and only Peru seemed very enthusiastic about actively collecting and trying out the genus. There have been some collections of *P. tamarugo* in Chile, but only for trees with good form and good seed production; the country has only a limited capacity for seed collection. The project suffered from the fact that only a small number of countries were involved. Because IBPGR acts only in a co-ordinating capacity there was no scope for a truly international exploration and collection programme able to encompass any of the four genera given high priority. Some breeding work for *P. cineraria* is being carried out at the Central Arid Zone Research Institute in Rajasthan, India. While eight species of *Prosopis* are widely distributed in Argentina, one local expert has reported that so far there have not been many collections of germplasm there, and called for an urgent programme of exploration and conservation before widespread destruction of natural stands makes it impossible (Hunziker 1985). Generally, little progress has been made in surveying and collecting seed from natural stands of *Prosopis*, and researchers have depended mainly upon seeds previously collected from a relatively small number of provenances.

The first, and still the only international comparative applied research programme for *Prosopis*, was carried out by a group of scientists led by Peter Felker, initially at the University of California, Riverside and more recently at Texas A & I University in Kingsville (Felker *et al*. 1981). The programme had the limited but pragmatic objective of screening *Prosopis* spp. for biomass (fuelwood or alcohol) production on semi-arid lands in the United States, so while the test areas were arid/semi-arid and subject to high temperatures, they were not truly tropical. Another limitation of the programme was that a number of *Prosopis* spp. with particular importance for the tropics (especially *P. juliflora* and *P. pallida*) experienced a high degree of mortality due to frost damage and so received only a limited evaluation. However, both species were recommended by Felker for further testing and development for use in tropical environments.

Felker observed high levels of biomass production (averaging 2.6 tonnes/ha per annum) for *P. alba*, *P. articulata* and *P. chilensis* in conditions equivalent to an annual rainfall of less than 500 mm, and *P. pallida* also gave encouraging results in the limited study period. The yields of the specimens of *P. tamarugo* tested were very low, but this could have been due to the choice of provenance. The yields of the highly drought-resistant *P. chilensis* and *P. tamarugo* did not decline very much as the level of irrigation was reduced. Felker and his colleagues demonstrated for the first time the practicability of vegetative propagation, necessary because high outcrossing results in up to 70% variation in biomass yields from seedlings derived from the same parent tree. The risk of *Prosopis* spreading as a weed was shown to be low if clones were selected for high biomass (and hence low pod) production, since spreading was achieved mainly by means of seeds eaten by animals and distributed in their faeces. In the case of clones selected for high pod production the risk of spreading could be reduced if pods were harvested soon after they had fallen or if animals were excluded from plantations.

Two groups in England have done pioneering research on multi-purpose trees. The Royal Botanic Gardens, Kew, established a small research group in 1983 to build a computerized database of economic plants for the world's drylands and held a major international conference on the subject in 1984 (Wickens *et al.* 1985). In 1982, the Oxford Forestry Institute of the University of Oxford set up a programme, under the direction of Brian Styles and Colin Hughes, to explore the native trees of drylands in Central America, collect seeds of multi-purpose trees and distribute them for trials in other regions, and lay the foundation for efforts to conserve threatened genetic resources. The Oxford Forestry Institute (formerly the Commonwealth Forestry Institute) has a wealth of expertise in seed collection, provenance surveys and trials for tropical pines in Central America, so this was a natural evolution of its activities. Twenty-five species were selected for initial testing, with potential for fuelwood production as the main criterion. These included species of genera like *Acacia, Albizia, Caesalpinia, Gliricidia, Leucaena, Mimosa, Parkinsonia* and *Prosopis*. Among those selected were little-known trees like *Acacia deamii*, a small tree common along the sides of roads in Honduras and highly regarded as a fuelwood, and *Leucaena shannoni*, which is far more drought-tolerant than *Leucaena leucocephala*. Hughes and Styles also regard as very promising *Parkinsonia aculeata*, a highly drought- and saline-tolerant tree, and *Guazuma ulmifolia*, a non-legume that produces excellent firewood, edible fruits and fodder, and is used as a living fence (Hughes and Styles 1984).

Development of multi-purpose trees requires far more than seed collection and trials. A deep knowledge of the physiology of these trees, including root growth and behaviour, water uptake, transpiration and adaptations to drought and salinity, is necessary to grow them successfully in extreme environments (Burley *et al.* 1984). The current level of physiological knowledge for different species is often very poor indeed. Since multi-purpose trees are commonly grown in combination with field and pasture crops, studies are also needed to determine if there are any possible conflicts between the root depth and nutrient requirements of trees and those of other plants with which they may be intercropped.

Many multi-purpose trees are legumes and therefore improve the fertility of degraded lands by fixing nitrogen, but to maximize this function their roots need to be inoculated with mycorrhizal bacteria. The US National Research Council, UNESCO, the French research institute ORSTOM (Office de la Recherche Scientifique et Technique Outre-Mer) and local government agencies have collaborated to establish two regional Microbiological Resources Centres, one for West Africa in Dakar, Senegal, and the other for East Africa in Nairobi, Kenya, to promote inoculation (UNEP 1985).

Technical Help for Farmers

One of the difficulties faced by farmers in their tree-planting activities is lack of technical support. They may receive free seedlings and visits from extension officers from government forestry departments, but these officers are likely to be mainly skilled in conventional forestry, rather than in agroforestry. Farmers therefore have to learn by trial and error, pushing back the frontiers of knowledge in agroforestry with every tree they plant. Often they need trees which forest department nurseries cannot supply, and information on agroforestry techniques and species which is not available in their own country.

Farmers would therefore benefit from a close relationship with a research-oriented organization with an extension programme and dedicated to serving their needs, in the same way that government forest departments have research divisions to advise their own foresters (Grainger 1984b). Such bodies are still few and far between, but a Kenyan programme dating back to 1982 may provide an example for other countries to follow. A network of six agroforestry research centres located in different parts of the country was established with support from USAID as part of its Renewable Energy Development Project. The main aim of each centre is to conduct trials to see which species are suited to the local climatic zone, and to experiment with different agroforestry systems by testing alternative harvesting systems and various combinations of trees and field crops. The centres train farmers and extension workers, and also have an extension programme to supply seeds and seedlings to community groups, schools and farmers. A wide range of species is stocked, currently numbering 127, with citrus, *Eucalyptus* and *Leucaena* the most popular. Recognizing the power of demonstration, seen so vividly in the farm forestry programme in Gujarat, the agroforestry centres invite volunteers to serve as "model farmers" by establishing demonstration plots which others can visit to see the potential of agroforestry realized in practice (Harrison 1986). In the Indian state of Gujarat, the centre of social forestry in that country, the International Tree Crops Institute (India), based in Baroda, works in a similar way, sharing knowledge and techniques with farmers.

Overcoming Legal Problems

Agroforestry and farm forestry programmes frequently

encounter legal obstacles, particularly regarding land tenure, which prevent them from being fully successful. In many countries farmers may not be able to fell their own trees without the permission of the government forest department, while in others it is against the law for individuals or non-state bodies to own forests or tree plantations and the government may hold rights to all trees even on private land. A project organized by a forest department could disregard such rules, but local people would still be aware of them and this could slow down a tree-planting project. It may therefore be necessary for laws to be changed so that social forestry and farm forestry are no longer illegal activities.

If freehold farmers are hesitant about planting trees which they think will become the property of the state, tenant farmers will be even more diffident if they fear that their landlord will demand a portion of the profits or increase the rent. They may also find it difficult to gain credit to finance tree growing because they do not have security of tenure. Studies by the World Bank have shown that land tenure problems have had an adverse effect on a number of forestry projects, and that designing projects in which the large number of landless people receive adequate supplies of wood and fodder presents a difficult challenge (Spears 1983). One of the reasons why farm forestry has been so successful in Kenya is that most smallholders own their lands and can plant what they like, even trees, in the knowledge that they can reap all the benefits. A survey of the Kakamega district of Kenya in 1983 showed that 80% of households had planted trees on their lands in the previous year, and 64% had woodlots (Harrison 1986).

MANAGING NATURAL WOODLANDS

Potential Yields Underestimated

Most strategies to increase wood production in the drylands have previously been concerned with establishing intensive plantations of fast-growing trees, on the assumption that they would give much higher yields than those from managed natural woodlands. However, recent data suggest that the potential yields from natural woodlands have been greatly underestimated. Thus annual increments of 4.0–4.8 cubic metres per hectare per year (m³/ha/yr) have been recorded for a protected stand of *Acacia senegal* in Chad; 0.8 m³/ha/yr for miombo woodland in Tanzania; and 0.67–2.35 m³/ha/yr for *Acacia senegal* in the Senegal's Bandia classified forest. This compares with 1 m³/ha/yr for a plantation of *Eucalyptus camaldulensis*. Even in favourable locations plantation production in the drylands rarely exceeds 5 m³/ha/yr (Freeman *et al.* 1983; FAO 1985b).

The time is therefore right for a re-evaluation of natural woodland management. A detailed study by Kenneth Jackson, George Taylor and Clémentine Condé-Wane concluded that: "By introducing protection and sound management practices, such as establishing optimum rotations, it should be possible to increase yields considerably at a fraction of the cost of establishing plantations" (Jackson *et al.* 1983). The cost of rehabilitating woodlands could be as low as $200 per hectare, compared with more than $1,000 to establish a hectare of plantation (Catterson 1984). Much research remains to be done, and this, together with the need to train local personnel in new techniques, means that it could take as long as 20–30 years for natural woodland management to become significant in most countries. On the other hand, the rewards from management could be immense, for even if in the initial stages, average yields per hectare are relatively low, the area of woodland involved is so large that the volume of wood produced would be substantial. As the study points out: "in Burkina Faso ... increasing the yield of the natural forest by 25% would be equivalent to the yield from 500,000 hectares of plantations." Exactly how much wood might become available will require extensive inventories to ascertain.

Trees and People

Another advantage of natural woodlands is that they produce a great variety of harvestable products, and this makes them far more attractive to local people than village woodlots. The range of products available from plantations is quite restricted, even if multi-purpose trees are used. Natural woodlands supply not only fuelwood, poles and fodder, but a whole range of foods, medicines, gums and other products that might not be available from intensive plantations. The strange-looking and fire-resistant baobab tree (*Adansonia digitata*), for example, is a common sight in African savanna lands and villages. Its leaves and fruit are used in cooking – fresh or powdered leaves are ingredients in sauces – and it is also a source of fodder, medicine, fibres, dyes and soap. The baobab and other trees such as néré (*Parkia biglobosa*) are planted and protected in villages, around fields, and elsewhere (Williams 1984b).

People in temperate countries do not generally realize how dependent people in the tropics are on the fruits of trees and shrubs, both as part of their regular diet and as emergency food reserves when crops fail. The fruit of *Zizyphus nummularia*, for example, is a valuable food in India and the Middle East, and the fruits of *Nitraria* spp. are consumed by people in the Central Asian desert. Africans eat the fruits, nuts, seeds, berries, flowers and leaves of many different trees and shrubs. The bark, roots, gum and young shoots are also used for cooking. Some are collected directly from wild trees or those growing in the house compound, while others are bought in local markets in fresh, dried or processed form.

At a time when food shortages in Africa continue to dominate the world's media, our ignorance of the composition of African diet can lead to serious misdirections in agricultural policy. According to Paula Williams:

> The "food problem" is generally considered to be a shortage of cultivated foods. In reading the popular press one can easily get the impression that Africans live on cereal crops, such as millet, alone. But it is difficult to meaningfully assess whether a population is receiving an adequate daily nutritional intake when no information exists on how many mangoes, how much soumbala and shea-nut butter, or how many leaf-based sauces enter the average diet. (Williams 1984a)

The neglect of tree-based foods is partly due to the way in which, based on the attitudes of scientists in the temperate world, trees are regarded as the province of foresters or horticulturalists rather than agriculturalists. Until recently, however, most foresters have been almost exclusively concerned with a single output from the forest: wood. Although the importance of tropical horticulture is acknowledged, statistics are generally available only for a fairly limited range of fruits grown in plantations and sold on commercial markets, and it is difficult to estimate the consumption of foods gathered from the wild or sold only in local markets. According to Paula Williams

If you ask a woman about all the things consumed during the course of a day and the ingredients of each meal, you may learn for example that she ate some shea-nut fruits (*Butyrospermum parkii*) in the morning when she was working in her field and brought home the nuts to cook shea-nut butter, a much utilized cooking fat. Perhaps also she had some dol, or local beer, that was fermented from the fruit of a local tree, such as *Sclerocarya birrea*. By asking about the ingredients that were used to make the tao and gumbo sauce you can learn that more than just cultivated millet and gumbos are used. The tao (a thick porridge made with millet) is often cooked with an acid to make it more digestible; commonly used acids are lemon juice, tamarind juice, tamarind fruit, tamarind leaves, or leaves of other tree species. The sauce probably contains, in addition to the gumbos, shea nut butter, soumbala (a condiment made from the fermented seeds of the néré tree *Parkia biglobosa*), salt, hot peppers and other seasonings. (Williams 1984a)

A wide variety of other products are collected from local vegetation. Gum arabic is extracted from the bark of planted and wild *Acacia senegal* and makes a major contribution to Sudan's economy. It is a common constituent in confectionery and is used in the manufacture of textiles and paper. Resins are another form of tree secretion but differ from gums in being insoluble in water. They are used as emulsifiers in medicine, binding agents in pill production, and in the manufacture of textiles, confectionery, water colours, perfumes and glues. A number of species of *Acacia* exude resins, including *A. arabica*, *A. senegal*, *A. decurrens* and *A. leucophlea*.

Some trees are the source of edible oils and fats. One such tree is *Butyrospermum parkii* (also known as *Vitellaria paradoxa*), referred to by Paula Williams. A hydraulic hand press was developed at the Royal Tropical Institute in Amsterdam to make it easier for women in Mali to make shea-nut butter from the fruits of that tree and so increase their income. Prototypes were tested in Mali and the final version is being produced by two competing local metal-working firms. The presses are bought by women's co-operatives and pay for themselves through increased profits in one to two years (Intermediate Technology Group, personal communication).

Oils are also important in cosmetics and medicine. Indian perfumiers use oil from *Cyperus scariosus* and *C. rotundus* as a fixative to flavour tobacco and scent soaps. Oil extracted from the leaves of various species of *Eucalyptus* is used as a treatment for colds, and in the manufacture of perfumes and as a solvent for tar, grease and paint. Traditional systems of medicine make extensive use of members of the plant kingdom. India, for example, has at least 2,500 medicinal plants. Depending upon the particular plant and medicine, parts of the plant may be consumed directly, or the active therapeutic agents are extracted from it first. Some plants are the source of commercial medicines, and a study of more than one hundred desert plants by scientists at the University of Arizona has identified two plants, *Caesalpinia gilliesii* and *Bursera microphylla*, which could be used in the treatment of cancer.

Fibres are obtained from members of the cactus genus *Agave*, particularly *A. lechugilla* and *A. sisalsana*, and *Bauhinia vahlii* in India. Tannin is still widely used for curing leather, and can be obtained from the bark of (among others) *Acacia cyanophylla, A. nilotica* and *Eucalyptus astringens*; from the fruit, leaves and bark of *Calotropis procera*; and from the bark and roots of *Zizyphus spina chisti* (FAO 1985c).

Given the diversity of products available from natural woodlands, it is perhaps understandable why plantations are not as attractive to rural people as they might have seemed to foresters, and why our whole approach to rural afforestation may need to be re-evaluated once again, both to broaden the range of products available from planted trees and to give a whole new emphasis to replenishing and managing remaining areas of natural woodlands.

The Guesselbodi Experiment

A pioneering experiment in natural woodland management is

being carried out in the Guesselbodi National Forest Reserve in Niger as part of the Forestry Land Use and Planning Project funded by USAID. The reserve, which covers 500 ha and is located 25 km east of the capital Niamey, has suffered from severe degradation and deforestation in the last thirty years. Comparison of aerial photographs taken in 1950 and 1979 showed a reduction of about a half in forest cover, caused by clearance for cropping and (especially) the effects of unauthorized grazing. Extensive erosion has left behind a crusted soil virtually devoid of organic matter, and in many cases consisting of hard, barren laterite. Urgent action was imperative for if current trends continued there would not be much woodland left in another thirty years. The project aims to restore the forest and protect it from further degradation, so that eventually it can supply the many needs of local people and provide wood to be sold in Niamey.

Foresters decided that local people held the key to the future of the forest and should be involved in the project from the start. So when it began in 1981 discussions were held with people from a number of surrounding villages to explain that the forest service wanted to try new management techniques at Guesselbodi and wished to obtain their ideas and suggestions. The villagers were happy to see that the project would provide employment for them, and during the discussions much valuable information was gained concerning past and present uses of the forest, the changes which have occurred in recent decades, and local forest customs. The villagers also told the foresters that they preferred to use for agriculture part of the reserve that was situated close to their fields. To show the good faith of the forest service and demonstrate its intention to integrate their suggestions into the design of the project, some 30 ha of this land was given to the villagers for growing crops. It was divided into thirteen fields, each about 2.5 ha in size, and allocated to those who applied by the drawing of lots. The land was given on the condition that villagers agreed to participate in an agroforestry system and plant crops between windbreaks of *Prosopis juliflora* and *Acacia holocerica*.

In the remainder of the reserve existing trees are being coppiced, and forest cover is being restored by both planting and natural regeneration. The dominant tree species remaining in the forest are *Combretum nigricans*, *C. micracanthum* and *Guiera senegalensis*. These are the three leading firewood species sold in

Niamey, so it was decided to coppice them on a ten-year rotation, although only trees above a certain minimum diameter are cut. The forest was divided into ten equal parcels, with each 50 ha parcel being cut in sequence every ten years. Because over much of the area the soil and vegetation were seriously degraded considerable restorative measures were needed. Tree seedlings raised in nurseries are being planted both within the remaining stands of trees and on more open land. Most of the species planted, selected after consultations with villagers, occur naturally in the forest but some exotics such as *Prosopis juliflora* and *Acacia holocerica* have also been included. Bunds made out of earth and stone have been built along contours to control soil erosion, and tree seedlings are planted behind the bunds to give extra stability. Other vegetation regenerates naturally on the moist land behind the bunds where water is trapped. Another technique to promote natural regeneration involves the laying of mulches, made of small branches and twigs, on the open stretches of barren land to improve the capacity of soil to absorb moisture and support plant growth. The control of grazing is obviously essential if tree regeneration is to be successful. Grazing is therefore forbidden for the first three years after planting, although local people can still cut hay for their animals and gather minor forest products as before. A forest guard is on hand to impound any animals which enter the reserve. When grazing resumes in each plantation, it is strictly regulated to stay in line with the carrying capacity of the pasture, so that the animals do not degrade the woodland (USAID 1979; Heermans 1986).

THE POTENTIAL OF AGROFORESTRY AND NATURAL WOODLAND MANAGEMENT

This chapter has looked at ways to increase or maintain tree cover which build on existing land uses instead of trying to create totally new ones. In some areas commercial farm forestry has been quite readily adopted, but in the longer term agroforestry systems which modify and enhance existing farming systems may well prove to be the most successful ways to promote the sustainable use of drylands. To realize fully the potential of agroforestry will require the development of multi-purpose trees and the conservation of their wild genetic resources. It is therefore encouraging to see that in 1983 the Algerian National Research Centre for Arid Zones and Houari Boumedienne University began a project to establish a gene bank of plant species of the Algerian Sahara at the El Golea Research Station. This should ensure the *in situ* conservation of important species (UNEP 1985).

One of the best ways to conserve valuable species may well be to encourage their continued use by managing natural woodlands in a sustainable way for a variety of products. The Guesselbodi project looks as though it will be successful and provide a model for projects in other areas. Despite the appeal of natural woodland management as the "latest" solution to forestry problems in the drylands, it is necessary to introduce a note of caution. There will be a substantial lead time before any significant results are seen, because much research is needed to counter previous neglect in this field, and it will take time for management policies to be introduced and for woodlands to respond to them. Further, most woodlands are already used quite extensively for grazing, fuelwood collection and the gathering of other products, so local people may not take kindly to regulation of this important resource.

The only really practical approach, according to Tom Catterson, appears to be the introduction of "local participatory management schemes which involve the people in adjacent villages in the activities being undertaken and including them as part of the ultimate destination of the benefits expected", as has been the case in Guesselbodi (Catterson 1984). Since this is contrary to the way in which most foresters have looked on these woodlands in the past, both sides will have difficulties in reorientating themselves. It is therefore likely that both the plantation and natural woodland management approaches will continue in the future, as appropriate to the conditions in particular areas and the needs of the people who live there.

9. Turning the Tide

The four previous chapters described ways of controlling desertification by making land uses more productive and sustainable. This chapter looks at projects which either reclaim land that has already been highly desertified, or prevent the spread of desertification by introducing soil conservation practices and stabilizing sand dunes that are encroaching on to farmland. Projects of this kind usually take place in areas where the social, economic and environmental impacts of desertification are already so serious that local people either request the help of governments and non-governmental organizations in saving their lands from further degradation or eagerly take advantage of such assistance when it is offered. Soil conservation and land reclamation, together with afforestation, account for the majority of current projects to control desertification.

PROGRESS IN SOIL CONSERVATION

Techniques for soil conservation are well known, having been intensively researched and tested over more than half a century in the Great Plains of the United States following the Dust Bowl tragedy of the 1930s. There are two main types of techniques: preventive cropping practices and tree planting which make the soil less vulnerable to erosive forces, and mainly "mechanical" curative techniques which reduce the freedom of water and soil to move over land surfaces.

Preventive cropping practices include growing crops in furrows that follow the contours of the land so that they do not provide channels for water to flow downhill and erode soil on its way; planting trees on watersheds or along the borders of fields to protect soil against erosion by water or wind; and restoring fallows to increase the soil's vegetative cover and organic matter content, and make it less prone to erosion. "Mechanical" techniques for soil conservation include terracing steep slopes to isolate soil and water flow; lining the edges of terraces with raised mounds or bunds; and establishing larger bunds around fields to capture rainwater so that it has time to infiltrate the soil instead of running off down the slope.

Soil conservation was enthusiastically promoted by colonial governments, like those of the British in Kenya and India, but because the approach was more dictatorial than consultative, these efforts were often resented by local people even if the results were beneficial. The techniques employed were largely mechanical, involving terracing and bunding which took up space and thus deprived some people of land. Consequently, when countries became independent, these soil conservation works were neglected, mainly because they were associated with the former colonial power rather than because they were not effective. The abandonment of soil conservation does not require anti-colonial spite, however: American farmers have been doing the same thing since the start of the 1970s simply to grow more food to make more money, even if the land was too poor to sustain such cropping.

Building a National Programme in Kenya

It takes time to restore soil conservation practices once they have been discontinued. Kenya now has an impressive programme, which grew from pilot projects in four districts in 1974 to cover twenty-two districts by 1977 and thirty-nine districts by 1986 (out of a total of forty-one districts in the country). The emphasis all along has been on ensuring that the programme grows steadily in scope without exceeding the administrative capacity of Kenyan personnel. The programme is funded by the government of Sweden, but while the role of Swedish advisers has been crucial, the number of expatriates has been restricted to three or four, and the programme operates within the structure of existing government ministries rather than as a separate organization. The growth of activities is therefore dependent upon the speed at which local agricultural officers are trained, either at home or abroad. The Kenyan character of the programme has been paramount, perhaps through sensitivity to local antipathy towards previous colonial impositions. President Daniel Arap Moi himself is a vigorous propagandist for the programme and regularly puts in a certain number of days of fieldwork every year.

Training farmers in practical techniques of soil conservation is at the heart of the programme. By 1986 more than 117,000 farmers had attended a one-day field workshop, peaking at 49,510 farmers per annum in 1984. However, even if that number could be trained each year it would still take another twenty years to reach all the small farmers in Kenya, so advice is also given during the regular farm visits of agricultural extension officers. Village chiefs and sub-chiefs, who have a major influence upon the adoption of soil conservation practices in their area, also attend the training sessions. School teachers are closely involved in the programme, since they can pass on techniques to the children in their care, encouraging them to establish tree nurseries and tell their parents about the programme. Associated with this programme is an East African regional training programme in which soil conservation officers from neighbouring countries, such as Tanzania, Ethiopia and Zambia, as well as Kenya, share experiences and techniques through seminars and tours of each other's countries.

Between 1974 and 1985, 490,000 farms were terraced in Kenya out of an estimated 1.1 million farms where terracing was needed. Because of Kenya's rapid population growth, the total number of farms (and therefore the number needing terracing) is increasing rapidly, having grown by more than half to 1.9 million over that period. Nevertheless, the programme is keeping pace with this expansion, and the proportion of targeted farms terraced by 1985 (48%) was larger than that in 1983 (39%). Assessing the impact of these activities on Kenyan agriculture is difficult, for although a number of scientific studies have shown that the terracing increases crop yields, it is not possible to distinguish the effect of terracing from that of improved seed and better management. Yield increases may also be biased by comparisons with previous low yields depressed by poor soil or rainfall. A report on the Kenyan programme by Keith Hudson states that: "Intangible benefits to the farmer may be as important as yield increases. There is evidence that conservation farming can lead to greater reliability of yields, to a better yield per unit of labour input, to better cash flow, and to better nutrition of rural farm families", although these too are difficult to quantify (Hudson 1987).

Swedish assistance for the programme averaged just under $2 million per annum between 1983 and 1985, although in 1986 it rose to $2.75 million, mainly due to an increased provision for the purchase and operation of transport facilities. A total of $15.35 million was donated between 1974 and 1987, equivalent to an average of just $28 per farm treated. The overall budget is much higher because it also includes specific funds from the government of Kenya and the cost of the regular agricultural services with which the soil conservation programme is integrated. Technical advice is free, but farmers are expected to pay for the installation of soil conservation works on their own land. However, the government pays for part or even all of the cost when a group of farmers get together to construct works that will benefit all of them.

The programme is not without its problems. Some experts argue that the programme could have grown faster if the government of Kenya had accepted more expatriate advisers, especially at the district level where the technical experience of local agricultural officers needs strengthening. There is also concern that farmers do not receive all the personal advice they need from agricultural extension officers, who normally use the contact-group technique, meeting with groups of farmers regularly once a fortnight, and relying on a selected "contact" farmer to pass on their advice to members of the group who cannot attend. Advice can therefore easily become garbled in being passed from farmer to farmer. Among other problems, there have not been enough staff to make school liaison a success, and farmers anxious to obtain trees have often found them unavailable because of the disappointing performance of the tree nurseries set up as an integral part of the project. Poor nursery management has resulted from inadequate staff training, a reliance upon casual labour, and the fact that soil conservation nurseries have to compete for labour with other government nurseries run by the forestry, agriculture and other departments. The emphasis in future will therefore be on fewer, more centralized and better-managed nurseries. Because of staff shortages and uncertainty about suitable techniques, the programme has so far tended to focus on higher-rainfall areas suited to cropping, rather than the semi-arid areas mainly used by nomadic pastoralists, but a pilot project is now under way at Mai Mahiu in the drier Nakuru district.

One of the main factors which has contributed to the success of the Kenyan programme is its long-term commitment to increasing the country's institutional capacity to conserve its soil resources. This means that projects can be planned thoroughly and implemented without undue haste, and officers can go abroad for training in the knowledge that the programme will still be in operation when they return. The Swedish International Development Agency (SIDA) supports the project by means of a grant, not a loan, but retains direct control over about a third of total expenditure so that it can, if necessary, withhold funds until administrative problems which may delay the programme are corrected. The fact that the expatriate advisers are small in number and integrated into the headquarters staff of the Kenya Soil and Water Conservation Branch of the Ministry of Agriculture and Land Development, with no separate project office, ensures close co-operation between the governments of Kenya and Sweden and gives less opportunity for friction to arise.

Lesotho's Campaign Against Soil Erosion

Institutional development is also at the heart of the Land Conservation and Range Development Project in Lesotho, one of the most eroded countries in Africa (USAID 1984a; Warren *et al.* 1985). This small mountain kingdom, about the size of the Netherlands, is entirely surrounded by the Republic of South Africa. Of its 1.4 million people, 90% are engaged in agriculture, but because of the difficult terrain and poor soils only 13% of national land area is suitable for growing crops. Livestock raising is therefore widespread and overgrazing is a prime source of erosion. The project, which began in 1980, aims to integrate soil conservation with the development of better range management techniques (described in Chapter 6).

Education at all levels is a key foundation of institutional development and by 1984 seven officers of the Ministry of Agriculture were either in training or had completed training at diploma level, fifteen were reading for Bachelor's degrees and seven for Master's degrees. District extension, conservation, range and agricultural officers, village chiefs and various farmers have also received training through meetings and field tours. The project, funded by USAID, includes nine expatriate advisers, four of whom work on the soil conservation component. The advisers are based at Ministry of Agriculture headquarters in Maseru, the capital of Lesotho, and (as in the Kenyan project) they are integrated with local staff within the ministry. In this project, however, all advisers have local counterparts for whom they often substitute during absences on foreign study leave.

Preparation of conservation plans was given a high priority in the project, and technology for interpreting aerial photography and satellite imagery was introduced to facilitate this. Data on soil types and agricultural potential are vital prerequisites for conservation plans; the goal was to complete a detailed survey of 150,000 ha, of which two-thirds was rangeland. By mid-1984, half of the target area had been surveyed, and conservation plans had been prepared for nine of the target of twenty-five areas (covering 19,657 ha) and for 209 farms covering 541 ha (18% of the target area). Preparation of plans was expected to speed up as personnel graduated from their training programmes and this allowed the number of conservation planning teams to increase.

Conservation plans cannot be implemented overnight, and although area plans can be prepared in six to twelve months it might take up to fifteen years to implement them. Thus, of the thirty-one area plans prepared under a preceding project only five had been fully implemented by 1984. Nevertheless, this project did aim to install soil conservation works on 64,000 ha of land, and by 1984 11,172 ha of cropland had been protected by terraces and water diversions, and another 200 ha by waterways, silt traps and the planting of grass and trees.

Reclaiming Degraded Hillsides in Ethiopia

The serious nature of soil erosion in the Ethiopian highlands has been referred to in earlier chapters. Seven-tenths of Ethiopia's population live in the highlands, and about half of the total area shows signs of accelerated soil erosion, with an estimated loss of 3.5 billion tonnes of topsoil every year as farmers cultivate ever steeper slopes. After extensive deforestation there is little vegetative cover remaining to prevent rain from flooding downhill and carrying the soil with it. Soils are now so thin that an estimated 20,000 sq km of land can no longer sustain cropping.

A large soil conservation project funded by the World Food Programme (WFP) has been under way in Ethiopia since 1973 in an attempt to reverse these alarming trends. The project now covers forty-four catchment areas. The people who work on the project receive payment in the form of food provided by the WFP. Where the slope is between 20% and 35% the land is terraced and may still be cropped. Where the slope is above 35% stone terraces are constructed and *Eucalyptus* trees planted. The plantations are later managed for fuelwood and pole production. Animals are not allowed to graze in the new plantations but local people can collect fodder from beneath the trees. Farmers cultivating lands which are to be treated by the scheme are allocated new croplands further down the slope, and given as much food as they would expect to grow in a single year to sustain them until they obtain the first harvest from their new lands.

By 1985, 700,000 ha of land had either been reforested or terraced. Despite the tremendous progress made by this project, the magnitude of the problem is so great that after twelve years only 4% of all seriously degraded lands had been treated; in 1985, the area of cropland reclaimed was only one-third of the area degraded in that year. Concentrating conservation work on the most critically affected lands allows degradation to continue elsewhere. The government is aware of the problem and determined to expand the current project into a national conservation programme covering the whole country (Harrison 1986; UNEP 1983; Milas and Asrat 1985).

The project has, however, run into problems over its food-for-work component. Farmers are often reluctant to build terraces or plant trees on their lands on their own initiative and prefer to wait until they receive food for doing so. Moreover, the huge shipments of free food for famine relief in 1984 undermined the scheme because priority was given by the government to transporting emergency relief food and there were no supplies to pay the soil conservation workers. Thus about 50,000 workers had to be laid off and, taking into account their families, this probably means that the total number of people receiving relief food rose by about 300,000. This is yet another example of how emergency short-term food aid can disrupt long-term agricultural development.

Lack of Maintenance: a Threat to Soil Conservation

Soil conservation demands continued vigilance on the part of farmers and a willingness to maintain soil conservation works and practices after a project has ended. This does not always occur, as exemplified in the last chapter by the removal of windbreaks in some parts of the US Great Plains. In the early 1960s USAID funded a project to construct bunds on half of a cultivated watershed in northwest Somalia. The 10,000 ha catchment was located near the town of Arabsiyo, 52 km west of Hargeysa. The aim of the three-year project, which ended in 1966, was to prevent soil erosion and concentrate the flow of water to improve its uptake by crops. Rainfall in the area is unpredictable and undependable, and when it does come it falls in short, fierce downpours lasting only a few hours. Much of the rain runs off along the surface, taking with it a lot of soil. The project therefore built bunds, or earth walls, to impound the water so that soil erosion would fall and soil moisture increase. Cropping was concentrated on the third of the treated area where most of the water was impounded. Results were immediate, and sorghum yields doubled without the use of fertilizer or improved seed. Yields of the stubble remaining after harvest also increased, leading to better nutrition for the livestock which grazed it.

Despite these remarkable gains the soil conservation works were allowed to fall into disrepair. A team sent back to Somalia by USAID in 1983 found that in the intervening seventeen years, "the farmers had failed to keep their end of the bargain and, except for repairing some of the more serious breaches, had not regularly repaired or maintained the bunds, which appeared considerably worn down and breached in many places" (McCarthy *et al.* 1985). The efficiency of water storage had declined as holes were allowed to develop in the bunds and their effective height was reduced because farmers had not removed the silt which accumulated behind them. Lack of control over water flows meant that gully formation, which had ceased for some years, had now resumed. Nevertheless the bunded land still yielded 50% more sorghum and maize than unbunded land.

There were three main reasons why the bunds were neglected. First, farmers' needs were not fully taken into account in the design of the project. Many farmers objected to the bunding of their lands, and although the layout of the bunds was modified so that they did not cross the boundaries of individual farms, some resentment still remained. The evaluation team concluded that farmers resented the amount of labour needed to maintain the bunds, and viewed cropping as a supplement to their livestock raising; hence their dislike of any additional hard work on the farm. The relationship between cropping and livestock raising was not considered during initial project design (McCarthy *et al.* 1985). Second, the government of Somalia did not fulfil its responsibilities after the project officially ended. Farmers received no ongoing support from the Ministry of Agriculture, and although the project was supposed to encourage the creation of a Soil and Water Conservation Department within the Ministry of Agriculture this hope was never realized. The ministry did not go on to complete the bunding of the other half of the watershed, despite a substantial training component in the USAID project. Further bunds were eventually built, however, as part of two separate FAO and World Bank projects. Of the four Ministry of Agriculture staff who received training in the USA only one was assigned to the project upon return – the other three were sent to entirely different regions of the country. Third, in 1969 the style of government changed, and a socialist republic replaced the former parliamentary democracy. This led to a new emphasis on large-scale farms, which contrasted with smallholdings like those at Arabsiyo.

A particularly serious and unpredicted side-effect of the project was that farmers abandoned their traditional rotation system in which only one-third of the land was cultivated at any one time and the remainder was left fallow for grazing. They now preferred to cultivate continuously those parts of their farms where water was held by bunds and yields were highest. The results of this change have been masked until now by the greater availability of water and the fertilizing effect of the silt deposited behind the bunds, but as the bunds deteriorate declining yields and increasing soil erosion are inevitable. Meanwhile, the continuous grazing of the other part of the farm has degraded pastures, removing the valuable species of grasses and shrubs and replacing them with less palatable species. The lesson of the Somali project is that similar failures are likely in the future unless projects take account of the needs of local people and provide sufficient funds for maintaining soil conservation works after the official end of the project.

PREVENTING SAND DUNE ENCROACHMENT

In the popular imagination desertification is still often associated with the relentless invasion of productive farmlands by sand dunes. Although this is not actually the main form of desertification, sand dune encroachment is nevertheless a worrying problem in a number of countries. There are two kinds of encroachment: the spreading of dunes on the fringes of deserts or in other desertified areas, and the movement inland of coastal dunes. Various techniques have been developed to stabilize or "fix" these dunes to prevent encroachment. The first step is to establish a temporary physical barrier against the movement of sand, usually by inserting micro-windbreaks in the form of small brushwood panels. These provide the tree seedlings or cuttings which are later planted in the dunes with vital protection against wind and sand. Seedlings are actually threatened by strong, desiccating winds more than by lack of water. Unstabilized dunes often contain substantial amounts of water which can be used by trees but, because the moisture is at low levels in the soil, initial top-watering should be avoided so that the trees develop deep roots (Kaul 1983).

Stabilizing Coastal Sand Dunes in Senegal

Senegal, on the western tip of Africa, is experiencing problems due to the encroachment of coastal sand dunes for up to 5 km inland on to one of its most productive agricultural areas, a fertile depression that runs parallel to the coast for 175 km from Dakar to Leonia and supplies 98% of all marketed vegetables in the country. About 9,000 people live in the area in sixteen major villages. The sand dunes are currently moving inland at a rate of about 13 metres every year.

The first experiments in dune fixation in Senegal took place as early as 1908, but work did not begin in earnest until the

1940s, and over the next twenty years about 20 km of dunes north of Malika were stabilized with *Casuarina equisetifolia*, an ideal tree for this purpose because it is resistant to both drought and salt spray. Activities expanded in the 1970s with the aim of fixing the remaining stretches of dunes between Dakar and St Louis. Three projects are currently under way. The first, funded by FAO, began in 1976 and covers 70 km of dunes halfway between Dakar and St Louis. The second project, funded by the government of Canada and started in 1979, is fixing dunes north of the FAO project to Gandial.

The most recent project started in 1981 with funding from USAID and aims to fix 73 km of dunes in an area immediately to the north of Dakar receiving between 200 and 500 mm of rainfall per annum. A band of *Casuarina* trees 250 metres wide is being planted parallel to the coast and 60 metres from the high water line; another parallel strip of *Eucalyptus camaldulensis* and *Acacia* trees are planted on "live" secondary dunes 3 km inland. Woven brushwood panels of the local shrub *Guiera senegalensis* were established on the seaward side of the dune to protect the trees from wind; the panels were anchored to sticks of *Euphorbia balsamifera* placed every 1.5 metres. Rows of trees were then spaced every 2–3 metres starting at a distance of 20 metres from the panels, and were given additional protection from wind and sand by further rows of brushwood windbreaks.

The government of Senegal has more than thirty years of experience in sand dune stabilization, so project staff are highly competent and the technical side of the project has proceeded relatively smoothly. By the end of 1986, 2,485 ha of trees had been established on coastal dunes, 858 ha on inland dunes, with another 550 ha on previously stabilized dunes. Additional windbreaks were established around villages and market gardens. The problems which arose were mainly financial and climatic. Funds for the project were delayed in 1982 and 1983 because of slow sales of rice. (One of the ways in which USAID supplies aid is in the form of food which is sold locally by the government so that the proceeds (in domestic currency) can be used for project funds.) Low rainfall in 1984–6 limited the periods when trees could be planted and those trees which were planted had low survival rates.

Despite such setbacks the project represents a substantial achievement in dune stabilization. The growth of trees on the dunes has been impressive and, according to Tim Resch of the US Forest Service's Forest Support Programme, it could well exceed that of all other plantations north of the Gambia River. Indeed, although originally planted for purely protective purposes, it could turn out to be one of the most productive plantations in the country (Resch 1987). The project was firmly managed and directed by government agencies who employed local people as labourers. In future, however, increased active participation of local people and arrangements for them to benefit from fuelwood and fodder production will be necessary to ensure that the new plantations are protected against incursions by grazing animals and fuelwood cutting.

Fixing Dunes in Somalia

Another country which has been suffering quite badly from coastal sand dune encroachment is Somalia in East Africa. An intermittent chain of sand dunes stretches from Eyl in the north of the country for 2,000 km along the Indian Ocean coast and for up to 25 km inland. The dunes encroach upon roads, villages, rangelands and fields. The moving sands disrupt communications and affect human and livestock health by irritating eyes, throats and skin. Work began as early as 1973 to fix dunes near Genale in lower Shabele province. With the help of 500 soldiers, 60 full-time workers and up to 5,000 part-time workers, brushwood, cactus and euphorbia plants were collected from nearby to stabilize the dunes. In 1977 a much larger project was started at Shallambot in the same province to erect windbreaks made of brushwood and commiphora bush, plant cactus, euphorbia and other plants to stabilize the acute face of moving dunes, and establish species of *Eucalyptus, Acacia, Tamarix* and other trees. By 1981 a 12 km stretch of dunes had been stabilized.

The Shallambot project was extended in 1982 to stabilize 1,200 ha around nearby towns and agricultural areas, with the assistance of funds from UNSO and the World Food Programme. The planting technique was changed to include far more trees and brushwood windbreaks instead of the cactus, euphorbia and commiphora used previously. The latter have to be collected from other areas which can themselves become vulnerable to desertification if too much material is removed.

The *Eucalyptus* and *Tamarix* trees could also be harvested for fuelwood in the future. This new project was slow to start because of administrative problems, and by 1984 only about 560 ha of dunes had been fixed, the tree nursery had not been established, the training and research components had not been implemented, and there were difficulties with project labourers because of low pay and non-delivery of food rations. However, the agencies supporting the project agreed to continue funding and thereafter progress was rapid. By 1986 1,310 ha of dunes had been planted and a further 450 ha revegetated by guarding them against grazing. The nursery was finally established and 160 ha of dunes were planted with trees in 1986 alone.

This project laid the foundation for other similar ones. Work began in 1982 at another site at Brava, about 19 km south of Shallambot, and in 1985 at Adele, north of the capital Mogadishu. Brava is a trading town and fishing port of about 25,000 people, but the only road into the town from the main coastal highway between Kismayo and Mogadishu passes over a large, 200-metre-high dune. Strong winds were causing the sand to move and threatening to engulf the town's lifeline, so the aim of the project, funded by the non-governmental organization Africare, was to stabilize a 30 ha area of the dune. The project relied entirely on voluntary labour, except for the National Range Agency technicians who were paid to dig trenches for windbreaks and holes for trees, and organize the collection of planting materials. Townspeople, mostly women, planted trees and stuck cuttings of cactus and euphorbia into the sand. The town of Brava was divided into six sections, each of which provided volunteers for one afternoon a week, competing with each other to see which could fix the largest area. With such vigorous action the dune was successfully stabilized and the town's main communications route was saved (UNSO, personal communication; Africare 1982, 1984).

Reclaiming Coastal Sand Dunes in China

China has reclaimed large areas of land that were formerly covered with sand dunes and shifting sands. Since the mid-1950s, 57 km of coastal windbreaks varying in width from 1 to 5 km have been established on Nanshan Island in Guangdong province. Saline-tolerant *Casuarina equisetifolia* trees are used to protect agricultural lands from the sands and tidal water which once threatened them, and crop yields have tripled. Agriculture in the Tungfanghung Production Brigade in Chifeng County had long been blighted by about 20,000 sand dunes each covering an area of between 0.1 and 0.5 ha. In the 1950s, using what can only be termed "mass mobilization", the brigade physically removed 2 million cubic metres of sand. After levelling, the land was flooded with silty water to form an alluvial soil. Once this had been completed, paddy fields were established, together with protective windbreaks to make sure the fields stayed sand-free in the future. Elsewhere, afforestation of sand dunes in the Taipingti People's Commune in Liaoning raised soil organic matter from 0.5 to 27%, increased nitrate levels from 5 to 45 parts per million, and doubled the levels of phosphorus and potassium, creating a soil 37 cm deep in thirty years (FAO 1978c).

Stabilizing Floodplain Dunes in Niger

At Assoro in the Bonza department of Niger fields were threatened by sand that was deposited on the banks of a seasonal river and grew to form a single dune 2 km long. Parts of the dune collapsed into the riverbed in the rainy season, but in the dry season sand blew over millet and other crops growing in the river's floodplain. Where the sand, accumulated it gave rise to many smaller dunes 3–5 metres high and 50–100 metres long. The area was extensively forested in the 1940s and 1950s, but the trees were felled for fuelwood and to provide additional land for growing crops. When trees became scarce the people burnt millet stalks, further exposing the soil to erosive forces, and reducing the organic matter content. As fertility declined, so did yields and it became necessary to crop even more intensively to maintain production. Eventually croplands became so degraded that they were just abandoned.

Work to fix the dunes began in 1978 with funding from CARE. Fences made of millet stalks were established in the dunes and behind them *Eucalyptus*, *Acacia* and other trees were planted. They were raised in five local nurseries that were set up as part of the project. The Forest Department organized the project and also managed the nurseries. People from eighty local villages worked voluntarily to make the fences and dig holes for seedlings, although they were paid for the millet stalks. Special guards were appointed to prevent goats from eating the young trees. When the project ended in 1982, 45,000 trees and 50 km of fencing had been established to fix seventeen dunes. As the trees grew taller the fences eventually rotted and improved soil organic matter. The *Eucalyptus* trees grew especially well, increasing in height by 1–2 metres every year.

The Forest Department continued to manage the project after external funding ceased. Any surplus trees from the nurseries were given free to local farmers, many of whom established woodlots for growing poles and fuelwood. The project produced tangible benefits for local people, reducing the threat of sand to their lands, providing cash income from the sale of millet stalks, and giving some paid employment. The costs which they had to bear were relatively low, since tree planting in the dry season did not conflict with regular farming activities, and the land being planted had already been lost to agriculture. In this particular case, initiation and direction of the project from the outside contributed in no small measure to its success, but could also pose problems in the future if the Forest Department withdraws from the project (Harrison 1986).

Dune Stabilization in India

Farmlands in the state of Rajasthan in the western part of India suffer very badly from the encroachment of sand dunes which cover about half of the state's total area (1.3 million ha). About 10,000 ha of dunes in Rajasthan are stabilized each year, and in a programme assisted by the World Bank and World Food Programme, an area of 50,000 ha was stabilized as part of the Rajasthan Canal Project between 1975 and 1979. The main constraints on dune stabilization, according to Dr C. N. Mathur, the Director of Desert Afforestation and Pasture Development in Rajasthan, are lack of finance, ineffective management and "above all, public apathy".

India's methods are less labour-intensive than those of China, and probably more typical of those used elsewhere in the world. The dunes are first fenced to protect them from grazing by livestock and to allow the natural regeneration of vegetation to proceed unimpeded. Then parallel strips of brushwood are fixed into the dunes, and with this protection it is possible to plant hardy trees, like *Acacia tortilis, Prosopis juliflora, Prosopis cineraria* and various species of *Zizyphus*, and to seed appropriate grasses. The dunes have to be carefully protected for at least two years, after which grass may be cut for forage, although no grazing is allowed for five years. Eventually, the trees can be cropped for fuelwood, with a typical yield of 38 tonnes/ha from *Prosopis juliflora* and 30 tonnes/ha from *Acacia tortilis*, both on a ten-year rotation (Kaul 1983). *Prosopis juliflora* has also been used in sand dune fixing programmes in Algeria where it proved to be the most successful of a variety of species tested (UNEP 1985).

New Techniques for Sand Dune Fixation in Libya and Egypt

Libya is currently fixing between 3,000 and 5,000 ha of dunes each year but is developing new capital-intensive techniques which it hopes will increase this rate by a factor of ten. Experiments with the use of a petroleum-based emulsion for sand dune fixation have been taking place since 1961. The substance is heated to 45°C and sprinkled on to sand dunes under pressure to form a thin black layer which is relatively permeable, prevents sand grains being blown away by wind, and helps to retain moisture near the dune surface so that the trees and shrubs used to fix the dune permanently can establish themselves successfully. In the Inshas zone of Egypt, other experiments are testing the effectiveness of a polymeric gel soil conditioner for a similar purpose. The gel consists of a polymer manufactured from a by-product of petroleum refineries (propylene gas) that is chemically combined with a cellulose substrate derived from agricultural waste. The experiments are proving successful and offer a low-cost technique for fixing dunes that considerably reduces the need for fertilizer and irrigation applications to trees like *Acacia, Prosopis* and *Atriplex* which are planted together with the gel (UNEP 1985).

THE PROSPECTS FOR SOIL CONSERVATION AND DESERT RECLAMATION

The sand dune stabilization programmes described in this chapter represent some of the best examples of collective action to control desertification in any sector. Perhaps it is because the threat of sand is so apparent, frightening and final that people are strongly motivated to take action. Certainly the task is not easy. In physical terms it would seem much less difficult to avoid soil degradation and ultimately desertification by devising sustainable forms of cropping, livestock raising and forest management, and improving existing practices by introducing soil conservation measures. This can take a long time, however, and "human nature" interposes itself, making targets difficult to attain. Thus the social aspects of controlling desertification often present a greater challenge than purely technical aspects, such as ensuring that trees planted in sand dunes are able to survive. For even if bunds, windbreaks and other soil conservation works are introduced on to farms, their effectiveness can soon decline if farmers fail to maintain them. These projects demonstrate what can be done to turn the tide of desertification, but it is very uncertain how widespread such action will be in the future and how long the beneficial effects of projects will last.

10. Controlling Desertification: Progress and Prospects

Desertification is one of the most serious environmental problems faced by humanity today. It is like a planetary cancer, and one which affects the northern and southern hemispheres and both rich and poor countries. But while the rich pay an economic price for their lack of diligence in taking care of the land, the poor (and their livestock) pay with their health and sometimes their lives. At the United Nations Conference on Desertification (UNCOD) in 1977, Mustafa Tolba, Executive Director of UNEP, called on the world to bring desertification under control by the end of the century (Tolba 1977). Despite the undoubted success of some projects described in this book, very little progress has been made in combating desertification, and it seems doubtful whether UNEP's target will be met. In 1981, four years after UNCOD, Tolba reported to UNEP's Governing Council that progress had been slow (Tolba 1981). Three years later, his report, based on a detailed two-year assessment of progress since 1977, was equally sombre: "Desertification, far from diminishing, has actually increased in severity", but national and international action to control it had been "patchy. ... The goal set by UNCOD to stop it completely by [the year 2000] has been shown to be no longer feasible" (Tolba 1984). This chapter assesses progress made so far, examines the reasons for the success of some projects and the overall dismal record, and suggests ways to improve the scale and effectiveness of future activities.

PROGRESS SO FAR

There are four main ways of assessing progress made in controlling desertification: the degree of fulfilment of the immediate objectives of the Plan of Action agreed at UNCOD; the amount of funding devoted to anti-desertification programmes on a regional and global basis; the number of successful projects and the area of land affected by them; and the development of institutions, policies and human resources to control desertification at national, regional and international levels. This section is largely based on the results of UNEP's 1982–3 global survey of desertification control activities, which remains the most recent source of data.

Progress in Fulfilling UNCOD Short-Term Priorities

The UNCOD Plan of Action to Combat Desertification (summarized in Appendix 1) had three main sections, dealing with: (1) priority actions to be taken by national, regional and international organizations (including non-governmental organizations); (2) practical measures to ensure improved land use in each country, along the lines of those described in Chapters 5–9; (3) ways of modifying national development planning to contribute to combating desertification.

Section 1 of the Plan of Action encouraged governments to take the lead in combating desertification in their countries. The top priority for each government was to nominate or appoint one of its ministries or agencies to co-ordinate anti-desertification activities. Initially, this co-ordinating body would collect data to assess the extent of desertification and monitor the rate at which it was spreading. Once some idea of the magnitude, character and location of desertification had been obtained, it would be possible to use this information to prepare national plans of action to deal with priority issues.

By 1983, six of the nineteen governments in the Sudano-Sahelian region had established special secretariats, ministries or interministerial committees to deal with environmental problems with an emphasis on desertification (Berry 1984a). Afghanistan had established a National Desertification Committee, Bolivia a Committee on Arid and Semi-Arid Lands, and Pakistan a Desertification Monitoring Unit (Dregne 1983b, 1985; Walls 1984). However, only two countries, Sudan and Afghanistan, had drawn up national plans of action; and in nine other countries plans were still in a draft stage. No national plan had yet been put into operation. According to a report prepared for UNEP by Sir E. R. Richardson: "Two national plans and nine draft plans constitute the meagre results of five years of effort on the part of the Desertification Branch" (Richardson 1984). Even the validity of such plans is open to question for, if they bore no relation to the rest of national development efforts, then what existed on paper might never be fulfilled in practice. Harold Dregne of Texas Tech University has suggested that the paucity of national plans after such a long time is a strong indication that they are not politically viable (Dregne 1983b).

Many desertification problems transcend national boundaries, so the Plan of Action encouraged governments to co-operate on matters of joint interest. The most spectacular short-term priorities were six giant transnational projects which were to result in: greenbelts on the northern and southern borders of the Sahara consisting of mosaics of shelterbelts, well-managed cropland and rangeland; the stratification of livestock production in the Sahel so that calves would be reared by nomadic herdsmen in the traditional manner, then moved south for fattening, either on farms in the semi-arid agricultural zone or on feedlots in more humid coastal areas where most would eventually be marketed; regional co-operation to manage aquifers underlying large areas of northeast Africa and Arabia; and satellite monitoring of desertification in West Asia (Afghanistan, India, Iran and Pakistan) and South America (Argentina, Bolivia, Chile and Peru). Almost no progress has been made since UNCOD in any of these projects, although tree planting has continued in some of the countries on the northern fringes of the Sahara, and in 1979 Algeria, Tunisia and Libya signed an agreement to collaborate on the northern greenbelt. Apart from the fact that donors may have been put off by the seeming impracticability of such grandiose ventures, one major reason for their failure to see the light of day was the distrust and even enmity which existed between a number of the countries that were intended to co-operate together. This was apparent even at the time of UNCOD and in most cases relations have worsened since.

The one clearly successful outcome of UNCOD, according to Leonard Berry of Florida Atlantic University (Berry 1984b), has been the greater awareness of desertification worldwide as a result of the scientific meetings, such as those in the USSR and China, which UNEP has sponsored on the subject both before and since UNCOD. These have helped to encourage administrators in affected countries to develop anti-desertification policies, and experts at development agencies to formulate programmes with a greater awareness of their impact on desertification. The UNEP Desertification Branch also publishes a valuable occasional bulletin, *Desertification Control*, to report on activities worldwide. However, it is instructive to note that in 1983–4, when the drought in Africa became even more intense than in the early 1970s, hardly one word was spoken about UNCOD or desertification in the popular media, so the extent of current awareness should not be overestimated.

Desertification has also become recognized as a subject for scientific investigation, as recommended in Section 3 of the Plan of Action. UNCOD was distinguished by the most extensive series of preparatory scientific investigations ever conducted into a global problem before a UN conference. This has continued with the establishment of two of the regional research centres proposed at UNCOD. The Sahel Institute, located in Bamako, Mali, is a specialized agency of CILSS. It undertakes, co-ordinates and disseminates research and proven technologies for the recovery and rehabilitation of degraded lands, and trains research workers and technicians. The Regional Centre for Agrometeorology and Hydrology (AGRHYMET) at Niamey in Niger trains technicians and engineers, carries out applied research into regional climatic and hydrological problems, and produces long-range forecasts. Two other institutions doing valuable research in the region are the International Institute of Tropical Agriculture (IITA) in Ibadan, Nigeria, which works on crop development, and the International Livestock Centre for Africa (ILCA) in Addis Ababa, Ethiopia, which undertakes multi-disciplinary studies of livestock raising and range management. Crop improvement and seed multiplication is also undertaken at the Institut National de Recherches Agronomiques du Niger (INRAN) in Niger. Further research into desertification is being conducted at the Institute for Development Studies at the University of Nairobi, with a focus on sociological studies; the Mazingira Institute, also in Nairobi; the Institute for Development Research in Addis Ababa; and the Institute of Environmental Studies in Khartoum, Sudan. The Environment Training Programme (ENDA) in Dakar, Senegal, provides training courses in ecosystem management and applied ecology and publishes a quarterly bulletin, *Africa Environment*. Despite these efforts, however, a substantial expansion is needed in the scale of scientific research into desertification.

Funding Devoted to Desertification Control

Another measure of progress is the amount of money spent on anti-desertification projects compared with UNEP's target budget for the developing world. This was $2.4 billion per annum at the time of UNCOD, but was later updated to $4.9 billion. Estimating the actual amount of money spent every year on desertification control is not at all easy since funds are channelled to dryland nations in numerous ways. UNEP has set up a Special Account to finance UN anti-desertification activities, but by 1983 it had attracted only $48,524 – subscribed entirely by developing nations (Walls 1984). Most governments have preferred instead to contribute to desertification control through traditional bilateral (country to country) channels and through multilateral agencies such as FAO and the European Development Fund. Between 1975 and 1980, total aid to the eight CILSS nations actually increased twice as fast as the average for all developing countries. By 1980 it had reached $1.5 billion per annum, and on a per capita basis was twice the African average. Altogether the region received $7.45 billion in aid during this period.

How much of this aid has been effective in controlling desertification? This raises a question about what is, and what is not, an anti-desertification project. Considerable funds have been directed to desertification-prone areas such as the Sahel since the mid-1970s, but it is difficult to say how much has had an impact on desertification as such. The UNCOD Plan of Action was extremely comprehensive, so many types of development projects conceivably fall within its scope, of which only a few can help to bring desertification under control, and some may actually exacerbate it.

It is known, however, that only 24% of all aid to the Sahel goes to agriculture and forestry. Two-fifths of this (10%) is for the development of irrigated agriculture, despite its poor performance, its potential for causing desertification, and estimates by the Club du Sahel that as much irrigated land goes out of productive use in the region every year because of poor management as comes into production. While 8% of all aid is directed to rainfed cropping, the majority of this (perhaps as much as 70%) is for cash crops rather than subsistence food crops which, as already shown, have been greatly neglected. Livestock development receives 5% of all aid, but without much success to show for it (Club du Sahel 1981).

In view of these reservations about the current use of funds in the cropping and livestock sectors, the ecology/forestry sector appears the one most likely to have an impact on desertification. Thus, the amount of funds devoted to it could, for want of anything better, be taken as a minimum estimate for funds spent on desertification control in the Sahel. However, this sector received only $104 million between 1975 and 1980 – a mere 1.4% of all aid to the Sahel. While success cannot be guaranteed by a large budget, the low priority accorded to the ecology/forestry sector in the Sahel is significant. The average afforestation rate in the Sahel during that period was a mere 4,800 ha per annum (Grainger 1986). The Club du Sahel has expressed its concern in no uncertain terms: "Do Sahelian governments and the donor agencies understand the gravity of the situation in the Sahel? International aid for reafforestation is still so low that to evaluate its impact would be an exercise in vain" (Club du Sahel 1980).

According to two estimates made for UNEP, annual world expenditure on desertification control is at least $150 million. Leonard Berry estimated that 3.5% of all development assistance to the nineteen Sudano-Sahelian countries, some $150 million per annum, is now allocated directly to desertification control programmes (Berry 1984a). Harold Dregne estimated that only about 10% of the $10 billion supposedly spent worldwide on projects with an anti-desertification component had in fact been directed towards desertification control (Dregne 1984b, 1985). This gives an upper figure of about $1 billion between 1977 and 1983 or $167 million per annum. Neither of these figures – nor the one in the previous paragraph – includes funds derived from NGOs, although they have a prominent role in controlling desertification, mostly in small-scale projects with much lower budgets than those funded by foreign governments and international agencies. Despite this, even Dregne's global estimate is equivalent to less than 4% of UNEP's target annual expenditure on desertification.

How is the aid money being spent? The largest share of development assistance to the Sahel – one-fifth of all aid between 1975 and 1980 – was used for building "infrastructure", mostly the construction and repair of roads. Roads are normally few and far between throughout the tropics; their absence impedes the marketing of agricultural produce and the transport of food aid. But should they consume so high a proportion of aid? By mid-1985 UNSO had spent more than $80 million in constructing 2,123 km of feeder roads in Burkina Faso, Gambia, Mali, Niger and Senegal. The Club du Sahel has criticized road construction as a "soft option": easy to justify and implement but not entirely necessary. Mauritania is building the "Road of Hope" between the capital Nouakchott on the coast and Nema near the eastern border. The aim is to transport food aid rapidly from the coast to the interior in times of famine and to export goods in the other direction. However, French consultant Michel Baumer has commented: "There does not exist a single product the value of which would be worth transport by road. As for famines, if not precisely predictable, they are nevertheless certain, and it would be wiser to build up stocks of food and fodder long in advance, using camel transport. This would give jobs to hundreds of nomads, who are particularly affected by exceptional drought" (Baumer, personal communication). Roads can also have detrimental effects by improving access to previously remote areas and facilitating the spread of economic activity. More roads make it easier for landless people to migrate into an area in search of land to clear and cultivate, and for gangs of woodcutters and their trucks to reach remaining stands of trees which they will then fell to supply the fuel needs of nearby urban areas. UNEP has identified rural roads as having a role in the destruction of the *Acacia nilotica* forests in the Chemama region of Mauritania (UNEP 1985).

Another major category of projects is concerned with water development. Some of these projects are fully justified. The Cape Verde Islands off the west coast of Africa, for example, receive very little rainfall and a $291,400 scheme has improved soil and water conservation in the San Jao Baptista Valley by reafforesting water-catchment areas and constructing flood-control dams, dikes, embankments and terraces (UNSO, personal communication). However, the majority of projects involve the sinking of boreholes for humans and livestock, and support for large-scale irrigation schemes. Little account seems to have been taken of the possibility that the new boreholes may themselves cause desertification. Mauritania has a $5.9 million project to equip thirty-six new deep boreholes with pumping stations. Officially launched in 1977, activities began in earnest in February 1979. Only in November 1979 did UNSO sponsor a visit by a consultant to report on ways of protecting the environment around the boreholes.

Thus the needs identified by UNCOD have not been given a high enough priority, either by donor nations or by Sahelian governments. Even the limited progress achieved in such fields as irrigation and afforestation may not be sustained unless sufficient funds are allocated to meet recurrent costs after project completion. Although donors provide funds to cover project establishment costs to purchase pumps, dig canals and plant trees, they rely on the host country to cover future maintenance. But if the government has no money to pay people to maintain canals and guard trees, then all that investment will be wasted. According to Ann de Lattre of the Club du Sahel: "New projects generate very heavy recurrent budget costs which cannot be met and so the project deteriorates. Phase 2 and Phase 3 of the project then have to be rehabilitation, not expansion. Since 1976 there has been a 35% increase in real aid flows to the Sahel but there is actually massive disinvestment taking place, because of those rehabilitation costs."

Progress in Implementing Projects to Control Desertification

Progress in combating desertification can also be assessed by looking at how successfully the different categories of projects have been implemented. Chapters 5–9 showed the paucity of successful anti-desertification projects. Improvement of rainfed cropping and rehabilitation of degraded irrigated cropland seem to have been largely neglected. Some efforts have been made to tackle rangeland management and livestock development but without much success, and whether ongoing projects will succeed is uncertain. Sand dune stabilization, soil conservation and afforestation projects have been more successful, but the cost has been high, and their extent and overall impact are unknown. It has also proved difficult to mobilize people at community level except in some sand dune stabilization projects. This lack of popular participation does not augur well for the future success of desertification control projects.

Harold Dregne, in his review for UNEP of progress in desertification control, also acknowledged the success of small-scale afforestation and sand dune stabilization projects. According to a survey carried out for UNEP by the Environmental Liaison Centre in Nairobi these account for half of all anti-desertification projects undertaken by NGOs (Dregne 1983b). However, degradation continued on lands that were marginally suitable for rainfed cropping, and there was no improvement in the situation in grazing lands and irrigated croplands. In the absence of data he based his conclusions on informed opinion, stating that: "The results of the assessment were discouraging. Desertification has been halted in only a few places. Unhappily it continues unabated in most of the land that was suffering at least moderate desertification in 1977." He observed that in the Sahel:

> As a consequence of the drought and increasing human and animal pressures, desertification continued as before. This occurred despite the hundreds of millions of dollars spent on what were called desertification projects .... In the remainder of Africa and other arid regions of the world, the same grim picture emerges: much money has been spent, much energy expended, but little to show for the effort. (Dregne 1984a)

The projects reviewed in this book are not entirely representative because, despite the hundreds of millions of dollars spent annually on development schemes, many agencies still have only rudimentary mechanisms for monitoring them and evaluating their achievements. Evaluation is often cursory, few detailed records are kept at headquarters, and projects tend to be assessed mainly with regard to the fulfilment of initial aims, neglecting broader aspects such as environmental impact. If it is difficult to assess the number of projects which have actually contributed to alleviating desertification, estimating the area affected by them is even harder.

Institutional Development

Tackling such a huge problem as desertification demands a considerable amount of organization. Since the mid-1970s a number of new organizations have been founded to support sustainable development in Africa. The Permanent Inter-State Committee for Drought Control in the Sahel (referred to as CILSS after its French name of Comité Permanent Inter-états de Lutte contre la Sécheresse dans le Sahel) was formed in 1973 by Burkina Faso, Mali, Mauritania, Niger and Senegal, primarily to represent the needs of the Sahelian region to donors. These countries were quickly joined by Chad, and later by Gambia and Cape Verde. CILSS works closely with the Club du Sahel, which was formed in Dakar in 1976 by representatives of a number of western development aid agencies to increase co-operation between the Sahelian countries and international donors. The Club's small co-ordinating secretariat is based at the Organization for Economic Co-operation and Development in Paris (Lattre and Fell 1984).

Within the United Nations system, a special UN Sudano-Sahelian Office (UNSO) was set up in 1973 to act as a central mechanism to co-ordinate efforts to aid the Sahelian countries in carrying out their drought recovery programmes, and to administer a UN Trust Fund for activities in the region under the direction of the UN Development Programme (UNDP). UNSO was later given responsibility for anti-desertification activities in the countries of the Sudano-Sahelian region (see Fig. 1.3), while UNEP is responsible for all other areas and for co-ordinating the implementation of the Plan of Action as a whole. UNEP established the Consultative Group for Desertification Control (DESCON) which comprises representatives of donor and recipient nations and UN and multilateral organizations. It acts as a preliminary screening body for potential anti-desertification projects, but so far it has had little influence upon project funding. A further body, the Interagency Working Group on Desertification (IAWGD), was established to share information and co-ordinate planning and action by the different UN agencies working on the problem.

Harold Dregne reviewed for UNEP the performance of the major international bodies responsible for desertification control. He found DESCON, IAWGD and UNEP all wanting:

DESCON has failed to live up to the hope that it would be a new source of funding for anti-desertification projects. Members of donor agencies have shown a notable lack of enthusiasm to finance projects that have been presented and there is no indication that the situation will change. ... As it now operates the IAWGD is ineffective and its meetings are a waste of time .... Finally, failure of the UNEP Desertification Branch to provide the leadership to mount an effective world effort has been most damaging ... [It has been] staffed at 50 percent or less of its proposed complement of professionals and has been managed by an acting director. (Dregne 1985)

Dregne recommended that IAWGD be abolished; the UNEP Desertification Branch be strengthened so it could actually fulfil its mandate to be the focal point of a global programme to combat desertification; and that DESCON be reorientated so that as well as providing a major forum for the exchange of information on desertification control activities it can commission studies on project planning, monitoring and evaluation, and recommend ways to improve implementation of the Plan of Action as a whole.

UNSO was criticized by Lee MacDonald and this author for its poor performance in the late 1970s and early 1980s (MacDonald 1981; Grainger 1983). UNSO has a dual mandate of assisting in economic rehabilitation and combating desertification. Unfortunately, the former aspect received most emphasis until recently, no doubt influenced by UNSO's close relationship with UNDP. Thus road construction accounted for more than two-thirds of all aid channelled to the CILSS countries by UNSO between 1975 and 1980. However, since that time UNSO has become a much more broadly based organization, and is now assisting projects concerned with afforestation, sand dune stabilization and soil conservation.

Actions Taken by Governments

No matter how many new regional and international organizations are established, progress in combating desertification ultimately depends upon the governments of countries in dry areas acknowledging desertification as a national priority and

initiating programmes to bring it under control. Many governments are now aware of the problem and have good intentions
to tackle it. Desertification is mentioned, for example, in the
national plans of Kenya, Somalia and some other countries, and
heads of state like President Moi of Kenya give leadership to
their peoples by continually referring in speeches to the
problems of soil erosion and desertification. Leonard Berry was
optimistic that "practically all ... countries in the [Sudano-
Sahelian] region have developed or will shortly develop national
strategies, plans or institutional machinery for combating
desertification and/or environmental degradation." At the same
time, he was cautious as to whether many countries had
sufficient funds and personnel to translate these policies into
real action, and he was critical of "a tendency in all the plans to
offer a sectoral approach to a problem that is basically multisectoral in nature" (Berry 1984a). Trying to achieve a balanced land
use is difficult when responsibilities for cropping, livestock
raising, forestry and water management are divided between
various ministries. Moreover, earlier chapters have shown how
government policies that favour urban areas rather than rural
areas, cash crops rather than subsistence crops, and cropping
rather than pastoralism, can increase desertification. Shortly
after the completion of UNEP's global survey of desertification
control activities, for which Dregne's and Berry's reports were
commissioned, the Brazilian government announced Projecto
Nordeste, a $12 billion programme which aims to increase the
drought resilience of the country's northeastern region. This
multisectoral programme, supported by the World Bank, covers
agriculture, education and health, and is set to last for fifteen
years. Although the programme was not specifically intended
for the control of desertification, its progress in this regard will
certainly be watched with interest (Magalhães and Rebouças
1988).

CONSTRAINTS ON PROGRESS

Social and Economic Development

That so little progress has been made in combating
desertification could be explained superficially by a lack of
interest or awareness on the part of governments or develop-

ment agencies. There is no doubt that this has been a significant factor, but other constraints have also been important. The most vulnerable countries are still in the early stages of social and economic development, and it is the poorest countries which experience the worst effects of desertification in terms of reduced food production, environmental degradation and the loss of both human and animal lives. Major characteristics of development in the tropics include a general decline in traditional land-use management systems as human numbers increase; the growth of cash cropping; and a breakdown in the cultural fabric which formerly underpinned land-use management. This breakdown results from factors including widespread migration to the cities and the disappearance of markets for traditional products (such as salt traded by nomadic pastoralists in northern Africa).

At the same time as developing countries were called upon to safeguard future food production by preventing environmental degradation, many were struggling with severe economic problems due to the general world recession which started in the late 1970s; falling commodity prices; and the economic impact of drought and desertification – reduced commodity exports, increased imports of food, and the high costs of drought and famine relief. These have all resulted in a lack of finance for fighting desertification and improving agricultural productivity. The situation is particularly bad in Africa, which was saddled with a foreign debt of $175 billion in 1986. This was only about 90% of the combined debt of Brazil and Mexico, but most African countries are much poorer than those of Latin America and find it totally impossible to manage debt repayments of about $12–14 billion a year (Smith 1987). As Michael Prowse of the *Financial Times* has put it: "the 12 African countries most plagued by debt now face scheduled repayments more than four times higher than the sums they could not manage in 1983–85" (Prowse 1987). Africa's earnings from commodity exports in 1986 fell by $19 billion from their 1985 level and the current account deficit tripled to $21 billion. With half their total export earnings needed to service their foreign debt, it is not surprising that African countries lack resources to invest in agricultural development or pay for imports of fuel, fertilizer and spare parts to keep vital equipment like pumps operational (Harden 1987).

Wars

Since UNCOD in 1977, there have been numerous wars, both within and between countries in dryland regions. These have diverted funds from long-term economic development needs. Ethiopia, Sudan and Chad have suffered civil wars. Libya's relationships with most of its neighbours have varied from friendly to hostile. Algeria and Morocco have been in conflict over the former Spanish territory of Western Sahara. Afghanistan has been under attack from guerrillas following its revolution and the entry of large numbers of Soviet troops. Relations between India and Pakistan have continued to be strained, while those between Iraq and Iran boiled over into full-scale war. The year 1988 saw a peace plan for Western Sahara agreed by Morocco and the Polisario Front (backed by Algeria); a cease-fire between Iraq and Iran; improved relations between India and Pakistan following the election of Prime Minister Benazir Bhutto; and the start of a phased withdrawal of Soviet troops from Afghanistan; but it is too soon to tell whether this means that dryland nations are now free to concentrate on peaceful pursuits. Such wars are in addition to the *coups d'état*, revolutions and other social upheavals that are customary in most of these countries and not only hinder the implementation of agricultural development and rehabilitation projects but also help to undermine further (and in some cases totally replace) traditional land-management systems. Between 1978 and 1982 alone, for example, there were five successful coups and three major attempted coups in African dryland countries.

Population Growth

The rapid growth in population offsets the impact of economic achievements in the developing world. The population of the Sudano-Sahelian region, for example, increased by almost a quarter between 1977 and 1984. Another mouth to feed for every four that were there seven years earlier would present a major challenge even without drought, desertification and depressed economic circumstances. But these factors turn the agricultural sector into a treadmill that only ever seems to move one way – backwards.

From the point of view of desertification, the problem is not only that overall populations are rising but that the movement of people leads to higher population densities in some areas; this in turn increases pressures on the land and exacerbates desertification. Drought, famine and desertification force people to flee from their homes to areas where the rains fall or where there are food relief supplies. People also migrate from country to country, fleeing in hundreds of thousands across unmarked borders to find refuge from famine and civil war. These refugees have to be fed and cared for in camps on a temporary basis, and then helped to establish new livelihoods. The relocation process can be traumatic for the inhabitants of areas where refugees are resettled; this causes further difficulties in managing land in a sustainable way. Other people may be moved by government edict against their will, whether starving Ethiopian peasants from degraded, drought-stricken highland areas or nomadic pastoralists forced to settle in villages.

Many people also migrate from the countryside to towns and cities, causing populations in urban areas to grow much faster than those in rural areas. Urban populations in the Sudano-Sahelian region have been growing on average at 5.5% per annum compared with 3.2% for the population as a whole (Berry 1984a). The rise in urban populations results in an increased demand for food, and often for "luxury" foods such as wheat and rice rather than staple foods. On the other hand, the migration of people from rural to urban areas depletes the supply of farm labour. Thus in the absence of sufficient economic growth to allow investment in a more productive agricultural sector, a country's capacity to produce food is reduced. Demand for charcoal and fuelwood leads to major pressures on woodlands, both adjacent to and far from towns and cities. The sheer concentration of people means that limited government funds have to be diverted away from rural development to satisfy the rapidly growing and seemingly limitless needs of urban areas. Governments usually want to keep food prices low in order to prevent opposition from urban dwellers. This also affects the rural economy because it removes incentives for farmers to improve food production.

Lack of Trained Personnel

One of the most crucial factors limiting the development of

desertification control activities is a shortage of qualified personnel. Educational systems are still embryonic in most developing countries, so it is only possible for them to train a relatively small number of graduates at home or abroad every year; this also applies to second- and third-level technical support staff. There may be a delay of up to five years as graduates continue their training and research, often overseas, before returning to assist in the national development effort. However, many are then assigned to high-level administrative posts, again because of the shortage of qualified personnel, so that their overall impact on technical activities is lower than it should be. There is no short-cut to solving this problem because, to be effective, resource management policy has to be developed by local people and not by expatriate advisers.

Complexity and Confusion

Another constraint is that desertification is a highly complex phenomenon, resulting from the impact on naturally fragile lands of a range of interdependent land uses and social and economic factors. To control desertification requires an interdisciplinary approach to land use and project planning and implementation, instead of the traditional sectoral approach that prevails in governments and most development agencies. Moreover, if projects are to be successful, those who plan and implement them should have a high level of social awareness and an ability to communicate, consult and work with local people; professional agronomists and foresters, however, have not been trained for this. Desertification is largely caused by the human misuse of land. Yet, says Harold Dregne, "Practically nothing has been done to improve understanding of the socio-economic factors underlying desertification, the human factor having been almost totally ignored" (Dregne 1984b). Thus, projects which are technically feasible fail because social factors are ignored or mismanaged.

Complexity and lack of knowledge cause confusion. There is still widespread misunderstanding about the relationship between drought, famine and desertification, so that short-term remedies are applied to solve not only short-term problems like drought and famine but also long-term problems like desertification. The way in which the seeds of desertification are sown by imbalanced land-use practices at great distances from deserts and degraded areas, is also not widely understood. Agricultural projects are therefore designed without any appreciation of their possible negative or positive impact on desertification. According to Harold Dregne, confusion about the meaning of desertification may be to blame for governments' general lack of interest in controlling it. They still think of desertification as only an "environmental" problem and do not realize the threat which it poses to their goal of increased food self-sufficiency (Dregne 1983b).

Part of the reason for our failure to control desertification is not confusion but a blatant neglect of past mistakes. Desertification has been occurring for millennia but has been largely ignored in agricultural and forestry training and research in temperate countries. Even when certain agricultural practices, like drilling boreholes for cattle, have been shown to be hazardous or likely to be unsuccessful in achieving their aims, they continue to be advocated by textbooks, and development agencies unwilling to change past habits still agree to fund projects in which they are included. Also inexcusable is poor planning: for example, water development projects which are begun without a detailed assessment of water availability or possible environmental impact. Whether for genuine ignorance or other reasons, development projects often make desertification worse instead of helping to control it.

SUCCESS OR FAILURE?

Identify People's Needs

The projects reviewed in Chapters 5–9 may not be entirely representative of anti-desertification schemes but they do suggest some ways of designing and implementing projects with a better chance of success than many previous ones. The most important lesson is that a project can fail almost before it has begun if, as a result of inadequate preliminary studies, its aims do not correspond to the needs and aspirations of the people who are supposed to implement it and benefit from it. Good examples of this are the woodlot projects in the Sahel where local people place far higher priority on obtaining fodder and food than the fuelwood which the plantations were established to produce. Managing natural woodlands for these and other products therefore stands a greater chance of being successful since it will supply key local needs.

Often what people need most are jobs or the opportunity to improve their standard of living. A number of projects, for example the Eastern Sudan refugee reafforestation project, have been successful because they offered jobs, while the Gujarat farm forestry project and the improved stoves projects in Niger and Gambia offered the chance for entrepreneurs to earn money. The success of social forestry projects in Gujarat and elsewhere, in which local people were employed to plant trees rather than organize the schemes themselves, shows that economic incentives can outweight the less tangible satisfaction of benefiting the wider "community". Identifying needs and capacities also applies to scientific research. The ICRISAT project to develop improved strains of sorghum initially spent much time and money in producing varieties which worked well on experimental stations but not on real farms with poor soils where farmers were not able to afford fertilizers, pesticides and other expensive inputs.

Improve Existing Practices and Institutions

Another key to success is to build on what already exists. Projects which require that existing land-use practices be totally replaced are unlikely to be successful. The emphasis should instead be on "working from within", improving the productivity and sustainability of existing practices in a way which capitalizes on, rather than denigrates, the skills and traditions of local people, and which does not present an economic risk because the value and timing of financial benefits are uncertain. One of the reasons for the success of the Majjia Valley windbreak scheme, for example, was that trees helped to improve the productivity of local agriculture.

The same philosophy also applies to institutional development. In Kenya the soil conservation scheme and the Baringo integrated agricultural development project worked through – and improved – existing local institutions instead of setting up separate project offices which might close after the project ended. The Senegal River irrigation scheme was able to take advantage of village institutions and experience with savings funds to manage the project and collect dues respectively. Some anti-desertification projects owe their success to the participation of NGOs with long-established links with particular villages. Committees already established by NGOs to run activities such as health schemes can also be used to plan and implement tree planting and other projects.

The institutional framework of agricultural societies is often informal and hidden but it is nevertheless deeply embedded in traditional culture. This applies especially to nomadic pastoralism which has relied for hundreds of years upon a complex form of social control to manage rangelands in a sustainable way. It is therefore vital to reinforce and sustain social control where it is seen to be positive, and rebuild it where it has degenerated, as is often the case in nomadic societies (due to outside influences). This is essentially what the pastoral associations in Mali, Senegal and Niger are attempting to do, although the new form of social control will not be identical to the traditional version. Respect for local cultures and their relationship with the land is vital since cultural factors can make theoretically attractive schemes totally unworkable and cause projects to fail.

Involve Local People

The need to consult local people is often stressed nowadays. Yet "consultation" may still mean in practice that officials merely impart information, instead of making a sincere attempt to involve people as full participants in a project. If, as in the Guesselbodi forestry project, people have the opportunity to play a part in designing a project so that it responds to some of their needs, then consultation can be really meaningful. The local people then feel like partners rather than employees.

Inadequate consultation before the start of a project can have serious repercussions later, for it can give rise to bad feeling which, inevitably, comes out into the open. In the Somali soil conservation project, for example, because farmers felt they had not been properly consulted initially they took no interest in maintaining soil conservation works after the project ended. Consultations should take place with all the people who are to participate in a project. Often women are not properly consulted even though they are the group most directly affected by a project. Women are primarily responsible for gathering fuelwood and collecting minor forest products and so will be important participants in the discussions that precede projects to establish woodlots or manage natural forests. Specially trained staff, able to listen and consult rather than just direct and inform, are needed to secure this kind of involvement. The Conservator of Forests in Gujarat established a special group of social foresters as early as 1969 for this purpose. The Majjia Valley windbreak scheme demonstrated that trust between officials and villagers can be vital in deciding whether a project starts at all.

The Power of Demonstration

The power of demonstration is enormous. It is much better to show a farmer an actual working example of the benefits of a new cropping technique than to spend hours describing it to him in words. A farm forestry project in Gujarat, for example, can start on a relatively small scale with a few farmers like Khalidas Patel, but once it becomes successful news spreads quickly by word of mouth, other farmers visit the area to learn about the techniques, and some of them may later apply to join the project. This was also found in other projects already mentioned, including small-scale irrigation on the Senegal River, soil conservation in Kenya, and the use of new cowpea varieties in Nigeria.

Decentralize Management

Trying to manage a resource by building a vast centralized bureaucracy is likely to end in failure. This has happened with a number of large-scale irrigation schemes, such as the Rahad project in Sudan. Better results will be obtained by decentralizing management, splitting a task or an area into smaller parcels which are easier to manage, and giving people participating in the scheme as much freedom as possible to make decisions, as in the Majjia Valley scheme where villagers were given the fuelwood harvested from the windbreaks and the responsibility for deciding how to distribute it. Small-scale irrigation projects such as those on the Senegal River are therefore likely to be more successful than large-scale projects.

Treat Technology With Care

Most dryland areas are still at a relatively low level of technological development. This means that the introduction of any technology should be treated with the utmost care. Irrigation schemes can break down completely if a simple pump goes out of service when there are no technicians to maintain it or foreign exchange to buy spare parts. The emphasis should therefore be on the kind of technology which is appropriate for the area, although this does not always have to be low technology. Nevertheless, project staff should always ask whether mechanical solutions are needed at all. Simply building lines of stones along contours in Burkina Faso, for example, has brought an enormous improvement in the supply of water to fields. Expensive pumps were not necessary. Those projects which do depend upon mechanical components should have a substantial training element so that local people are shown how to maintain and repair the machinery on the spot and farmers must become aware of the importance of maintenance.

Gain Political Support

Finally, strong political support is vital to ensure the success of any project. For a small-scale project this means the support of local village councils and chiefs. For a national project like the Kenya Soil Conservation Programme, active and visible leadership by the head of state can make the difference between success and failure. Even more importantly, political support at the highest level for desertification control can modify agriculture and forestry policies, so that projects which might cause desertification are given less emphasis and those which might control it are given a higher priority. Such support can also transcend divisions between different ministries and help to change government policies in economic, social and other areas which could indirectly prejudice the successful implementation of anti-desertification projects.

PRIORITIES FOR ACTION

In addition to improving the design and implementation of projects generally, there are four main priorities for long-term action that will lay a sound foundation for controlling desertification: better monitoring of desertification; changes in rural policies; improved training of personnel; and the development of the institutional capacity of countries to control desertification.

Better Monitoring of Desertification

One of the top priorities is to increase our knowledge of the extent and rate of increase of desertification. Obtaining reliable data on one of the world's most serious environmental problems is surely reason enough for a major effort to monitor the physical, biological and social aspects of desertification but, as Harold Dregne has pointed out, there is a much more pragmatic reason: "Few decision makers will be convinced of the need to allocate scarce resources to combat desertification in the absence of good data on desertification damage" (Dregne 1985). Thus, while some could object that the main priority ought to be practical action rather than data collection, the ability to undertake that action depends directly upon how well we can justify it by our detailed knowledge of the problem.

How are these data to be collected? Monitoring should take place on a continuous basis, because desertification is a dynamic phenomenon and therefore needs to be monitored regularly. Chapter 4 showed that it is much harder to monitor desertification than, say, deforestation in the humid tropics, but a programme involving satellite monitoring, aerial photography, side-looking airborne radar and widespread field checks would undoubtedly give us much of the information we need. Why have UN agencies not taken action to do this? For many reasons reviewed elsewhere (Grainger 1984a), UNEP and FAO, the two agencies mandated to work in this area, are hesitant to engage in direct global monitoring of natural resources in the tropics. UNEP, despite having a Global Environment Monitoring System (GEMS), has to rely largely on other UN agencies, particularly FAO, to collect data on tropical forest areas, but FAO is apparently more interested in continuing to obtain data as it always has in the past, by requesting it from member countries. Since few developing countries have the proper facilities to monitor vegetation and land-cover changes the result is that few data are collected. "Desertification monitoring is a matter which ... everyone thinks is needed but no one is prepared to do", says Harold Dregne (Dregne 1983b).

The only way to overcome this impasse, in this author's opinion, may be for the United Nations to create a wholly new agency with the express purpose of monitoring vegetation and land use worldwide on a continuous basis. Ideally, this agency would be allocated sufficient resources to have its own purpose-built remote-sensing satellite, ground-receiving stations and central computer systems for storing and analysing data. Unless an effective, continuous monitoring system is established, it is unlikely that there will be progress in developing a methodology for categorizing and classifying desertification in rigorous quantitative terms. Such a methodology is absolutely essential for the analysis of the results of monitoring. It would also help to identify the areas under severe threat of desertification which should have priority for reclamation and protection.

Changes in Government Policies

In many countries, governments will have to rethink their policies which relate to rural areas. Concentrating on urban and industrial development and keeping food prices artificially low has had severe repercussions on the viability of the agricultural sector in many countries. Exasperation at such policies led tens of thousands of Indian farmers to stage a sit-in protest in Delhi in October 1988 (Brown 1988). Ways must be found to restore the balance between urban and rural economies so that it is in the best interests of farmers to make agriculture as sustainable and productive as possible. If rural communities are prosperous then people will think twice before migrating to towns and cities in search of jobs. This would help to control urban expansion and therefore reduce the rate of increase on government spending on urban areas and the spread of environmental degradation around them.

Related to this is the need for a change in the emphasis which governments give to different types of agriculture. At present, irrigated cropping generally benefits at the expense of rainfed cropping, and cash crops at the expense of subsistence food crops and pastoralism. While more irrigation may appear to be the logical way to increase agricultural productivity in dryland areas, experience and the considered opinion of bodies like the Club du Sahel teaches otherwise. Growing more cash crops is an attractive way to increase agricultural earnings, but too much emphasis on cash crops can lead to a decline in the production of food for domestic consumption. Countries then have to continue expanding cash-crop exports to pay for ever-increasing food imports. Nomadic pastoralism may appear outdated, and nomads may seem to be a population effectively outside the orbit of the law and the taxman. But nomadic pastoralism may well be the only long-term sustainable use for large areas of marginal lands and so should not be lightly discarded. Factors like these demand a major change in land-use policies, towards a more integrated approach to natural-resource management, and away from the present fragmented attitude that completely neglects the way in which the expansion of one land use can impact on others and lead to desertification.

Improved Training of Personnel

Since lack of trained personnel is one of the most crucial constraints on effective action to control desertification, there should be a new emphasis on human resource development at every level. People are the most precious asset of any country, so the best hope is to continue to expand the education systems of developing countries so that the young people of today will become the skilled farmers, agronomists, foresters and planners of tomorrow. The people of the dryland regions must themselves take the main part in developing new and more appropriate cropping systems which are sustainable even under extreme climatic conditions. This is preferable to continuing to try to adapt the agricultural techniques used in developed nations where conditions are different. When experts in tropical countries make discoveries and develop land-use policies for themselves, their governments are far more likely to heed their advice than that of outsiders. United Nations and other development agencies, as well as donor nations generally, should therefore step up their support for the education and training of the people of developing countries, both at home and overseas. Many development projects now include training components and require the appointment of local staff to serve as counterparts to expatriate advisers, but projects are usually too short for such training to make a meaningful addition to their accomplishments. Experience described earlier in this book suggests that longer-term projects can improve this situation since trainees would then know that there will be a project to return to after they finish their courses of instruction or research.

The retraining of existing professionals is also needed to overcome two major obstacles to desertification control – sectoral insularity and a purely technical approach to project design and implementation, disregarding social aspects. Sectoral insularity refers to the traditional tendency for agronomists, foresters and other professionals to carry out their work in isolation from other sectors. Agronomists, for example, would regard it as strange to plant trees as part of a cropping project, and many foresters would be uncomfortable about opening up forest reserves to partial cultivation, even if at the same time new trees were being planted by the cultivators. This book has shown how short-sighted such approaches are, and how they result in unprotected farmlands suffering from terrible soil erosion, and in "forest reserves" without trees because there are not enough people to plant them.

All too often projects fail because they have been designed in a social vacuum, and people living in the project area are regarded as labourers rather than as participants. Meetings before the start of projects are considered to be occasions for officials to inform local people of what is to happen, rather than to listen to villagers' advice and modify their project design accordingly. An unwillingness to listen to such advice can cause projects to fail since they may contain components that are quite infeasible on the basis of local knowledge. Social and cultural factors, such as traditional usage rights or products which are important to local people, may be entirely overlooked and this may not merely make a project unattractive but also incite active opposition to it.

Institutional Development

Shortages of trained personnel slow down the development of

institutions that are capable of actually implementing government policies to combat desertification and develop sustainable land uses. The lack of a sufficiently broad institutional base is a serious constraint on progress, and there is little which can be done quickly to overcome this problem. One strategy would be for governments to encourage the development of, and create close links with, indigenous NGOs, since these have at their disposal a vast national network of voluntary workers, educators and fund raisers able to (for example) plant trees or improve health care. Bearing this in mind, it is no surprise that in some countries more trees are planted by NGOs than by government forest departments. NGOs have been referred to by Harold Dregne as perhaps the most effective agents in the campaign to combat desertification (Dregne 1983b). However, they generally suffer from a lack of technical support, and closer collaboration with government agencies could help to alleviate this problem. More recognition of the importance of NGOs by senior government officials is also needed. Officials at local level will then be more willing to give their time freely to help NGOs realize national aims.

Institutional development is also needed at the international level. Leonard Berry, for example, has pointed out that there is no regional organization in East Africa to match CILSS in West Africa (Berry 1984a). While there are historical reasons for the distrust of such organizations in the region, for example the abortive East Africa Federation whose failure still affects relations between Kenya and Tanzania, it must surely be possible to overcome differences at a relatively informal level, bearing in mind not only the shared threats to lands, agriculture and forestry in the region, but also the very serious refugee problem. Commenting on the fact that UNEP, the main co-ordinator and instigator of anti-desertification programmes, has been unable to staff its Desertification Branch fully, Harold Dregne has said that: "Failure of the UNEP Desertification Branch to provide the leadership to mount an effective world effort has been most damaging. Strengthening the Desertification Branch, as UNEP Executive Director Mustafa Tolba has proposed, is vital" (Dregne 1984a). Unless UNEP has sufficient personnel to provide a competent focal point for the collection of data on trends in desertification and progress in combating it, donor nations and development agencies could be forgiven for thinking that desertification is not as serious a problem as they are often told it is. Hopefully, the revival of interest shown by governments in the world environment, as demonstrated at the seven-nation economic summit in Paris in July 1989, and the UNEP Governing Council two months earlier, will lead to a strengthening of UNEP generally and thereby improve its capacity to co-ordinate the global campaign against desertification.

Priorities for Developed Countries

The governments of developed countries can help to control desertification by supporting the governments of dryland countries in their efforts to improve rural policies, develop institutional capabilities, train more skilled personnel, and generally develop local solutions to desertification problems. They can insist that all development projects which they fund in dry areas, either directly or through multilateral agencies, be evaluated (both before and after implementation) for their likely impact on desertification. They should also be more selective in the way in which they allocate development assistance. More funding should be given in future to projects which improve the productivity of rainfed cropping, rehabilitate degraded irrigated cropland, and reclaim other desertified land; to small-scale forestry and irrigation projects; and to projects which bring natural woodland under local management. A greater proportion of funding should be channelled through NGOs or to projects in which they are involved. Governments should greatly increase funding to scientific programmes which monitor desertification on a global basis. They should also ensure that UNEP has sufficient funds to fulfil its crucial coordinating role. Finally, the governments of developed countries which include drylands should ensure that these are well managed.

CONCLUSIONS

This book has been partly optimistic and partly pessimistic. Desertification is a tremendous and disturbing phenomenon, about which we know something but not a great deal. The UNCOD Plan of Action listed many of the techniques which could in principle control desertification but, as this book has shown, we seem to lack the ability to apply these techniques in the real world where social and economic factors and misguided government policies hve such a great influence. The almost total neglect since UNCOD of the role of social and economic factors in desertification does not give cause for hope that such constraints will be overcome in the ñear future (Dregne 1983b).

Confusion about the nature, scale and significance of desertification is also hampering efforts to learn more about it and bring it under control. Desertification is one of the most serious threats to the global environment, yet for some reason we are fearful of tackling it as the major scientific problem which it undoubtedly is, and hesitant about allocating sufficient technical and financial resources to monitoring its extent and rate of change. Many governments in dryland areas, even now, consider desertification to be just an "environmental problem", rather than something which is undermining their agriculture and their aspirations for national economic development. They therefore do not give priority to programmes explicitly designed to control desertification, and fail to see the link between the economic policies which they adopt and poor land use in rural areas. They stumble from drought to drought and from famine to famine, responding to such crises with short-term solutions, instead of adopting a long-term approach to controlling the underlying problem of desertification. This would help to reduce the impact of climatic fluctuation, and lay the foundations of a sustainable future for the drylands and the people who inhabit them.

Future hopes for controlling desertification therefore lie in the hands of both scientists and government leaders. We urgently need to obtain more reliable data on the scale and rate of change of desertification, and to look far more deeply at the human processes which contribute to it. Information, as Harold Dregne has said so forcefully, is absolutely crucial to the perceptions of governments. Scientists can find that information and bring it to the attention of governments who would then have a strong justification to take action. (The astonishing speed at which the governments of developed nations reacted to the discovery of a hole in the earth's atmospheric ozone layer and agreed on a sharp reduction in the use of chlorofluorocarbons is a good example of what can be achieved.) It is now twelve years since UNCOD. Without comprehensive and detailed scientific studies to fill these vital gaps in our knowledge of desertification, it is likely that by the end of the century

– the UNCOD deadline for bringing it under control – the world's governments will still not even have become convinced of the seriousness of the problem and the need for concerted action.

One factor which could bring an end to this apathy is the imminence of global climatic change forecast to occur over the next 50 to 100 years as a result of the greenhouse effect. The results of research into desertification, and the experience gained in controlling it, could be crucial if we are to devise practical strategies to cope with the effects of impending changes in world temperature and precipitation patterns. We cannot be sure whether or not the prolonged drought in the Sahel and East Africa is one of the first major consequences of the greenhouse effect and, as was argued at length earlier in this book, drought is still only regarded as an indirect cause of desertification. Nevertheless, if the forecasts are true, farmers all over the world will have to learn to adapt their agricultural practices to cope with greater extremes of climate. In those areas where climatic shifts will make crop production a marginal activity, there is a very real danger that overcultivation will degrade the land. The parallels and possible linkages between desertification and the greenhouse effect are therefore apparent, even if not clearly defined or proven, and we should do well to place far more emphasis on desertification control activities now in order to have a sound basis for wider action in the future.

Desertification will be with us for a long time to come, and like a chronic disease it demands continued medication. Governments north and south, and the international agencies which serve them, must stop ignoring the problem – and the feelings of millions of ordinary men and women across the planet that something must be done to bring desertification under control. This will require vision to realize that it is a priority problem, leadership to show the way forward, statesmanship to bridge gaps between different nations and different sections inside nations, and imagination to grasp the immense scale of the problem and identify strategies capable of solving it. Also needed will be a sense of humility, so that sectional interests, whether among scientists, administrators, communities or nations, can be transcended. Given the prevalence of armed conflict between and within dryland nations, perhaps this may be too much to hope for, but a start must be made soon.

Appendix: Summary of the UNCOD Plan of Action to Combat Desertification

A. RECOMMENDATIONS FOR NATIONAL AND REGIONAL ACTION

(1) Evaluation of desertification and improvement of land management

- Recommendation 1: Countries should assess the magnitude and extent of desertification, its causes and effects, by monitoring dryland dynamics, including the human condition. Satellite imagery should be used where appropriate. Atmospheric processes, soil cover, vegetation, shifting dunes, wildlife, crop yields, demography, human health and nutrition are among the items to be monitored. Desertification maps should be prepared.

- Recommendation 2: Techniques of land-use planning and management based on ecologically sound methods should be introduced into areas affected or likely to be affected by desertification, in conformity with social equity and geared to fostering social and economic development. In any area a pilot project should be used to test plans, which should allocate within the area such uses as crops, livestock, wildlife, biosphere reserves, mining, industry, recreation, tourism, urbanization and roads.

- Recommendation 3: Public participation should be made an integral part of action to prevent and combat desertification and account should be taken of the needs, wisdom and aspirations of people. Special educational packages and the full use of extension services are needed to increase general public awareness of the problem. Public discussion should be maximized, by making use of mass media and encouraging the expansion and strengthening of community organizations.

(2) The combination of industrialization and urbanization with the development of agriculture and their effect on ecology in arid areas

- Recommendation 4: UNEP, in conjunction with UNDP, UNIDO and other UN organizations, are recommended to study and publicize positive and negative local and world-wide experience of the role which industrialization and urbanization play under different social and economic conditions in changing the ecological status of the environment and intensifying, preventing or eliminating the processes of desertification in arid areas.

(3) Corrective anti-desertification measures

- Recommendation 5: Environmentally-sound management and development of water resources should be introduced as part of measures against desertification, and the UN Water Conference recommendations should be endorsed. Countries should develop local and appropriate water technologies, using local materials. Various appropriate technologies are mentioned, including rainwater collecting from roofs, sand filters, water harvesting, low-cost reservoirs, solar and wind pumps, and composting. There should be regional data banks, flood control and river pollution schemes. Revegetation of watersheds and other measures to reduce erosion, flood hazards and siltation are recommended. Appropriate methods for water purification, recycling, pollution control and desalinization should be introduced.
- Recommendation 6: Degraded rangelands should be improved, with better range management, wildlife management, and improved welfare for pastoral communities. Specific techniques suggested include deferred or rotational grazing; managed water points; new breeds and species of livestock; forage reserves; drought grazing reserves; use of crop residues and agro-industrial wastes as fodder; removing stock from drought areas; better livestock marketing; mobile abattoirs; irrigated production of forage; price stabilization and marketing schemes for nomads; better land tenure and water rights; control of tourism; use of rangeland plants for fibre, alcohol, beverages, medicine; health, education and

other social services for dispersed populations; assistance with resettlement or partial sedentarization. Regional centres should develop drought-resistant, high-yield and nutritious forage plants. Systems which combine forestry, agriculture and livestock raising should be studied, designed and implemented so as to give rise to economically and socially stable systems.

- Recommendation 7: Rain-fed farming should be improved so as not to cause desertification. This could include cover crops, rational use of organic and chemical fertilizers, terracing, strip cropping, shelter belts, sand stabilization, diversification of farming systems and better land tenure. Watersheds, upland pastures and woodlands should be revegetated and protected from grazing, cultivation or cutting for fuel where this entails a risk of desertification.

- Recommendation 8: Although new arid lands should be irrigated and each nation must determine its own priorities, "at the global scale the greater urgency is for the rehabilitation and improvement of existing (irrigation) schemes". Measures should be taken to combat waterlogging, salinization and alkalinization. Detailed suggestions are given as to how this should be done, including the development of salt-tolerant crops.

- Recommendation 9: Restore, maintain and protect vegetative cover so as to stabilize and protect soils in denuded areas, especially on watershed and mountain slopes, dunes, and where villages, roads and farms are threatened. The establishment of shelter belts and other tree plantations can be an essential part of the reclamation of degraded areas. Specific suggestions include restoring damage by mines, industry and tourism; fenced reserves around settlements; traffic control, especially of off-road vehicles; stabilizing sand via matting, mulches, bitumen coatings, dune revegetation; restrictions on firewood-gathering, and establishment of fuelwood plantations; greenbelts – "mosaics of revegetated areas". Regional co-operation will be needed for the production and distribution of plants for large revegetation campaigns.

- Recommendation 10: Governments should take all necessary steps to conserve the flora and fauna in areas subject or likely to be subject to desertification. Regulations should be adopted governing the import, sale or use of threatened animal and plant species. Regional co-operation in conservation is desirable, especially on the fringes of national parks and where a certain ecosystem extends across international boundaries.

- Recommendation 11: National or intra-regional systems for monitoring climatic, hydrological or pedological conditions and the ecological conditions of land, water, plants or animals should be established or strengthened, as appropriate, in areas affected or likely to be affected by desertification.

(4) Socio-economic aspects

- Recommendation 12: The social, economic and political factors which have an important bearing on desertification should be analysed and evaluated, particularly inequitable relationships at national, regional, international levels, and methods to equalize them. New educational programmes are needed, that will allow the adult population especially to become fully aware of the ecological aspects of development, and to participate more fully in programmes aimed at combating desertification.
- Recommendation 13: Countries which so wish should adopt demographic and economic policies that will ameliorate problems arising from population growth, e.g. increasing pressure on rural lands, and migration to urban areas which increases problems there as well as leading to labour shortages and poorer land use in rural areas.
- Recommendation 14: Peoples affected by desertification should be provided with health services, including family planning, comparable to the rest of the population. Services could include mobile health teams, paramedical personnel and air services. The resources of the local pharmacopoeia (i.e. local drugs) should be fully utilized. It should be remembered that "the problems of metabolism among rural dryland peoples ... are quite different from those in urban areas."
- Recommendation 15: Human settlement in drylands should not conflict with land productivity, and should take full account of local climate, local building materials, local architectural traditions and social habits. Buildings should com-

bine modern design with the principles that "made traditional abodes comfortable"; they should minimize energy demand for heating and cooling; use solar and wind energy where possible. Regional co-operation is needed to share the traditional experience and knowledge of design and materials which often crosses political boundaries.

- Recommendation 16: In lands subject to desertification, national systems should be established for monitoring the human condition. Such variables as population, health, food, settlements, education, production, should be recorded. This applies to migration and movements across borders during disasters.

(5) Insurance against the risk and the effects of droughts

- Recommendation 17: There should be insurance schemes to cope with drought disaster: official relief action should recognize the indigenous strategies which people already possess to cope with drought, "rather than ignoring or damaging them". Techniques should include crop and livestock insurance; fodder, food, fuel and seed reserves; permanent breeding stocks of animals; refugee pasture areas; stockpiles of tools for relief employment. Regionally, movements of peoples across frontiers should be encouraged during severe drought, as this is a form of insurance against further loss.

(6) Strengthening science and technology at national level

- Recommendation 18: Indigenous scientific and technical capabilities should be strengthened, imported technologies should be adapted to suit local conditions. Programmes to revive traditional techniques for combating desertification should be encouraged.
- Recommendation 19: Alternative and unconventional energy sources should be encouraged. "Collection of woody plants and charcoal manufacture are at present, and are likely to remain, the main source of energy for many arid land dwellers." Woodlots, forest reserves, new woody species, improved charcoal manufacture, could increase the supply of fuel; solar energy should be used for water pumps, water

heaters, cookers, refrigerators, food dryers; wind energy should be used for pumping and energy production; biogas could be made from animal wastes, and charcoal, gas and oil via pyrolysis from grain husks, peanut hulls, crop stalks and palm leaves. Such devices should be subsidized and distributed.

- Recommendation 20: Training and education should be undertaken, especially via the mass media. This would include out-of-school education, communication with remote areas, radio, TV, cinema, pamphlets, posters, demonstration centres for farmers and nomads and in courses at schools and universities.

- Recommendation 21: Co-ordinated national machinery to combat desertification and drought should be established where none exists. This national body would analyse, evaluate and disseminate existing information on desertification; prepare a national plan of action and arrange the financing for its implementation; monitor the progress of anti-desertification measures; and liaise with regional and international organizations.

(7) Integration of anti-desertification programmes into comprehensive development plans

- Recommendation 22: Programmes to combat desertification should be formulated, whenever possible, in accordance with the guidelines of comprehensive development plans at the national level.

B. RECOMMENDATIONS FOR INTERNATIONAL ACTION AND CO-OPERATION

(1) International action

- Recommendation 23: UN agencies and other international bodies (UNDP, UNEP, UNIDO, UNCTAD, FAO, WHO, WMO, UNESCO, World Bank and the UN regional commissions are specifically mentioned) should revise and adapt their activities to incorporate the Plan of Action. The UN system should, before 1980, develop a methodology for assessing, monitoring and forecasting desertification; make available technical assistance to governments; publish a desertification atlas by 1978–82; promote rational management of arid and semi-arid rangelands; provide financial and technical support for the sedentarization of nomads; increase the effectiveness of disaster relief organizations; research and develop simple, cheap, efficient alternative energy devices for drylands.

• Recommendation 24: The UN General Assembly should endorse the activities of the World Meteorological Organization, the International Council of Scientific Unions and interested UN agencies that are directed at understanding and resolving climate problems, and urge governments, international agencies, and other interested bodies to support and participate in the planning and execution of the World Climatic Conference, and the Global Atmospheric Research Programme.

• Recommendation 25: The UN should invite intergovernmental and non-governmental organizations to participate in the Plan of Action.

(2) International co-operation

• Recommendation 26: The processes of desertification at times transcend national boundaries. While recognizing national sovereignty, it is recommended that countries should co-operate in the sound and judicious management of shared water resources as a means of combating desertification effectively.

C. RECOMMENDATIONS FOR IMMEDIATE INITIAL ACTION

(1) At national level

Governments are recommended to:

a. Establish or designate a national body to combat desertification (Recommendation 21).
b. Assess the extent of desertification in the country (Recommendation 11).

c. Establish national priorities for actions against desertification.

d. Prepare a national plan of action against desertification.

e. Select among national priorities those actions which could be taken nationally, with the support of regional and international organizations or other foreign sources, in the framework of regional or international co-operation, or only with foreign aid.

f. Prepare and submit requests for international support for specific projects within the above priorities, if required.

g. Implement actions in accordance with national plans of action.

(2) At regional level

a. Convene regional post-conference technical workshops and seminars.

b. Organize inter-regional consultations and studies for selecting sites for the establishment of regional centres such as:

- six regional anti-desertification centres, for Latin America, S. Asia, W. Asia, Africa south of the Equator, Sahel, N. Africa, based on existing institutes or on new sites;
- a number of international rangeland and livestock management centres;
- five international sand dune fixation centres;
- management-training irrigation farms;
- rainfed agriculture demonstration farms;
- revegetation/afforestation demonstration stations;
- regional networks of biosphere reserves.

c. Organize and implement six transnational projects:
- an integrated scheme for management of livestock and rangelands in the Sudano-Sahelian belt;
- management of regional aquifers in N.E. Africa and Arabia;
- satellite monitoring of desertification in S.W. Asia;
- satellite monitoring of desertification in Latin America;
- establish a "greenbelt" – a mosaic of forest plantations and sustainable rangeland and cropland – on the northern border of the Sahara from Algeria to Egypt;
- establish a similar green belt in the Sahel from Cape Verde to Sudan.

(3) At international level

a. UN agencies are requested actively to associate themselves with the implementation of relevant parts of the Plan of Action.
b. Governments are requested to put forward their needs for international support for their national plans of action.
c. Formulate specific actions by appropriate joint programming.
d. Mobilize the necessary financial resources.
e. Arrange, co-ordinate and design specific projects and strategies for financing and implementing anti-desertification programmes.

D. RECOMMENDATIONS FOR IMPLEMENTATION OF THE PLAN OF ACTION

- Recommendation 27: UNEP with its Governing Council and Environment Co-ordination Board should be responsible for following up and co-ordinating the implementation of the Plan of Action. Regional commissions of the UN should be responsible for co-ordinating, catalysing and executing intra-regional programmes adopted by member states. The regional commissions should in this regard actively participate in the Environment Co-ordination Board.
- Recommendation 28: The following forms of financing are recommended for consideration:

a. sub-regional co-operation;
b. bilateral, multilateral and multi-bilateral assistance;
c. consultative group/club or group-type financing;
d. special account;
e. additional measures such as funds in trust, fiscal measures entailing automacity, and an international fund.

Bibliography

Abdou, N., M. Keita, G. Kaka, H. Hamadou, J. Eriksen, C. Beal, B. Skapa and G.B. Greenwood, 1985. *The Niger Integrated Livestock Project. A Mid-Term Evaluation* (Washington DC: USAID).

Abernethy, C.L., 1984. "Irrigation in Africa: present situation and prospects", in D.L. Hawksworth (ed.) 1984: 342–6.

Adams, R., 1975. "Comments", in International Geophysical Union, 1975, Working Group on Desertification, *Proceedings of a Meeting on Desertification*, Cambridge University, 22–26 September: University of Cambridge, Department of Geography: 138.

Adams, W.M. and A.T. Grove (eds), 1983. *Irrigation in Tropical Africa. Problems and Problem Solving*, Cambridge African Monographs No. 3 (University of Cambridge: African Studies Centre).

Adefolalu, D.K., 1983. "Desertification of the Sahel", in Ooi Jin Bee (ed.), 1983: 402–38.

Africare, 1982. "A proposal for sand dune stabilisation, Brava, Somalia, April, 1983" (Washington DC: Africare).

Africare, 1984. "Brava Sand Dune Stabilisation Project. Democratic Republic of Somalia, Second Progress Report" (Washington DC: Africare).

AgRISTARs, 1983. *Research Report – Fiscal Year 1982* (Houston: NASA).

Ali, S.H., 1980. "Practical experience of irrigation reform, Andhra Pradesh, India", IDS Discussion Paper No. 153 (University of Sussex: Institute of Development Studies).

Anon., 1977. "Desertification: an overview", UN Conference on Desertification, Nairobi, 29 August – 9 September 1977, Document A/CONF.74/1/Rev.1 (Nairobi: UNEP).

Anon., 1982a. *The State of India's Environment 1982* (New Delhi: Centre for Science and Environment).

Anon., 1982. "Women of the trees". *Development Forum*, Jan–Feb. 1982: 3–4 (UNDP, Geneva).

Anon., 1983. *Tunisia: The Wheat Development Programme*, AID Project Impact Evaluation Report No 48 (Washington DC: USAID).

Anon., 1987a. "Wetter weather linked to greenhouse effect", *New Scientist*, 9 July: 27.

Anon., 1987b. "Neither rain nor famine", *Economist*, 15 August: 27.

Anon., 1987c. "Sahara sand falls over Britain", *Sunday Telegraph*, 18 August: p. 1.

Anon., 1988a. "The blessed rain", *Economist*, 2 July: 54.

Anon., 1988b. "Magma flows blamed for El Nino events", *New Scientist*, 10 December: 17.

Anon., 1988c. "Twin plagues of war and famine", *Time*, 28 March: 39–40 (US edition).

Anon., 1988d. "Sudan counts the cost of a disastrous war", *New Scientist*, 12 November: 24.

Anson-Meyer, M., 1983. "Les illusions de l'autosuffisance alimentaire: exemple du Benin, du Ghana, du Nigéria et du Togo", (Illusions of food self sufficiency: the example of Benin, Ghana, Nigeria and Togo), *Mondes en Développement* 11: 51–79.

Anteneh, A., 1984. "Trends in sub-Saharan Africa's livestock industries", in D.L. Hawksworth (ed.), 1984: pp. 224–32.

Arizo-Nino, E.J., L. Herman, M. Makinen and C. Steedman, 1981. "Livestock and meat marketing in West Africa. Vol. 1 synthesis, Upper Volta" (Washington: USAID).

Arnold, J.E.M., 1983. "Community forestry and meeting fuelwood needs", *Commonwealth Forestry Review* 62: 183–9.

Arnold, J.E.M., 1984. "Forestation for local community development" in K.F. Wiersum (ed.), *Strategies and Designs for Afforestation, Reforestation and Tree Planting*: 48–62 (Wageningen, Netherlands: Pudoc).

Arnold, J.E.M. and J.J. Jongma, 1978. "Fuelwood and charcoal in developing countries: an economic survey", *Unasylva* 29 (118): 2–9.

Arnold, R.G., 1986. "Opening remarks", in D.L. Hintz and J.R. Brandle (eds), 1986: 1–3.

Aronson, D.R., 1982. "Toward development for pastoralists in Central Niger. A synthesis report of the work of the Niger Range and Livestock Project" (Niamey: USAID/Republic du Niger).

Aubréville, A., 1949. *Climats, Forêts et Désertification de l'Afrique Tropicale* (Paris: Société d'Edition Géographiques, Maritimes et Coloniales).

Bach, W., 1979. "Short-term climatic alterations caused by human activities: status and outlook", *Progress in Physical Geography* 3: 55–83.

Baker, R. St B., 1954. *Sahara Challenge* (London: Lutterworth Press).

Baker, R., 1984. "Protecting the environment against the poor. The historical roots of the soil erosion orthodoxy in the Third World", *The Ecologist* 14: 53–60.

Bansil, P.C., 1983. "A critique of Indian irrigation policy. Managerial factors in water use", *Quarterly Economic Report* 27(4): 21–34 (Indian Institute of Public Opinion).

Barnett, T., 1984. "Small-scale irrigation in sub-Saharan Africa: sparse lessons, big problems, any solutions?", *Public Administration and Development* 4: 21–47.

Barry, M.A., 1982. "Cap-Vert: halte à la sécheresse" (Cape Verde: a stop to drought), *Jeune Afrique* 22: 41–4.

Barry, R.G. and R.J. Chorley, 1987. *Atmosphere, Weather and Climate*, 5th edn (London: Methuen).

Basu, N.G., 1984. "Community forestry and the local community", in K.F. Wiersum (ed.), 1984: 193–204.

Baumer, M., 1981. "Livestock stratification: help or hindrance to arid land pastoralism", *Mazingira* 5: 72–80.

Benge, M., 1987: "Information memorandum", 12 January, USAID, Washington DC, unpublished.

Bennett, K., 1984. "Gambia National Improved Cookstoves Project. Stove technologist, end of mission report, August 1983–February 1984" (London: Intermediate Technology Consultants Ltd).

Benoit, M., 1984. "Le Séno Mango ne doit pas mourir: pastoralisme, vie sauvage et protection au Sahel" (The Séno Mango must not die: pastoralism, wildlife and the protection of the Sahel), Mémoires de l'ORSTOM No. 103 (Paris).

Bernard, F.E., 1985. "Planning and environmental risk in the Kenyan drylands", *Geographical Review* 75(1): 58–70.

Bernus, E., 1977. "Case study on desertification. The Eghazer and Azawak Region, Niger", United Nations Conference on Desertification, Nairobi, 29 August–9 September 1977, Document A/CONF.74/14 (Nairobi:UNEP).

Berry, L., 1984a. "Assessment of Desertification in the Sudano-Sahelian Region 1978–1984", United Nations Environment Programme Governing Council, 12th Session, 16 May 1984, Item 9 of the provisional agenda, UNEP, Nairobi.

Berry, L., 1984b. "Desertification in the Sudano-Sahelian Region 1978–1984", *Desertification Control Bulletin* 10: 23–8.

Berry, L. and R.B. Ford, 1977. "Recommendations for a system to monitor critical indicators in areas prone to desertification", Clark University, Massachusetts, unpublished.

Beshai, A.A., 1984. "The economics of a primary commodity: gum arabic", *Oxford Bulletin of Economics and Statistics* 46: 371–81.

Beudot, F. (ed.), 1982. "Elements for a bibliography of the Sahel drought" (Paris: OECD Development Centre).

Bhuiyan, S., 1985. "Irrigation technology for food production: expectations and realities in South and Southeast Asia", paper presented at the Conference on Water and Water Policy in World Food Supply, Texas A & M University, 26–30 May 1985.

Bills, N.L. and R.E. Heimlich, 1984. "Assessing erosion on US cropland: land management and physical features", Agricultural Economic Report No. AER-513 (Washington DC: Economic Research Service, USDA).

Bjerknes, J., 1966. "A possible response of the atmospheric Hadley circulation to equatorial anomalies of ocean temperature", *Tellus* 18: 820–9.

Bjerknes, J., 1969. "Atmospheric teleconnections from the equatorial Pacific", *Monthly Weather Review* 97: 163–72.

Bjerknes, J., 1972. "Large-scale atmospheric response to the 1964–65 Pacific equatorial warming", *Journal of Physical Oceanography* 2: 212–17.

Blackie, M.J. (ed.), 1984. *Proceedings of African Regional Symposium on Small Holder Irrigation* (Harare: University of Zimbabwe).

Blaikie, P., 1985. *The Political Economy of Soil Erosion in Developing Countries* (London: Longman).

Blench, R., 1985. "Pastoral labour and stock alienation in the sub-humid and arid zones of West Africa", Pastoral Network Paper No. 19e

(Cambridge: Agricultural Administration Unit, Overseas Development Institute).

Bliss, F., 1983. "Probleme der Neulandentwicklung im Wadi al-Gadid. Westliche Wüste Agyptens" (Development problems in newly reclaimed areas in the Wadi al-Gadi in the Egyptian western desert), *Tropenlandwirt* 84 (April): 63–78.

Bliss, F., 1984. "Wüstenkultivierung und Bewässerung im 'Neuen Tal' Agyptens" (Cultivating and irrigating the desert in the "New Valley" in Egypt), *Geographische Rundschau* 36: 256–62.

Blom, P.S., 1981. "*Leucaena*, a promising versatile leguminous tree for the tropics", *International Tree Crops Journal* 1: 221–36.

Bothomani, I.B., 1984. "The food crisis in East and Central Africa with special reference to Kenya, Malawi, Tanzania and Zambia", *Journal of African Studies* 11: 148–55.

Bottrall, A., 1982. "Management – irrigation's soggy centre", *International Agricultural Development* July/August 1982: 20–21.

Boukhobza, M., 1982. "Pastoralisme et processus de désertification de la steppe algérienne" (Pastoral farming and erosion in the High Plateaux steppe lands of Algeria), *Production Pastorale et Société* 10: 64–67.

Bowonder, I.B. and Ravi C., 1984. "Waterlogging from irrigation projects: an environmental management problem" (Hyderabad: Centre for Energy and Technology, Administrative Staff College of India).

Breed, C.S. and J. F. McCauley, 1986. "Use of dust storm observations on satellite images", *Climatic Change* 9: 243–58.

Breman, H. and C.T. de Wit, 1983. "Rangeland productivity and exploitation in the Sahel", *Science* 221: 1341–7.

Brown, D., 1988. "Farmers lift siege of Indian capital", *Guardian*, 1 November.

Brown, M.M. (ed.), 1985. *Famine. A Man-Made Disaster?*, A report for the Independent Commission on International Humanitarian Issues. (London: Pan).

Buckoke, A., 1988. "Eritrean group outwits MiGs to hold off famine", *The Times*, 24 November: 11.

Budyko, M.I., 1958. *The Heat Balance of the Earth's Surface*, translated by N.A. Stepanova (Washington DC: US Dept of Commerce).

Budyko, M.I., 1974. *Climate and Life*, edited by D.H. Miller (New York: Academic Press).

Burley, J., C.E. Hughes and B.T. Styles, 1985. "Genetic systems of tree species for arid and semi-arid lands", paper presented to Symposium on Establishment and Productivity of Tree Planting in Semi-arid Regions, Texas A & I University, Kingsville, Texas.

Burley, J., P.A. Huxley and F. Owino, 1984. "Design, management and assessment of species, provenance and breeding trials of multipurpose trees", in R.D. Barnes and G.L. Gibson (eds), *Provenance and Genetic Improvement in Tropical Forest Trees*: pp. 70–80 (Oxford: Oxford Forestry Institute).

Bush, R., 1985. "Drought and famines", *Review of African Political Ecnomy* 33: 59–64.

Caldwell, J.C., 1984. "Desertification, demographic evidence 1973–1983",

Occasional Paper No. 37 (Canberra: Development Studies Centre, Australian National University).

Campbell, D.J., 1984. "Response to drought among farmers and herders in Southern Kajiado District, Kenya", *Human Ecology* 12: 35–64.

Carl Bro International, 1982. "An evaluation of livestock management and production in Botswana with special reference to Communal Areas" (Gaborone: European Development Fund/Government of Botswana).

Carruthers, I. (ed.), 1983. *Aid for the Development of Irrigation* (Paris: OECD).

Carruthers, I., 1985. "Protecting irrigation investment: the drainage factor", *Ceres* 18(4): 15–21.

Carter, T.R. and N.T. Konijn, 1988. "The choice of first-order impact models for semi-arid regions", in Parry *et al.*, 1988: 61–83.

Catterson, T., 1984. "AID forestry activities in Africa 1976–1983. A summary assessment", in USAID, 1984b: Appendix A.

Catterson, T.M., F.A. Gulick and T. Resch, 1985. "Desertification – rethinking forestry strategy in Africa: experience drawn from USAID activities", paper prepared for FAO Expert Consultation on the Role of Forestry in Combating Desertification, Saltillo, Mexico.

Charney, J., 1974. *Symons Memorial Lecture*, (London: Royal Meteorological Society).

Charney, J., 1975. "Dynamics of deserts and drought in the Sahel", *Quarterly Journal of the Royal Meteorological Society* 101: 193–202.

Charney, J., P.H. Stone and W.J. Quirk, 1975. "Drought in the Sahara: a biophysical feedback mechanism", *Science* 187: 434–5.

Charney, J., W.J. Quirk, S.H. Chow and J. Kornfield, 1977. "A comparative study of the effects of albedo change on drought in semi-arid regions", *Journal of Atmospheric Schiences* 34: 1366–85.

Christiansson, C., 1981. *Soil Erosion and Sedimentation in Semi-Arid Tanzania. Studies of Environmental Change and Ecological Imbalance* (Uppsala: Scandinavian Institute of African Studies).

CIDA, 1986. *Development*, June 1986: 40 (Hull, Quebec: Canadian International Development Agency).

Cloudsley Thompson, J., 1977. *The Desert* (London: Orbis Publishing).

Club du Sahel, 1979. *Development of Irrigated Agriculture in the Sahel* (Paris: OECD).

Club du Sahel, 1980. "The Sahel Drought Control and Development Programme", 1975–1979: A Review and Analysis (Paris: OECD).

Club du Sahel, 1981. "Official development assistance to CILSS member countries 1975–1980" (Paris: OECD).

Club du Sahel, 1983. "Development of rainfed agriculture in the Sahel. Overview and prospects". Fifth conference of the Club du Sahel, 26–28 October 1983. OECD, Paris.

Cooke, H.J., 1983. "The struggle against environmental degradation – Botswana's experience", *Desertification Control* 8: 9–15.

Cooley, M.E. and R.M. Turner, 1982. "Application of LANDSAT products in range- and water-management problems in the Sahelian zone of Mali, Upper Volta and Niger", Geological Survey Professional Paper No. 1058 (Washington DC: US Department of the Interior).

Cottingham, R., 1987. "Dry season gardening projects. Niger, West

Africa", (Washington DC: Lutheran World Relief.

Coughlin, C., 1988. "Silent death in the Sudan's siege towns", *Sunday Telegraph.*

Courel, M.F., R.S. Kandel and S.I. Rasool, 1984. "Surface albedo and the Sahel drought", *Nature* 307: 528–31.

Creek, M.J., 1984. "Stratification of beef industry in Kenya", in B. Nestel (ed.): pp. 365–77.

Darkoh, M.B.K., 1982. "Population expansion and desertification in Tanzania", *Desertification Control* 6: 26–33.

Darkoh, M.B.K., 1986. "Combating desertification in Zimbabwe", *Desertification Control Bulletin* 13: 17–28.

Davy, E.G., 1974. "Drought in West Africa", *WMO Bulletin* 123: 18–23.

Delehanty, J.M., J.T. Thomson an M. Hoskins, 1985. "Majjia Valley Windbreak Project. Sociological evaluation. Draft report" (Niamey: CARE International).

Delgado, C. and J. Staatz, 1981. *Livestock and Meat Marketing in West Africa Vol. III: Ivory Coast and Mali* (Washington: USAID).

Delgado, C.L. and C.P.J. Miller, 1985. "Changing food patterns in West Africa", *Food Policy* 7: 55–62.

Dennett, M.D., J. Elston and J.A. Rodgers, 1985. "A reappraisal of rainfall trends in the Sahel", *Journal of Climatology* 5: 353–61.

Dennison, S., 1986. *The Majjia Valley Windbreak Evaluation Study. An Examination of Progress and Analyses to Date* (Niamey: CARE International).

Diemer, G. and E.C.W. van der Laan, 1983. "Small scale irrigation along the Senegal river", paper presented at the lunchtime meeting of the Irrigation Management Network, Overseas Development Institute, London.

Dougrameji, J.S. and M.A. Clor, 1977. "Case study on desertification. Greater Mussayeb Project, Iraq", United Nations Conference on Desertification, Nairobi, 29 August–9 September 1977, Document A/ CONF.74/10 (Nairobi:UNEP).

Downing T.E., D.N. Mungai and H.R. Muturi, 1988. "Drought climatology and development of the climatic scenarios", in Parry *et al.* 1988b: 149–73.

Dregne, H.E., 1976. *Soils of the Arid Regions* (Amsterdam: Elsevier).

Dregne, H.E., 1977. "Map of the status of desertification in the hot arid regions", United Nations Conference on Desertification, Nairobi, 29 August–9 September 1977, Document A/CONF.74/31 (Nairobi: UNEP).

Dregne, H.E. 1982. *Dryland Soil Resources* (Washington DC: US Agency for International Development).

Dregne, H.E., 1983a. *Desertification of Arid Lands* (Chur, Switzerland: Harwood Academic Publishers).

Dregne, H.E., 1983b. "Evaluation of the implementation of the Plan of Action to Combat Desertification", report for UNEP.

Dregne, H.E., 1984a. "Guest editorial. Desertification – present and future", *International Journal for Development Technology* 2: 255–9.

Dregne, H.E., 1984b. "Combating desertification: evaluation of progress", *Environmental Conservation* 11: 115–21.

Dregne, H.E., 1985. "Aridity and land degradation", *Environment* 27(8): 16, 18–20, 28–33.

Dunne, N., A. Harris, D. Hargreaves, T. Dickson and D. Owen, 1988. "The US drought. Special report", *Financial Times*, 14 July: 8.

Eckholm, E., 1976. *Losing Ground* (New York: W.W. Norton).

Eckholm, E., 1982. *Down to Earth* (London: Pluto Press).

Eckholm, E., G. Foley, G. Barnard and L. Timberlake, 1984. *Fuelwood: The Energy Crisis that Won't Go Away* (London: Earthscan).

Edwards, C. and L. Vincent, 1983. "Irrigation in Africa and Asia politics and problems", Development Studies Reprint No. 143, School of Development Studies, University of East Anglia.

Ejeta, G. and D. Woods Thomas, 1984. "Chronology of a new sorghum hybrid for the Sudan", unpublished.

El Baz, F., 1983. "A geological perspective of the desert", in S.G. Wells and D.R. Haragan (eds), *Origin and Evolution of Deserts*: 163–83. (Albuquerque: University of New Mexico Press).

El-Lakany, M.H., 1986. "The importance of shelterbelts in Egyptian agriculture", in D.L. Hintz and J.R. Brandle (eds), 1986: 133–4.

Elwell, H.A., 1979. "Modelling soil losses in Zimbabwe-Rhodesia", Institute of Agricultural Engineering, Harare, unpublished.

Elwert, G., 1983. *Bauern und Staat in West Africa. Die Verflechtung sozioökonomischer Secktoren am Beispel Benin* (Farmer and state in West Africa. The inter-weaving of socio-economic sectors, as exemplified by Benin) (Frankfurt: Campus Verlag).

Eyre, S.R., 1968. *Vegetation and Soils. A World Picture* (London: Edward Arnold).

Falvey, J.L., 1982. "*Gliricidia maculata* – a review", *International Tree Crops Journal* 2: 1–14.

Famoriyo, S., 1984. "Land acquisition and irrigation in Nigeria", *Land Use Policy* 1: 55–63.

FAO, 1975. *Production Yearbook 1974* (Rome: FAO).

FAO, 1977. "Report of the fourth session of the FAO Panel of Experts on Forest Gene Resources. Canberra, 9–11 March 1977" (Rome: FAO).

FAO, 1978a. *Report on the Agro-ecological Zones Project. Vol. 1. Methodology and Results for Africa* (Rome: FAO).

FAO, 1978b. *Forestry for Rural Community Development*, FAO Forestry Paper No. 7 (Rome: FAO).

FAO, 1978c. *China: Forestry Support for Africa*, FAO Forestry Paper No. 12 (Rome: FAO).

FAO, 1981. *Map of the Fuelwood Situation in the Developing Countries* (Rome: FAO).

FAO, 1982. *Production Yearbook 1981* (Rome: FAO).

FAO, 1983. "Keeping the land alive: soil erosion – its causes and cures", *FAO Soils Bulletin* 50 (Rome: FAO).

FAO, 1985a. *Irrigation in Africa South of the Sahara. A Study with Particular Reference to Food Production* (Rome: FAO).

FAO, 1985b. "Arid Zone Forestry Programmes. State of knowledge and experience. An overview", Expert Consultation on the Role of Forestry in Combating Desertification, Saltillo, Mexico, 24–28 June 1985,

Document FOR 85/1.

FAO, 1985c. "Arid zone forestry programmes. Processing and utilization. An overview", Expert Consultation on the Role of Forestry in Combating Desertification, Saltillo, Mexico, 24–28 June 1985, Document FOR 85/5 (Rome: FAO).

FAO, 1985d. "Arid zone forestry programmes. International programmes. FAO activities for controlling desertification", Expert Consultation on the Role of Forestry in Combating Desertification, Saltillo, Mexico, 24–28 June 1985, Document FOR 85/2 (Rome: FAO).

FAO, 1987. *Production Yearbook 1986* (Rome: FAO).

FAO/UNEP, 1984. "Provisional methodology for assessment and mapping of desertification" (Rome: FAO).

FAO/UNESCO/WMO, 1977. "World map of desertification", United Nations Conference on Desertification, Nairobi, 29 August–9 September 1977, Document A/CONF.74/2.

Farmer, G. and T.M.L. Wigley, 1985. "Climatic trends for tropical Africa", a research report for the Overseas Development Administration. Climatic Research Unit, University of East Anglia, Norwich.

Felker P., G.H. Cannell, P.R. Clark, J.F. Osborn and P. Nash., 1981. *Screening Prosopis (Mesquite) Species for Biofuel Production on Semi-Arid Lands*", (Kingsville: Texas A & I University).

Fishwick, R.E., 1965. "Neem plantations in N. Niger", quoted in Spears (1983).

Fleer, H., 1981. "Teleconnections of rainfall anomalies in tropics and subtropics", in J. Lighthill and R.B. Pearce (eds), *Monsoon Dynamics*: pp. 5–8 (Cambridge: Cambridge University Press).

Fleuret, P. and A. Fleuret, 1983. "Socio-economic determinants of child nutrition in Taita, Kenya: a call for discussion", *Culture and Agriculture* 19: 8–20.

Floor, W. and J. Gorse, 1987. "Household energy issues in West Africa", (Washington DC: World Bank).

Floret, C., M. Le Floch, R. Pontanier, F. Romane, M.S. Hadjej, T. Ionesco, T. Jalel, N. Gaddas, A. Ben Saleh and M. Dupuy, 1977. "Case study on desertification. Oglat Merteba Region, Tunisia", United Nations Conference on Desertification, Nairobi, 29 August–9 September 1977 (Nairobi: UNEP).

Foley, G., 1987. "New analytic approaches to the woodfuel crisis in the Western Sahel and other African countries", Panos Institute, London, unpublished.

Foley, G. and G. Barnard, 1984. *Farm and Community Forestry*, Technical Report No. 3 (London: Earthscan).

Foley, G. and G. Barnard, 1985. "Farm and community forestry", Network Paper 1b, Social Forestry Network, Overseas Development Institute, London.

Forse, B., 1987. "Staring starvation in the face", *New Scientist*, 3 December: 30–1.

Franke, R. and B.H. Chasin, 1981. "Peasants, peanuts, profits and pastoralists", *The Ecologist* 11: 156–68.

Freeman, P., R.F. Damon, O. Fugalli, T. Resch, G.F. Taylor and T. Wood

Stervinou, 1983. "Third year evaluation of the Senegal Fuelwood Production Project (Washington DC: USAID).

Freeman, P.H., 1986. "Natural resources in sub-Saharan Africa. Review of problems and management needs", Bureau for Africa, US Agency for International Development, Washington DC (Draft).

Gadgil, S., A.K.S. Huda, N.A. Jodha, R.P. Singh and S.M. Virmani, 1988. "The effects of climatic variations on agriculture in dry tropical regions of India", in Parry et al., 1988b: pp. 495–578.

Gardner, R.L., 1984. *Economics and Cost Sharing of Salinity Control in the Colorado River Basin*, Dissertation, Colorado State University, 1983. (Diss. Abstr. Int. A 45(1) 252.)

Garnett, N., 1988. "Drought stunts US farm income", *Financial Times*, 11 October.

Gautier, E.F., 1970. *Sahara: the Great Desert*, translated by D.F. Mayhew (New York: Octagon Books).

Glantz, M.H. (ed.), 1977. *Desertification. Environmental Degradation in and around Arid Lands* (Boulder, Colorado: Westview Press).

Glantz, M.H., 1987a. "Drought, famine and the seasons in sub-Saharan Africa", in R. Huss-Ashmore and S. Katz (eds), *Anthropological Perspectives on the African Famine* (New York: Gordon and Breach Science Publishers).

Glantz, M.H., 1987b. "Drought and economic development in sub-Saharan Africa", in M.H. Glantz (ed.), *Drought and Hunger in Africa: Denying Famine a Future* (Cambridge: Cambridge University Press).

Glantz, M.H. and R.W. Katz, 1985. "Drought as a constraint to development in sub-Saharan Africa", *Ambio* 14: 334–9.

Glantz, M.H. and N.S. Orlovsky, 1986. "Desertification: anatomy of a complex environmental process", in K.A. Dahlberg and J.W. Bennett (eds), *Natural Resources and People. Conceptual Issues in Interdisciplinary Research*: 213–29 (Boulder, Colorado: Westview Press).

Goldschmidt, W., 1981. "The failure of pastoral development projects in Africa", in J.G. Galaty, D. Aronson, P.C. Saltzman and A. Chouinard (eds), *The Future of Nomadic Peoples* (Ottawa: International Development Research Centre).

Goudie A., 1978. "Dust storms and their geomorphological implications", *Journal of Arid Environments* 1: 291–310.

Goudie, A., 1984. *The Nature of the Environment* (Oxford: Basil Blackwell).

Government of Botswana, 1980. "A human drought relief programme for Botswana", (Gaborone: Ministry of Local Government).

Government of Mali, 1982. "La stratégie alimentaire du Mali" (Mali's food strategy) (Paris: OECD for Club du Sahel).

Grace, J.R., 1986. "Plant response to wind", in D.L. Hintz and J.R. Brandle (eds), 1986: 39.

Grainger A., 1983. *Desertification* (London: Earthscan).

Grainger, A., 1984a. "Quantifying changes in forest cover in the humid tropics: overcoming current limitations", *Journal of World Forest Resource Management* 1: 3–63.

Grainger, A., 1984b. "Increasing the effectiveness of afforestation projects in the tropics involving non-governmental organisations", *International Tree Crops Journal* 3: 33–47.

Grainger A., 1986a. *The Future Role of the Tropical Rain Forests in the World Forest Economy*, D.Phil. thesis, Department of Plant Sciences, University of Oxford (Oxford: Oxford Academic Publishers).

Grainger A., 1986b. "Deforestation and progress in afforestation in Africa", *International Tree Crops Journal* 4: 33–48.

Grainger, A., 1988. "Estimating areas of degraded tropical lands requiring replenishment of forest cover", *International Tree Crops Journal* 5: 31–61.

Grainger, A. and H.W. Esbenshade, 1978. "The development of tree crops for agroforestry systems", *Proceedings. Eighth World Forestry Congress, Jakarta*: pp. 709–18 (Rome: FAO).

Grassl, H., 1979. "Possible changes of planetary albedo due to aerosol particles", in W. Bach, J. Pankrath and W.W. Kellogg (eds), *Man's Impact on Climate*: pp. 229–41, (Amsterdam: Elsevier).

Gregersen, H.M., 1982. *Village Forestry Development in the Republic of Korea – A Case Study*. Forestry for Local Community Development Programme Series (Rome: FAO).

Gribbin, J., 1979. *Climate and Mankind* (London: Earthscan).

Griffin, K. and Hay, R., 1985. "Problems of agricultural development in socialist Ethiopia: an overview and a suggested strategy", *Journal of Peasant Studies* 13: 37–66.

Griffiths, G.H. and W.G. Collins, 1981. "The development of LANDSAT systems for monitoring desert encroachment in northern Kenya", in D. Lynn and J.A. Allen (eds), *Matching Remote Sensing Technologies and Their Applications*. Proceedings of an International Conference, London, December 1981: 53–64 (Reading: Remote Sensing Society).

Grove, A.T., 1973. "Desertification in the African environment", in D. Dalby and R.J. Harrison Church (eds), *Drought in Africa*. Report of the 1973 Symposium, School of Oriental and African Studies, University of London.

Grove, A.T., 1977. "Desertification", *Progress in Physical Geography* 1: 296–310.

Grove, A.T., 1978. *Africa* (Oxford: Oxford University Press).

Gupta D.K., 1982. "Impact of lining of watercourses on agricultural output – a case study", *Wamana* 2 (4): 1, 8–16.

Gupta, T., 1983. "The economics of tree crops on marginal agricultural lands with special reference to the hot arid region in Rajasthan, India", *International Tree Crops Journal* 2: 155–94.

Gupta, T., 1985. "Protection of social forestry plantations in Madhya Pradesh and Uttar Pradesh", unpublished.

Haldeman, J.M., 1985. "Problems of pastoral development in Eastern Africa", *Agricultural Administration* 18: 199–216.

Hamid, S., M.A. Qayyum, C. Ata-ur-Reham, H.A. Chaudry, S.M.H. Zaidi, G. Haider, M.A. Ahmad, S.B. Hasan and M.A.R. Farooqu, 1977. "Case study on desertification. Mona Reclamation Experimental Project, Pakistan", United Nations Conference on Desertification. Nairobi, 29 August–9 September 1977 (Nairobi: UNEP).

Hansen, J., D. Johnson, A. Lacis, S. Lebedeff, P. Lee, D. Rind and G. Russell, 1981. "Climate impact on increasing atmospheric carbon dioxide", *Science* 213: 957–66.

Harden, B., 1987. "Africa's poor on the brink", *Washington Post*, 7 June: A1, A25.

Hardin, G.J., 1972. "The tragedy of the commons", in G.H.E. Daly (ed.), *Towards a Steady State Economy*: 133–48 (San Francisco: W.H. Freeman).

Hare F.K., 1977. "Climate and desertification", UN Conference on Desertification, Nairobi, 29 August–9 September 1977, Document A/CONF.74/5 (Nairobi: UNEP).

Hare, K., 1983. *Climate and Desertification: A Revised Analysis*, World Climate Applications Programme Report No. 44. (Nairobi: World Meteorological Organization/UNEP).

Hare, F.K., 1984. "Recent climatic experiences in the arid and semi-arid lands", *Desertification Control Bulletin* 10: 15–22.

Harrison, P., 1986. *The Greening of Africa* (London: Earthscan).

Hawksworth, D.L. (ed.), 1984. *Advancing Agricultural Production in Africa*. Proceedings of CAB's First Scientific Conference, Arusha, Tanzania, 12–18 February 1984 (Farnham Royal: Commonwealth Agricultural Bureau).

Heathcote, R.L., 1983. *The Arid Lands: Their Use and Abuse* (London: Longman).

Heermans, J., 1986. "The Guesselbodi experiment with improved management of brushland in Niger", *Development Anthropology Network* 4(1): 11–15.

Hegmar, E., Pa-Momandou N'jie and A. Närman, 1982. "Agricultural development in Kenya with special reference to Eastern Division, Kitui District", *Choros* No. 8.

Helldén, U., 1978. *Evaluation of LANDSAT-2 Imagery for Desertification Studies in Northern Kordofan, Sudan, Rapporter o. Notiser* No 38 (Department of Physical Geography, University of Lund, Sweden).

Helldén U., 1980. *Satellite Data for Regional Desertification and its Control*, Rapporter o. Notiser No. 50 (Department of Physical Geography, University of Lund, Sweden).

Helldén, U., 1984. *Drought Impact Monitoring. A Remote Sensing Study of Desertification in Kordofan, The Sudan*, Rapporter o. Notiser No. 61 (Department of Physical Geography, University of Lund, Sweden).

Helldén, U., 1988. "Note", *Desertification Control Bulletin* 17: 8–12.

Helldén, U. and M. Stern 1980. *Monitoring Land Degradation in Southern Tunisia*, Rapporter o. Notiser No. 48 (Department of Physical Geography, University of Lund, Sweden).

Henning, D. and H. Flohn, 1977. "Climate aridity index map, United Nations Conference on Desertification, Nairobi, 29 August–9 September 1977, Document A/CONF. 74/31 (Nairobi: UNEP).

Herath, H.M.G., 1985. "Economics of salinity control in Sri Lanka: some exploratory results", *Agricultural Administration* 18: 191–7.

Higgins, G.M., A.H. Kassam, L. Naiken and M.M. Shah, 1982. "Africa's agricultural potential", *Ceres* 14(5): 13, 18–21.

Higgins, G.M., A.H. Kassam and M.M. Shah, 1984. "Land, food and population in the developing world", *Nature and Resources* 20(3): 2–10.

Hilsum, L., 1987. "Camels. New potential for pastoralism in Kenya", *Development International* Jan./Feb. 1987: 44–5.

Hintz, D.L. and J.R. Brandle (eds), 1986. *Proceedings. International Symposium on Windbreak Technology, Lincoln, Nebraska, June 23–27 1986*, Great Plains Agricultural Council Publication No. 117 (Lincoln, Nebraska).

Hobbs, J., 1988. "Climatic patterns and variability in the Australian wheatbelt", in Parry *et al.* 1988b: 687–717.

Hobbs, J., H. Harris and J.R. Anderson, 1988. "Introduction: policy and planning issues for the Australian wheat industry", in Parry *et al.* 1988b: 673–86.

Hooper, J., 1988. "Saharans join against threat of locust plague", *Guardian*, 4 October: 13.

Horowitz, M.M., 1979. "The sociology of pastoralism and African livestock projects", AID Program Evaluation Discussion Paper No. 6 (Washington DC: USAID).

Hoskins, M.W., 1982. "Observations on indigenous and modern agro-forestry activities in West Africa", paper presented at the Workshop on Agroforestry, Freiburg, Germany, 31 May–5 June 1982, United Nations University and Albert Ludwigs University, Freiburg.

Hudson, N.W., 1987. "Soil conservation programme in Kenya", paper presented to Sustainable Development Conference, 28–30 April 1987, (International Institute for Environment and Development London).

Hughes, C.E. and B.T. Styles, 1984. "Exploration and seed collection of multipurpose dry zone trees in Central America", *International Tree Crops Journal* 3: 1–31.

Hulme, M., 1987. "Rainfall in central Sudan: an asset or a liability?" *Geoforum* 18: 321–31.

Hulme, M., 1989a. "Monitoring and predicting Sahelian rainfall: a review of progress and an assessment of data needs", in *Proceedings of Conference on Arid Lands: Prospects and Problems in the Sahel, 18–29 May 1986* (London: Earthscan Publications).

Hulme, M., 1989b. "Is environmental degradation causing drought in the Sahel? An assessment from recent empirical research", *Geography*.

Hunziker, J.H., 1985. "Studies on the taxonomy, hybridization and genetic variation of Argentine species of *Prosopis*", paper presented to Symposium on Establishment and Productivity of Tree Planting in Semi-arid Regions, Texas A & I University, Kingsville, Texas.

Huxley, P.A. (ed.), 1983. *Plant Research and Agroforestry*, proceedings of a consultative meeting held in Nairobi, 8 to 15 April 1981, ICRAF, (Nairobi: International Council for Research in Agroforestry).

IBPGR, 1977. *Report of 3rd Session* (Rome: International Board for Plant Genetic Resources/FAO).

IBPGR, 1980. *Genetic Resources of Tree Species in Arid and Semi-Arid Areas* (Rome: FAO).

Ibrahim, F., 1980. *Desertification in Nord-Darfur*, Hamburg Geographical Studies No 35 (Hamburg).

Indian Planning Commission, 1973. *Integrated Agricultural Development in Drought Prone Areas. Report of Task Force on Integrated Agricultural Development* (New Delhi: Government of India).

Jackson, J.K., G.F. Taylor and C. Condé-Wane, 1983. "Management of natural forest in the Sahel Region", Forestry Support Program

(Washington DC: USDAFS/USAID).

Jackson, R.D. and S.B. Idso, 1975. "Surface albedo and desertification", *Science* 189: 1012–13.

Jerve, A.M., 1981. "Pastoralists, peasants or proletarians? An analysis of the role of livestock production in the economy of Tsabong Area, Southern Kgalagadi", DERAP Working Papers No. A226 (Norway: Chr. Michelsen Institute).

Jodha, N.S., 1988. "Introduction", in Parry *et al.* 1988: 503–21.

Johnson, S.H. III, 1982. "Large-scale irrigation and drainage schemes in Pakistan: a study of rigidities in public decision making", *Food Research Institute Studies* 18: 149–80 (Stanford University, California).

Jones, P.D., T.M.L. Wigley, C.K. Folland, D.E. Parker, J.K. Angell, Lebedeff S. and J.E. Hansen, 1988. "Evidence for global warming in the past decade", *Nature* 332: 790; 333: 122.

Joseph, S. and J. Loose, 1983. "Design and laboratory testing of portable metal briquette burning cooking stoves for The Gambia" (London: Intermediate Technology Consultants Ltd).

Joshi, P.K., 1983. "Benefit-cost analysis of alkali land reclamation technology – an ex-post evaluation", *Agricultural Situation in India* 38: 467–70.

Joshi, P.K. and A.K. Agnihotri, 1984. "An assessment of the adverse effects of canal irrigation in India", *Indian Journal of Agricultural Economics* 39: 528–36.

Joyce, C., 1988. "American politicians warm to greenhouse effect", *New Scientist*, 8 September: 30; *New Scientist*, 30 June: 35.

Junge, C., 1979. "The importance of mineral dust as an atmospheric constituent", in C. Morales (ed.), *Saharan Dust: Mobilization, Transport, Deposition*, SCOPE Report No. 14 (Chichester: John Wiley).

Karamchandani, K.P., 1982. *Extension Components of Social Forestry in Gujarat* (Vadodara: Gujarat State Forest Department).

Kates, R.W., D.L. Johnson and K. Johnson Haring, 1977. "Population, society and desertification", UN Conference on Desertification, A/ CONF.74/8 (Nairobi: UNEP).

Kaul, R.N., 1983. "Some silvicultural aspects of sand dune afforestation", *International Tree Crops Journal* 2: 133–46.

Keeling, C.D., R.B. Bacastow, A.E. Bainbridge, C.A. Ekdahl Jr, P.R. Guenther, L.S. Waterman and J.F.S. Chin, 1986. "Atmospheric carbon dioxide variations at Mauna Loa Observatory, Hawaii", *Tellus* 28: 538–51.

Kelemen, P., 1985. "The politics of the famine in Ethiopia and Eritrea", Occasional Paper No. 17, Department of Sociology, University of Manchester.

Kernan, H.S. and T.M. Resch, 1984. "Forestry for Chad and the USAID role" (Washington DC: Forestry Support Program, USDAFS/USAID).

Khogali, M.M., 1983. "The grazing resources of the Sudan", in Ooi Jin Bee (ed.), 1983: 326–45.

Kilahama, J.B.R., 1980. "Village afforestation in Tanzania", Forestry Division, Ministry of Natural Resources and Tourism, unpublished.

Kort, J., 1986. "Benefits of windbreaks to field and forage crops", in D.L. Hintz and J.R. Brandle (eds), 1986: 53–4.

Kousky, V.E., M.T. Kagano and F.A. Cavalcanti, 1984. "A review of the Southern Oscillation: oceanic-atmospheric circulation changes and related rainfall anomalies", *Tellus* 36A: 490–504.

Krings, T.F., 1980. *Kulturgeographischer Wandel in der Kontaktzone von Nomaden und Bauern in Sahel von Obervolta: am Beispiel des Oudalen (Nordost-Obervolta)* (Social and geographical change in the zone of contact between nomads and sedentary farmers: the case of Oudalan (north-east Upper Volta), Hamburg Geographical Studies 36 (Hamburg).

Krings, T., 1985. "Viehalter contra Ackerbauern. Eine Fallstudie aus dem Nigerbinnendelta (Republik Mali)" (Pastoralists against peasants – a case study of the Inner Niger Delta, Mali), *Erde* 116: 197–206.

Lamb, H.H., 1973. "Some comments on atmospheric pressure variations in the Northern Hemisphere", in D. Dalby and R.J. Harrison Church (eds), 1973: pp. 27–8.

Lamb, H.H., 1979. *Climate, Present, Past and Future* (London: Methuen).

Lamb, P.J., 1978. "Large scale tropical Atlantic surface circulation patterns associated with subSaharan weather anomalies", *Tellus* 30: 240–51.

Lamb, P.J., 1982. "Persistence of subSaharan drought", *Nature* 299 (5878): 46–8.

Lamb, P.J., 1985a. "SubSaharan drought", paper prepared for Department of State Conference on African Environmental Issues, 8 October 1985, Washington DC.

Lamb, P.J., 1985b. "Rainfall in subSaharan West African during 1941–1983", *Zeitschrift-für Gletscherkunde und Glazialgeologie* 21: 131–9.

Lamb, P.J., 1986. "Waiting for rain", *The Sciences*, May/June 1986: 31–5.

Lamb, P.J., 1987. "On the development of regional climatic scenarios for policy-oriented climatic-impact assessment", *Bulletin of American Meteorological Society* 68: 1116–23.

Lamb, P.J., R.A. Peppler and S. Hastenrath, 1986. "Interannual variability in the tropical Atlantic", *Nature* 322: 238–40.

Lamprey, H.F., 1975. "Report on the desert encroachment reconnaissance in Northern Sudan, October 21–November 10 1975" (Khartoum: National Council for Research, Ministry of Agriculture, Food and Resources). Reprinted in *Desertification Control Bulletin* 17 (1988): 1–7.

Lamprey, H.F. and H. Yussuf, 1981. "Pastoralism and desert encroachment in northern Kenya", *Ambio* 10 (2/3): 131–4.

Landsberg, H.E., 1975. "Sahel drought: change of climate or part of climate", *Archiv für Meteorologie, Geophysik und Bioklimatologie* B23: 193–200.

Lanly, J.P., 1982. *Tropical Forest Resources*, FAO Forestry Paper No. 30 (Rome: FAO).

Latremolière, J., 1981. "Pour une seule agriculture africaine: cultures vivrières et culture de rente" (Towards a single African agriculture: subsistence and cash crops), *Marchés Tropicaux et Méditerranéens* 37: 3281–3.

Lattre, A. de, *et al.* 1981. *Le dossier Sahel: tome 3. Les cultures irriguées et les aménagements hydroagricoles dans le cadre des 2 premiers programmes de génération (1977–1985)* (Sahel dossier: Volume 3. Irrigated crops and hydroagricultural schemes within the framework of the first two programmes 1977–85) (Paris: Ediafric-Documentation Africaine).

Lattre, A. de and A.M. Fell, 1984. "The Club du Sahel. An experiment in international cooperation" (Paris: OECD).

Laval, K., 1986. "General circulation model experiments with surface albedo changes", *Climatic Change* 9: 91–102.

LeBlond, R., 1982. "Remote sensing and development. Report on IDRC-supported projects in the Sudan, Bolivia, Tanzania, Bangladesh and Mali". Publication No. IDRC-174e (Ottawa: International Development Research Centre).

Lee, J.A.C. (ed.), 1985. *Conservation in Sub-Saharan Africa. An Introductory Bibliography for the Social Sciences*, Cambridge African Monographs No. 5 (Cambridge: School of African Studies, University of Cambridge).

Le Houérou, H.N., 1977. "The nature and causes of desertisization", in M.H. Glantz, 1977: 17–38.

Leng, G., 1982. *Desertification: a Bibliography with Regional Emphasis on Africa*, Fachbereich, Schwerpunkt Geographie, Universität Bremen No. 1.

Lindley, D., 1988. "Once-great lake facing a slow, salty death", *The Times*, 2 September.

Lockwood, J.G., 1985. *World Climatic Systems* (London: Edward Arnold).

Lockwood, J.G., 1988. "Climate and climate variability in semi-arid regions at low altitudes", in Parry *et al.* 1988b: 85–120.

Long, S., 1985. "Interim report. Influence of neem (*Azadirachta indica*) shelterbelts on microclimatic factors, water use, and yield of intercropped millet in the Majjia Valley, Niger. 1984 growing season" (Niamey: CARE International).

Lyles, L., 1976. "Wind patterns and soil erosion on the Great Plains", in R.W. Tinus (ed.), *Shelterbelts on the Great Plains*, Great Plains Agricultural Council Publication No. 78: 22–30 (Lincoln, Nebraska).

Mabbutt, J.A., 1983. "General assessment of progress in the implementation of the Plan of Action to Combat Desertification", report for UNEP.

Mabbutt, J.A., 1984. "A new global assessment of the status and trends of desertification", *Environmental Conservation* 11: 100–13.

Mabbutt, J.A., 1985. "Desertification of the world's rangelands", *Desertification Control Bulletin* 12: 1–11.

Mabbutt, J.A., 1986. "Desertification indicators", *Climatic Change* 9: 113–22.

Mabbutt, J.A and C. Floret (eds), 1980. *Case Studies on Desertification*, Natural Resources Research No. 18 (Paris: UNESCO).

McCarthy, J.W., C. Clapp-Wincek, S. Londner and A. Thomas, 1985. "A soil and water conservation project in two sites in Somalia: seventeen years later", AID Project Evaluation Report No. 62 (Washington DC: USAID).

MacDonald, L.H., 1981. "Natural resources development in the Sahel: the role of the United Nations system", M.S. thesis (Ann Arbor: School of Natural Resources, University of Michigan).

McDowell, R.E., D.G. Sisler, E.C. Schermerhorn, J.D. Reed and R.P. Bauer, 1983. "Game or cattle for meat production on Kenya rangelands?" Cornell International Agricultural Development Mimeograph No. 101 (Ithaca: Department of Economics, New York State College of Agriculture, Cornell University).

McElvoy, A., 1988. "Infants main victims as hunger and disease grip southern Sudan", *The Times*, 18 October: 8.

McGahuey, M., 1985. "Assessment of the *Acacia albida* extension projects in Chad" (Washington DC: Chemonics International).

MacGregor, M.T.G. de and C.V. Valverde, 1975. "Evolution of the urban population in the arid zones of Mexico 1900–1970", *Geographical Review* 65: 214–28.

MacKenzie, D., 1987. "Ethiopia's grains of hope", *New Scientist*, 20 August: 20–1.

McMillan, D.E., 1984. "A resettlement project in Upper Volta", dissertation, Northwestern University, Evanston.

McTainsh, G., 1980. "Harmattan dust deposition in northern Nigeria", *Nature* 286: 587–8.

McTainsh, G., 1985. "Desertification and dust monitoring in West Africa", *Desertification Control Bulletin* 12: 26–33.

Magalhaes, A.R. and O.E. Rebouças, 1988. "Introduction: drought as a policy and planning issue in Northeast Brazil", in Parry *et al.* 1988b: 279–304.

Makadho, J.M., 1984. "A review of some factors affecting the viability of small holder irrigation schemes in Africa", in M.J. Blackie (ed.), 1984: 209–19.

Makinen, M. and E. Ariza-Niño, 1982. "The market for livestock from the pastoral zone of Central Niger", Discussion Papers Series No. 7 (Tahona, Niger: Niger Range and Livestock Project, Ministère du Developpement Rural and USAID).

Marchal, J.Y., 1983. "Yatenga, Norde Haute Vaulta. La dynamique d'un espace rural soudano-sahelien" (Yatenga, northern Upper Volta. The dynamics of a rural area in the Sudan-Sahel region), Travaux et Documents de l'ORSTOM No. 167 (Paris: Office de la Recherche Scientifique et Technique Outre-Mer).

Mathai, W., 1985. "The Green Belt Movement", unpublished.

Maydell, H.J. von, 1983. *Arbres et arbustes du Sahel: leurs charactéristiques et leurs utilizations*, Schriftenreihe der GTZ No. 147 (Eschborn: GTZ).

Maydell, H.J. von, B. Becker, S. Klug, T. Lutke and K.F. Panzer, 1983. *Agroforstliche Landnutzung im Einzugsbereich zentraler Orte im Sahel. Fallbeispiel Nord-Senegal* (Agroforestry land use in the catchment area of central places in the Sahel. A case study from North Senegal), Forschungsberichte des Bundesministeriums für Wirtschaftliche Zusammenarbeit No. 47 (Hamburg: Bundesforschungsanstalt für Forst-und-Holzwirtschaft).

Mayer, H. *et al.*, 1983. "Der Pastoral-nomadische Subsistenzsector in Africa-Ursachen und Folgen ökonomischer und sozialer Transformationspozesse" (The pastoral-nomadic subsistence sector in Africa: causes and consequences of economic and social changes), *Afrika Spectrum* 18: 295–304.

Mbwala, P.P., 1980. "Irrigation development and peasant participation in Tanzania", *Eastern Africa Journal of Rural Development* 13: 68–91.

Mechergiu, M. and H.J. Mellouli, 1986. "Competition between a windbreak and an irrigated crop", in D.L. Hintz and J.R. Brandle (eds), 1986: pp. 263–9.

Meigs, P., 1953. "World distribution of arid and semi-arid homoclimates", in *Arid Zone Hydrology*, UNESCO Arid Zone Research Series No. 1: 203–9 (Paris: UNESCO).

Melamed-Gonzalez, R. and L. Giasson, 1987. *A Directory of NGOs in the Forestry Sector*, 2nd Africa edition (New York: International Tree Project Clearinghouse, Non-Governmental Liaison Service, United Nations).

Metral, F., 1984. "State and peasants in Syria: a local view of a government irrigation project", *Peasant Studies* 11: 69–90.

Meuer, G., 1984. "Food and politics in Africa. How to hit the hunger list", *Development and Cooperation* 5: 4–9.

Meyer, G., 1982. "Promotional measures and present state of development in the field of nomadism in Syria", *Applied Geography and Development* 19: 97–107.

Middleton, N.J., 1985. "Effect of drought on dust production in the Sahel", *Nature* 316: 431–4.

Middleton, N.J., 1987. 'Wind erosion in the Sahel", *Geography Review*, November 1987: 26–30.

Middleton N.J., 1989. "Desert dust", in D.S.G. Thomas (ed.), *Arid-Zone Geomorphology*. In press.

Milas, S. and Asrat, M., 1985. "Eastern Africa's spreading wastelands", *Desertification Control Bulletin* 12: 34–40.

Miles, M.K. and Folland, C.K., 1974. "Changes in the latitude of the climatic zones of the Northern Hemisphere", *Nature* 252: 616.

Miller, J., 1988. "Price of corruption: a vanishing Soviet Sea", *Sunday Telegraph*, 30 October: 11.

Mnzava, E.M., 1980. "Village afforestation: the lessons of experience in Tanzania" (Rome: FAO).

Monnier, Y., 1981. *La poussière et la cendre: paysages, dynamique des formations végétales et stratégies des sociétés en Afrique de l'Ouest* (Dust and ashes: countryside, vegetation cover and strategies of societies in West Africa) (Paris: Agence de Coopération Culturelle et Technique).

Mooley, D.A. and B. Parthasarathy, 1983. "Indian summer monsoon and El Niño", *Pure and Applied Geophysics* 121: 339–52.

Mooley, D.A. and B. Parthasarathy, 1984. "Fluctuations in all India summer monsoons during 1871–1978", *Climatic Change* 6: 287–301.

Morales, C., 1977. *Saharan Dust: Mobilisation, Transport, Deposition. Review and Recommendations from a Workshop, Gothenberg, Sweden, April 1977* (Stockholm: Ecological Research Committee, Swedish Natural Science Research Council).

Moris, J.R., 1984. "Managing irrigation in isolated environments: a case study of Action Blé-Diré, Mali", in M.J. Blackie (ed.), 1984: 245–56.

Morse, B., 1987. "Africa beyond the famine: new hope", foreword in M.H. Glantz (ed.), 1987b: xiii–xx.

Mrema, G.C., 1984. "Development of small holder irrigation in Tanzania: problems and prospects", in M.J. Blackie (ed.), 1984: 307–16.

Munslow, B., 1984. "State intervention in agriculture: the Mozambican experience", *Journal of Modern African Studies* 22: 199–221.

Murthy, K.R.S., 1985. "An analysis of the objectives and performance of the Uttar Pradesh Social Forestry Project", unpublished.

NAS, 1974. *More Water For Arid Lands. Promising Technologies and Research*

Opportunities (Washington DC: US National Academy of Sciences).

NAS, 1975. *Underexploited Tropical Plants with Promising Economic Value* (Washington DC: US National Academy of Sciences).

NAS, 1977. Leucaena. *Promising Forage and Tree Crop for the Tropics* (Washington DC: US National Academy of Sciences).

NAS, 1979. *Tropical Legumes* (Washington DC: US National Academy of Sciences).

NAS, 1980. *Firewood Crops, Vol. 1* (Washington DC: US National Academy of Sciences).

NAS, 1983a. *Changing Climate* (Washington DC: Carbon Dioxide Assessment Committee, US National Academy of Sciences).

NAS, 1983b. *Firewood Crops, Vol. 2* (Washington DC: US National Academy of Sciences).

Neftel, A., E. Moor, M. Oeschger, and B. Stauffer, 1985. "Evidence from polar ice cores for the increase in atmospheric CO_2 in the last two centuries", *Nature* 315: 45–7.

Nestel, B. (ed.), 1984. *Development of Animal Production Systems* (Amsterdam: Elsevier Science Publishers).

NGLS, 1986. *Tree Project News* 1(4) (New York: UN Non-Governmental Liaison Service).

Nicholson, S.E., 1983a. "Sub-Saharan rainfall in the years 1976–80: evidence of continued drought", *Monthly Weather Review* 111 (Aug.): 1646–54.

Nicholson, S.E., 1983b. "The climatology of sub-Saharan Africa", in NRC, 1983: 71–92.

Nickum, J.E., 1982. *Irrigation Management in China: A Review of the Literature*, World Bank Staff Working Paper No. 545 (Washington DC: World Bank).

Nobre, C.A. and L.C.B. Molion, 1988. "The climatology of droughts and drought prediction", in Parry *et al.* 1988: 305–23.

Norton, C.C., F.R. Mosher and B. Hinton, 1979. "Investigation of surface albedo variations during the recent Sahel drought", *Journal of Applied Meteorology* 18: 1252–62.

Novikoff, G. and M. Skouri, 1981. "Balancing development and conservation in pre-Saharan Tunisia", *Ambio* 10: 135–41.

NRC, 1983. *Environmental Change in the West African Sahel* (Washington DC: US National Research Council).

O'Keefe, P., P. Raskin and S. Bernow, 1984. *Energy, Environment and Development in Africa, Vol. 1* (Uppsala: Biejer Institute).

Ochieng, E.O., 1981. "The problem of food imports and the need to eliminate it: the case of Botswana", *Eastern Africa Journal of Rural Development* 14: 61–104.

OECD, 1981. *Directory of Non-Governmental Organisations in OECD Member Countries Active in Development Cooperation* (Paris: OECD).

Okaeme, A.N., 1981. "Significance of migrant Fulani cattle and present developments around Kainji Lake Basin", *Nigerian Journal of Agricultural Extension* 1: 38–43.

Olayide, S.O., 1983. "The focus on smallholder agricultural production in West Africa", *West African Journal of Agricultural Economics* 3: 39–60.

Olsson, K., 1984. *Long Term Changes in the Woody Vegetation in N. Kordofan, The Sudan,* Rapporter o. Notiser No. 60 (Department of Physical Geography, University of Lund, Sweden).

Olsson, K., 1985. *Remote Sensing for Fuelwood Resources and Land Degradation Studies in Kordofan, The Sudan,* Avhandlingar No. 100 (Department of Physical Geography, University of Lund, Sweden).

Olsson, L., 1985. *An Integrated Study of Desertification* (Department of Physical Geography, University of Lund, Sweden). Avhandlingar No. 98.

Ooi Jin Bee (ed.), 1983. *Natural Resources in Tropical Countries* (Singapore: Singapore University Press).

Otterman, J., 1974. "Baring high-albedo soils by overgrazing: an hypothesised desertification mechanism", *Science* 186: 531–3.

Otterman, J., 1975. "Reply to Jackson R.D. and Idso S.B. Surface albedo and desertification", *Science* 189: 1013–15.

Owen, D., 1988. "Rain washes out Canada's remaining grain hopes", *Financial Times,* 8 September.

Oxby, C., 1982. "Collective ranches in Africa", *World Animal Review* 42: 11–18.

Oxby, C., 1984. "Settlement schemes for herders in the subhumid tropics of West Africa: issues of land rights and ethnicity", *Development Policy Review* 2: 217–33.

Ozanne, J., 1988. "Massive locust swarms blot out Sudan's hopes of recovery", *Financial Times,* 18 October: 42.

Parry, M.L. and T.R. Carter, 1988. "The assessment of effects of climatic variation on agriculture. A summary of results from studies in semi-arid regions", in Parry *et al.* 1988b: 9–60.

Parry, M.L., T.R. Carter and N.T. Konijn, 1985. "Climate change. How vulnerable is agriculture?", *Environment* 27: 4–5, 43.

Parry, M.L., T.R. Carter and N.T. Konijn (eds), 1988a. *The Impact of Climatic Variations on Agriculture. Vol. 1. Assessment in Cool Temperate and Cold Regions* (Dordrecht: Kluwer Academic Publishers).

Parry, M.L., T.R. Carter and N.T. Konijn (eds), 1988b. *The Impact of Climatic Variations on Agriculture. Vol. 2. Assessments in Semi-Arid Regions* (Dordrecht: Kluwer Academic Publishers).

Patel, A.R., 1983. "Irrigation schemes and small farmers", *Kurukshetra* 31(21): 7–11.

Paylore, P. and J.R. Greenwell, 1980. "Fools rush in: pinpointing the Arid Zones", *Arid Lands Newsletter* 10: 17–18.

Pearce, F., 1988. "Cool oceans caused floods in Bangladesh and Sudan", *New Scientist,* 8 September: 31.

Peck, A.J., J.F. Thomas and D.R. Williamson, 1983. "Effects of man on salinity in Australia", in *A Perspective on Australia's Water Resources to the Year 2000* (Canberra: Department of Resources and Energy).

Peck, R.B., 1984. "Traditional forestation strategies of local farmers in the tropics", in K.F. Wiersum (ed.), 1984: 205–16.

Pedersen, B., 1980. "A note on the genus *Prosopis*", *International Tree Crops Journal* 1: 113–23.

Pedersen, B. and A. Grainger, 1981. "Bibliography of *Prosopis*", *International Tree Crops Journal* 1: 273–86.

Persaud N., S. Long, M. Gandah and M. Quathara, 1986. In D.L. Hintz and J.R. Brandle (eds), 1986: 211–12.

Powell, J.M. and E. Taylor Powell, 1984. "Cropping by Fulani agropastoralists in central Nigeria", *ILCA Bulletin* 19: 21–7.

Powell, R., 1988. "Ethiopia's biggest camp devastated by floods", *Guardian*, 4 October: 13.

Prakash, I., 1977. "Case study on desertification. Luni Development Block, India", United Nations Conference on Desertification, Nairobi, 29 August–9 September 1977, Document A/CONF.74/11 (Nairobi: UNEP).

Prowse, M., 1987. "Sub-Saharan Africa's agony", *Financial Times*, 20 March: 25.

Rapp, A., 1974. *A Review of Desertization in Africa – Water, Vegetation and Man* (Stockholm: Secretariat for International Ecology).

Rapp, A., 1982. "Processes of desertification and dust storms in African drylands", unpublished.

Rapp, A., 1986. "Introduction to soil degradation processes in drylands", *Climatic Change* 9: 19–31.

Rapp, A. and U. Helldén, 1979. "Research on environmental monitoring methods for land use planning in African drylands", *Rapporter o. Notiser* 42 (Department of Physical Geography, University of Lund, Sweden).

Rapp, A., H.N. Le Houerou and B. Lundholm (eds), 1976. "Can desert encroachment be stopped?", *Ecological Bulletin* 24 (Swedish Natural Science Research Council).

Rasool, S.I., 1984. "On dynamics of deserts and climate", in J.T. Houghton (ed.), *The Global Climate*: 107–20. (Cambridge: Cambridge University Press).

Reining, P., 1978. *Handbook on Desertification Indicators* (Washington DC: American Association for the Advancement of Science).

Reis, M.M., M. Ribon, E. Paniago and S.C. de Alvarenga, 1982. "Planos óptimos de cultivos no Projeto de Irrigaçao de Bebedouro, Municipio de Petrolina, Pernambuco" (Optimum cultivation plans for the Bebedouro Irrigation Project, Petrolina Municipality, Pernambuco), *Revista Ceres* 29: 242–58.

Repetto, R., 1986. *Skimming the Water. Rent Seeking and the Performance of Public Irrigation Systems*. Research Report No. 4 (Washington DC: World Resources Institute).

Republic of Kenya, 1984. "Baringo Pilot Semi-Arid Area Project. Interim report" (Nakuru).

Resch, T., 1981. "Somalia forestry and natural resources sector analysis", Forestry Support Program (Washington DC: Forestry Support Program, USDAFS/USAID).

Resch, T., 1987. "Senegal PL 480 Title III Report: 47–52" (Washington DC: Forestry Support Program, USDAFS/USAID).

Resch, T., A.G.S. el Din, D. de Treville and A.A. Saleem, 1985. "Mid-term evaluation. Eastern Refugee Reforestation Project" (Washington DC: Forestry Support Program, USDAFS/USAID).

Rezig, I., 1982. "L'état présent du nomadisme dans la région de Djelfa" (Nomadism at the present day in the central Algerian region of Djelfa)

Production Pastorale et Société 10: 47–53.

Richardson, Sir E.R., 1984. "Evaluation of institutional and financial arrangements", report for UNEP.

Risser, J., 1985. "Soil erosion problems in the USA", *Desrtification Control Bulletin* 12: 20–5.

Roose, E.H. and F. Lelonmg, 1976. "Les facteurs de l'érosion hydrique en Afrique tropicale. Etudes sur petites parcelles expérimentales de sol", *Revue de Géographique Physique et de Géologie Dynamique* 18: 365–74.

Rorison, K.M. and S.E. Dennison, 1986. "Majjia Valley windbreak evaluation. Windbreak and windbreak harvesting influences on crop production. 1985 growing season" (Niamey: CARE International).

Sales, T., 1982. *Agreste, Agrestes, Transformaciones Recentes na Agricultura Nordestina* (Agreste, region or regions? Recent changes in agriculture in the Northeast) (Rio de Janeiro: Paz e Terra).

Sandford, S., 1983. *Management of Pastoral Development in the Third World* (London: John Wiley).

Santoir, C., 1983. "Raison pastorale et politique du développement. Les problémes des Peul sénégalais face aux aménagements" (Pastoral rights and development. The problems of the Peul in Senegal confronted by development projects), Travaux et Documents de l'ORSTOM No. 166 (Paris: Office de la Recherche Scientifique et Technique Outre-Mer).

Schlesinger, M.E. and J.F.B. Mitchell, 1985. "Model projections of the equilibrium climatic response to increased carbon dioxide", in M.C. MacCracken and F.M. Luther (eds), *Projecting the Climatic Effects of Increasing Carbon Dioxide*, Report DOE/ER-0237: 83–147 (Washington DC: Office of Energy Research, Carbon Dioxide Research Division, US Department of Energy).

Schmidt-Wulffen, W., 1985. "Dürre- und Hungerkatastrophen in Schwarzafrika-das Fallbeispiel Mali" (Drought and famine in sub-Saharan Africa: a case study of Mali), *Geographische Zeitschrift* 73: 46–59.

Seetheraman, S.P. and P.M. Singh, 1985. "Community forestry. A case study of Samrer village", unpublished.

Sen, A., 1981. *Poverty and Famines. An Essay on Entitlement and Deprivation* (Oxford: Clarendon Press).

Sen, B., 1982. "Some aspects of fertilizer use by small farmers: a review", *AID Research and Development Abstracts* 10(3/4): 33.

Sene, E.H., 1984. "Rapport introductif. Conference Préparatoire à la Campagne Nationale de Reboisement 1984", Ministère de la Protection de la Nature, Direction des Eaux, Forêts et Chausesses, Directeur des Eaux et Forets, Dakar, Senegal.

Shah, S.A., 1985. "Eucalyptus – friend or foe?" *International Tree Crops Journal* 3: 191–5.

Sharma, K.K., 1989. "India reaps record harvest", *Financial Times*, 12 July: 34.

Shen-Chang-Jiang, 1982. "Pastoral systems in arid and semi-arid zones of China", Pastoral Network Paper No. 136 (London: Agricultural Administration Unit, Overseas Development Institute).

Shepherd G., 1985. "Social forestry in 1985. Lessons learnt and topics to be addressed", Network Paper 1a (London: Social Forestry Network,

Overseas Development Institute).

Sindiga, I., 1984. "Land and population problems in Kajiado and Narok, Kenya", *African Studies Review* 27: 23–40.

Singh, G., 1985. "Impact of social forestry. A case study in Badaun Division", unpublished.

Sircoulon, J., 1983. *Contribution to WMO, 1983*. "Report of the Expert Group Meeting on the Climatic Situation and Drought in Africa, Geneva, October 6–7" (Geneva: World Meteorological organization).

Skarpe, C., 1981. "A Report on the Range Ecology Project, Western Botswana" (Uppsala: Animal Production Research Unit).

Skutsch, M., 1985. "Forestry by the people for the people – some major problems in Tanzania's village afforestation programme", *International Tree Crops Journal* 3: 147–70.

Smith, P., 1987. "Africa's unforgiving debt hinders promised reform", *Financial Times*, 15 June: 2.

Snowdon, B., 1985. "The political economy of the Ethiopian famine", *National Westminster Bank Review*, November 1985: 41–55.

Spears, J.S., 1983. "Tropical reforestation: an achievable goal?" *Commonwealth Forestry Review* 62: 201–17.

Spears, J., 1985. "Deforestation issues in developing countries – the case for an accelerated investment programme", *Commonwealth Forestry Review* 64: 313–43.

Speece, M. and M.J. Wilkinson, 1982. "Environmental degradation and development of arid lands", *Desertification Control Bulletin* 7: 2–9.

Spooner, B. and H.S. Mann, 1982. *Desertification and Development: Dryland Ecology in Social Perspective* (London: Academic Press).

Staples, R.R., 1939. "Runoff and soil erosion tests in semi-arid Tanganyika Territory", *East African Agricultural Journal* 7: 153–63; 189–95.

Stebbing, E.P., 1935. "The encroaching Sahara", *Geographical Journal* 86: 509–10.

Stiles, D.N., 1983. "Camel pastoralism and desertification in northern Kenya", *Desertification Control* 8: 2–8.

Stryker, J., and C.H. Gotsch, 1981. "Investments in large scale infrastructure: irrigation and river management in the Sahel", *AID Research and Development Abstracts* 10 (3/4): 34–5.

Stubbs, A.T., 1977. "The tribal trust lands in transition: land use", *The Rhodesia Science News* 11 (8): 181–4.

Stürzinger, U., 1983. "Tchad: 'mise en valeur' – coton et développement (Chad: cotton cultivation and "development"), *Revue Tiers-Monde* 24 (95): 643–52.

Sud, Y.C. and M. Fennessy, 1982. "A study of the influence of surface albedo on July circulation in semi-arid regions using the GLAS GCM", *Journal of Climatology* 2: 105–25.

Sudzuki, H.F., 1969. "Absorcion foliar de humedad atmosférica en tamarugo", *Boletin Tecnico Estado Facultad de Agronomia* 30: 1–23 (University of Chile).

Sundar, A. and P.S. Rao, 1981. "Farmers' organizations for efficient water use in irrigated agriculture – an overview", *Wamana* 1: 1–13.

Sur, H.S., 1986. "Role of windbreaks and shelterbelts on wind erosion, moisture conservation and crop growth – an Indian experience", D.L. Hintz and J.R. Brandle (eds), 1986: 247–8.

Swift, J., 1977. "Pastoral development in Somalia. Herding cooperatives as a strategy against desertification and famine", in M.H. Glantz (ed.), 1977: 275–305.

Swift, J. (ed.), 1984. *Pastoral Development in Central Niger. Report of the Niger Range and Livestock Project* (Niamey: UDAID/Republic of Niger).

Swindale, L.D., 1981. "A time for rainfed agriculture (I & II)", *Eastern Economist* 77: 1120–7; 1178–82.

Thakur, J. and P. Kumar, 1984. "A comparative study of different irrigation systems in Western Uttar Pradesh", *Indian Journal of Agricultural Economics* 39: 521–7.

Thirgood, J.V., 1986. "The struggle for sustention on the island of Cyprus", *Journal of World Forest Resource Management* 2: 21–41.

Thomas, G.A. and A.J. Jakeman, 1985. "Management of salinity in the River Murray basin", *Land Use Policy* 2: 87–103.

Thomson, J.T. and M. Hoskins., 1984. "Majjia Valley Windbreak Project. Sociological evaluation. Interim report" (Niamey: CARE International).

Thomson, R., 1987. "Peking admits errors on forest fires", *Financial Times*, 2 June: 20.

Thornthwaite, C.W., 1948. "An approach towards a rational classification of climate", *Geographical Review* 38: 55–94.

Tiffen, M., 1984. "Human resources in African irrigation", in M.J. Blackie (ed.), 1984: 57–73.

Tiffen, M., 1985. *Land Tenure Issues in Irrigation Planning Design and Management in Sub-Saharan Africa*, Working Paper No. 16 (London: Overseas Development Institute).

Tolba, M.K., 1977. "Closing statement", UN Conference on Desertification, Nairobi, 29 August–9 September 1977 (Nairobi: UNEP).

Tolba, M.K., 1981. "Coordination and follow-up of the implementation of the Plan of Action to Combat Desertification", Report by the Executive Director. Governing Council. 13–26 May 1981 (Nairobi: UNEP).

Tolba, M.K., 1984. "Harvest of dust", *Desertification Control* 10: 2–4.

Toulmin, C., 1983. "Herders and farmers or farmer-herders and herder-farmers?" Pastoral Network Paper No. 15d (London: Agricultural Administration Unit, Overseas Development Institute).

Toulmin, C., 1988. "Smiling in the Sahel", *New Scientist*, 12 November: 69.

Trewartha, G.T., 1968. *An Introduction to Climate*, 4th edn (New York: McGraw-Hill).

Tucker, C.J., J.R.G. Townshend and T.E. Goff, 1985. "African land-cover classification using satellite data", *Science* 227: 369–75.

Tucker, C.J. and C.O. Justice, 1986. "Satellite remote sensing of desert spatial extent", *Desertification Control Bulletin* 13: 2–5.

Tucker, C.J. and B.J. Choudhury, 1987. "Satellite remote sensing of drought conditions", *Remote Sensing of Environment* 23: 243–51.

Tueller, P.T., 1987. "Remote sensing science applications in arid environments", *Remote Sensing of Environment* 23: 143–54.

UK Meteorological Office, 1988. "Russian sun, Indian rain", *Guardian*, 6

September.

UN, 1977. "Draft Plan of Action to Combat Desertification", UN Conference on Desertification, Nairobi, 29 August–9 September 1977, Document A/CONF.74/L.36 (Nairobi: UNEP).

UN, 1978. *United Nations Conference on Desertification 29 August–9 September 1977. Round-Up, Plan of Action and Resolutions* (New York: United Nations).

Underhill, H.W., 1984. "Small-scale irrigation in Africa in the context of rural development" (Rome: FAO).

UNDP/World Bank, 1986a. "Niger Improved Stoves Project. Mid-term progress report" (Washington and New York: UNDP/World Bank Joint Energy Sector Management Assistance Program).

UNDP/World Bank, 1986b. "Test results on charcoal stoves from developing countries" (Washington and New York: UNDP/World Bank Joint Energy Sector Management Assistance Program).

UNEP, 1983. "Stone deserts on Africa's roof", *Desertification Control* 8: 20–3.

UNEP, 1985. "Promotion of exchange of information and expertise on desertification control and technology in Africa. Vol. 1. Actions" (Nairobi: UNEP).

UNESCO/UNEP/FAO, 1978. *Tropical Forest Ecosystems. A State of Knowledge Report* (Paris: UNESCO).

UNSO, 1986. "Quadripartite review. Restocking of Gum Belt for Desertification Control Phase II. Mission report." (New York: UN Sudano-Sahelian Office).

USAID, 1979. "Niger Forestry and Lands Use Planning Project. Project paper" (Washington DC: USAID).

USAID, 1980. "Sahelian livestock industry status and development strategy" (Washington DC: USAID).

USAID, 1983a. "Senegal Fuelwood Production Project. Special evaluation report, May 1983" (Washington DC: USAID).

USAID, 1983b. "Sudan. Eastern Refugee Reforestation Project. Operational program grant proposal" (Washington DC: USAID).

USAID, 1984a. "Lesotho Land Conservation and Range Development Project. Project evaluation statement, June 20th 1984" (Washington DC: USAID).

USAID, 1984b. "Report of workshop on forestry program evaluation" (Washington DC: Bureau for Africa, USAID).

US Department of Agriculture, 1984. "Sub-Saharan Africa: outlook and situation report" (Washington DC: USDA Economic Research Service).

US Soil Conservation Service, 1980. *Field Windbreak Removals in Five Great Plains States 1970–1975* (Washington DC).

USSR, Ministry for Reclamation and Water Management, 1977. "An associated case study. USSR, Golodnaya Steppe", United Nations Conference on Desertification, Nairobi, 29 August–9 September 1977, Document A/CONF.74/23 (Nairobi: UNEP).

van der Laan, E.C.W., 1984. "Small scale irrigation schemes along the Senegal River", *Agricultural Administration* 17: 203–13.

Vermeer, D.E., 1981. "Collision of climate, cattle and culture in Mauritania during the 1970s", *Geographical Review* 71: 281–97.

Verstraete M.M., 1986. "Defining desertification: a review", *Climatic*

Change 9: 5–18.

Vincent, L., 1982. "Welling doubts about India's expansion plans", *International Agricultural Development* July/Aug. 1982: 14–15.

Wade, R., 1984. "Managing a drought with canal irrigation: a south Indian case", *Agricultural Administration* 17: 177–202.

Walach, B., 1984. "Irrigation developments in the Krishna Basin since 1947", *Geographical Review* 74: 127–44.

Walker, G.T., 1923. "Correlations in seasonal variations of weather VIII: a preliminary study of world weather (World Weather I)", *Memoirs of the India Meteorological Department* 24: 75–131.

Walls, J., 1984. "A summons to action. The background of the first general assessment of progress in implementing the Plan of Action to Combat Desertification", *Desertification Control* 10: 5–14.

Walter H., 1973. *Vegetation of the Earth in Relation to Climate and Ecophysiological Conditions* (London: English Universities Press).

Warren, M., G. Honadle, S. Montsi and B. Walter, 1985. *Development Management in Africa: The Case of the Land Conservation and Range Development Project in Lesotho*, AID Evaluation Special Study No. 31 (Washington DC: USAID).

Weber F., 1982. Review of CILSS Forestry Sector Program Analysis Papers (Washington DC: Forestry Support Program, USDAFS/USAID).

Wenyue Hsiung, 1983. "Forestry progress in China", *Commonwealth Forestry Review* 62: 191–3.

Whittaker, R.H., 1975. *Communities and Ecosystems* (New York: Macmillan).

Wickens, G.E., J.R. Goodin and D.V. Field (eds), 1985. *Plants for Arid Lands* (London: George Allen and Unwin).

Wiersum, K.F. (ed.), 1984. *Strategies and Designs for Afforestation, Reforestation and Tree Planting* (Wageningen, Netherlands: PUDOC).

Williams, O.B., H. Suijdendorp and D.G. Wilcox, 1977. "Associated case study. Australia, Gascoyne Basin", United Nations Conference on Desertification, Nairobi, 29 August–9 September 1977, Document A/CONF. 74/15 (Nairobi: UNEP).

Williams, P., 1984a. "A 'minor' forest product?", Letter No. 9 to Institute of Current World Affairs, Hanover, New Hampshire.

Williams, P., 1984b. "Traditional agroforestry", Letter No. 12 to Institute of Current World Affairs, Hanover, New Hampshire.

Williams, P., 1985a. "(No longer) blowing in the wind", Letter No. 15 to Institute of Current World Affairs, Hanover, New Hampshire.

Williams, P., 1985b. "Mobilizing popular tree planting efforts", Letter No. 16 to Institute of Current World Affairs, Hanover, New Hampshire.

Wilson, R.T., 1982. "The economic and social importance of goats and their products in the semi-arid arc of northern tropical Africa", in *Proceedings of the Third International Conference on Goat Production and Disease, January 10–15, Tucson, Arizona*: 1986–95.

Winer, N., 1980. "The potential of the carob tree (*Ceratonia siliqua*)", *International Tree Crops Journal* 1: 15–26.

Winrock International, 1985. "World agriculture: review and prospects into the 1990's; a summary" (Morrilton, Arkansas).

Winstanley, D., 1973. "Recent rainfall trends in Africa, the Middle East and India", *Nature* 243: 464–5.

WMO, 1986. *Report of the International Conference on the Assessment of the Role of Carbon Dioxide and of Other Greenhouse Gases in Climatic Variations and Associated Impacts, Villach, Austria, 9–15 October 1985*, Report No. 661 (Geneva: World Meteorological Organization).

World Bank, 1978. *Forestry Sector Policy Paper* (Washington DC: World Bank).

World Bank, 1981. *World Development Report 1981* (Oxford: Oxford University Press).

World Bank, 1982. *World Development Report 1982* (Oxford: Oxford University Press).

World Bank, 1984. "Pakistan. Subsector report: public and private tubewells. Emerging issues and options" (Washington DC: World Bank).

World Bank, 1986a. *Sudan Forestry Sector Review* (Washington DC: World Bank).

World Bank, 1986b. "World Bank financed forestry activity in the decade 1977–86. A review of key policy issues and implications of past experience to future project design" (Washington DC: Agriculture and Rural Development Department, World Bank).

World Bank, no date. "Mexico: irrigation subsector survey" (Washington DC).

World Commission on Environment and Development, 1987. *Our Common Future* (Oxford: Oxford University Press).

Wyatt-Smith, J., 1982. "The agricultural system in the hills of Nepal: the ratio of agricultural to forest land and the problem of animal fodder", Occasional Paper No. 12 (Kathmandu, Nepal: Agricultural Project Service Centre).

Index

Sahel Institute, 296
saline seepage, 93, 122
saline tolerant crops, 33
Salinity Control and Reclamation
    Projects, 182–3
salinization:
    causes, 90–95
    hardpan and, 33
    irrigation and, 33, 90–92, 129
    reclamation, 182
    reversibility of, 20
Samburu people, 194, 205
sand dunes, 31, 80, 256, 290–91, 300,
    301
San Jao Baptista Valley, 299
San Joaquin Valley, 92, 132
satellites, 133, 144–5, 147, 148,
    150–56, 157, 181, 295, 315:
    ground stations, 153
savannas, 25
Schlabathebe area, 212–13
Sclerophyll vegetation, 26
seawater, 93
seeds: costs of, 163
semi-arid zones, 13
Sen, Amartya, 117
Senegal:
    groundnuts in, 74
    tree plantations in, 216, 236
    woodland clearance, 97
Senegal River, 44, 185, 310, 312, 313
Séno Mango, 87
sesame, 69
shadoof, 189
Shah, S. A., 247
Shallambot, 287, 288
shea-nuts, 95, 271
sheep, 72, 78, 79, 81, 85, 89, 193
Shepherd, Gill, 227
silvopasture, 245, 258–9
Sinai Desert, 20
Sind province, 248
Sine-Saloum area, 173
Skutsch, Margaret, 232
social forestry, 217–26:
    awareness generation, 227
    benefit, distribution of, 227–8
    benefits, time delays and, 233–4
    economic incentive, 229–30
    future of, 226–34, 244
    land conflicts and, 230–31
    political support for, 230
    products other than firewood, 226
    social structure and, 228, 229

trust and, 231–2
soil:
    compaction, 33, 34, 77
    conservation, 121, 122, 158, 170:
        farmer training in, 278
        maintenance and, 283–5
        progress in, 276–85
        prospects for, 291–2
    degradation, types of, 31–2
    fertility, 27, 68, 76, 211
    moisture, 57
    structure, 65
    see also following entry; desertification
soil erosion:
    devegetation and, 34
    extent of, 7
    mechanized farming and, 72, 73
    overcropping and, 34
    trampling and, 34
    trees and, 158
Somalia:
    dune fixation in, 287–8
    soil conservation and, 283–5, 311
Sonoran Desert, 51
sorghum:
    drought resistance, 66
    improved varieties, 171
    mechanized cultivation, 223
    yield, 40
Souss, 193
South America:
    desertification, 13, 132
    drylands and, 9, 11
Southern Oscillation, 49
South Korea: tree plantation in, 217,
    218
soybeans, 73
Spain, 11, 259
Spreece, Mark, 93
sprint wheat, 41
Sri Lanka, 175
"steppes", 25
Stern, Mikail, 151
Stiles, Daniel, 194
stone lines, 169–70
stoves: improvement of, 238–43
Styles, Brian, 265, 266
Stylosanthes hemata, 197
Stylosanthes scabra, 197
sub-humid areas, 14, 140
sub-Saharan Africa, 14, 17, 209, 215:
    soils in, 30
    see also under names of countries
subsistence cropping, 115, 125, 172

KING ALFRED'S COLLEGE
LIBRARY